Lecture Notes in Computer Science

Edited by G. Goos, J. Hartmanis and J. van Leeuwen

Advisory Board: W. Brauer   D. Gries   J. Stoer

Lecture Notes in Computer Science

Edited by G. Goos, J. Hartmanis and J. van Leeuwen

Advisory Board: W. Brauer   D. Gries   J. Stoer

Terence C. Fogarty (Ed.)

Evolutionary Computing

AISB Workshop
Leeds, U.K., April 11-13, 1994
Selected Papers

Springer-Verlag
Berlin Heidelberg New York
London Paris Tokyo
Hong Kong Barcelona
Budapest

Series Editors

Gerhard Goos
Universität Karlsruhe
Postfach 69 80, Vincenz-Priessnitz-Straße 1, D-76131 Karlsruhe, Germany

Juris Hartmanis
Department of Computer Science, Cornell University
4130 Upson Hall, Ithaka, NY 14853, USA

Jan van Leeuwen
Department of Computer Science, Utrecht University
Padualaan 14, 3584 CH Utrecht, The Netherlands

Volume Editor

Terence C. Fogarty
Faculty of Computer Studies and Mathematics, University of the West of England
Frenchay Campus, Coldharbour Lane, Bristol BS16 1QY, United Kingdom

CR Subject Classification (1991): F.1, F.2.2, I.2.6, I.2.8-9, I.5.1, I.5.4, J.3

ISBN 3-540-58483-8 Springer-Verlag Berlin Heidelberg New York

CIP data applied for

© Springer-Verlag Berlin Heidelberg 1994
Printed in Germany

Typesetting: Camera-ready by author
SPIN: 10479154   45/3140-543210 - Printed on acid-free paper

Preface

This volume contains the post-workshop proceedings of a workshop on evolutionary computing sponsored by the Society for the Study of Artificial Intelligence and Simulation of Behaviour (AISB) and held at the University of Leeds, England, 11–13 April 1994. The workshop brought together most of the people doing research on evolutionary computing in the UK and some colleagues from abroad. Thirty-seven papers were presented at the workshop selected on the basis of abstracts refereed by the organising committee. This was composed of the following people:

Terry Fogarty, University of the West of England, Bristol,
Ray Paton, University of Liverpool,
Nick Radcliffe, University of Edinburgh,
Phil Husbands, University of Sussex,
Colin Reeves, Coventry University,
Dave Corne, University of Edinburgh,
Peter Fleming, University of Sheffield.

After the workshop a full paper version of each presentation was reviewed by three referees drawn from the organising committee and the following additional referees:

Jonathan Shapiro, Manchester University,
Geoff Miller, University of Sussex,
Peter Hancock, University of Stirling,
Marco Dorigo, Free University of Brussels,
Peter Ross, University of Edinburgh,
Hugh Cartwright, Oxford University,
Stuart Flockton, University of London.

On the basis of the advice of the referees, which took into consideration scientific progress made at the workshop, twenty-two papers were selected and revised for publication in this volume. In addition two authors, Ray Paton and Colin Reeves, were invited to contribute papers.

This volume only gives a partial, though polished, view of the proceedings of the workshop and would not be complete without a report on the workshop itself.

In June 1993 Steve Greig of Napier University suggested the holding of a European evolutionary computation conference. The "powers that be" did not see the need for one and so, despite widespread support in the UK, he did not pursue the idea. At about the same time I was filling out a questionnaire and suggested that the AISB hold more workshops in my area of research, i.e., evolutionary computation in general and classifier systems in particular. When the AISB asked me to organise a workshop on evolutionary computation I put the idea to the other organisers, who all agreed to help.

I drew up a call for papers and posted it to all UK researchers who had expressed an interest in Steve Greig's original proposal and on all the relevant electronic bulletin boards.

I received next to nothing up to the day the call for papers expired. Then over 40 abstracts arrived. These were refereed by the organisers - eighteen were selected for presentation as talks and twelve as posters. I invited the organisers to present papers and the show was on the road.

The workshop turned out to be more like a conference at which the various people working in the area in the UK and some guests from abroad presented their work. Some of this was familiar and built on work already published but much of it was novel.

The first day was taken up with presentations of work on the application of evolutionary computation to problems in robotics, signal processing and control. These are particularly active areas for the application of evolutionary computing.

There were two papers demonstrating the use of genetic algorithms to design control systems for real robots. Marco Dorigo of the Free University of Brussels told us about his work on behaviour based control implemented as a set of interacting classifier systems. He demonstrated, with the help of a video, a small robot that uses his techniques to learn some simple behaviours. Adam Fraser of Salford University explained how two genetically programmed robots have learned to co-operate with each other in transporting a shared load. He showed us one of the robots and proposed a general architecture for intelligent control.

The IEEE recently organised a whole workshop on adaptive computing, including genetic algorithms, for signal processing applications, and there were three papers in this area presented at this workshop. Stuart Flockton from Royal Holloway, University of London, described a genetic algorithm for optimising the coefficients of recursive digital filters that guarantees filter stability. David Beasley from University of Wales College of Cardiff presented his method of expansive coding that reduces the complexity of algorithms used in digital signal processing using a genetic algorithm. Julian Miller of Napier University showed how genetic algorithms can be used to optimise Reed-Muller expressions.

The next three papers were on classifier systems for control. Classifier systems, for the uninitiated, are rule-based classification or control systems that are optimised with genetic algorithms. The genetic algorithm is used to optimise either the whole rule-base or individual rules. I presented my paper comparing the use of genetic algorithms and classifier systems in optimising rule-sets for controlling a cart-pole simulation. Tony Pipe of the University of the West of England, Bristol compared the perfor-mance of a classifier system and a genetically optimised neural implementation of an adaptive heuristic critic. Andy Fairley of Liverpool University examined the evolutionary stability of the bucket brigade in a classifier system using evolutionary game theory.

The second day was less specialised and covered biological foundations of evolutionary computation, the theory of genetic algorithms, selection, and applications in scheduling. All of the posters were also presented on this day.

There are a number of psychologists and biologists active in the area of evolutionary computation and two of them presented papers at this workshop. Geoffrey Miller of the University of Sussex reviewed the biological theory of sexual selection and some of its possible applications in search, optimisation, and diversification. He used simulation results to illustrate some key points. Ray Paton of the University of Liverpool presented some insights from biological systems including mechanisms acting on strings both as independent identities and within ecologies.

There were two papers on the theory of genetic algorithms and one on selection methods in various evolutionary algorithms. The first theory paper presented a novel and interesting formulation of genetic algorithms while the second extended some well established work in the field of Walsh coefficients. Jonathan Schapiro of Manchester University presented a statistical mechanical formulation of the dynamics of genetic algorithms. He used this theory to make some predictions for a test problem - search for the low energy states of a random spin chain - and confirmed it with experiments. Paul Field of Queen Mary and Westfield College presented a generalisation of partition coefficients and Walsh coefficients to nonbinary alphabets. Peter Hancock of the University of Stirling compared selection methods in evolutionary algorithms. His experiments covered genetic algorithms, evolution strategies and evolutionary programming. A polite discussion on the pros and cons of the different selection methods was sparked off by this talk.

A lot of research has been done on the use of genetic algorithms for scheduling and the next two scheduled papers demonstrated the application of the techniques developed to solve real scheduling problems. Unfortunately the second of them had to be rescheduled to Wednesday but I will mention them both here. Dave Corne of Edinburgh University described the optimisation of timetables using his genetic algorithm based system in various universities in the UK. Hugh Cartwright of Oxford University discussed how a genetic algorithm can be implemented to handle a full industrial flowshop taking into account that the flowshop operates continuously and must be able to adjust the order in which products are made as new requests are received.

The poster session in the afternoon of the second day consisted of the presentation of the following twelve papers:

A Genetic Algorithm for University Timetabling, by Edmund Burke, David Elliman and Rupert Weare, University of Nottingham,

Learning Anticipatory Behaviour Using a Delayed Action Classifier System, by Brian Carse, University of the West of England, Bristol,

The Co-evolution of Classifier Systems in a Competitive Environment, by Keith W. Chalk and George D. Smith, The University of East Anglia,

Genetic Algorithms and Directed Adaptation, by John Coyne and Ray Paton, University of Liverpool,

Evolving Go Playing Strategy in Neural Networks, by Paul Donnelly, Patrick Corr, Danny Crookes, The Queen's University of Belfast,

Walsh and Partition Functions Made Easy, by Paul Field, Queen Mary and Westfield College,

The last day was taken up with talks on the use of local search techniques, multi-objective optimisation and applications of genetic programming.

The use of the genetic algorithm for global search has long been established but the use of genetic operators which do local search is gathering momentum. This is demonstrated in the four papers at the workshop in this area. Lawrence Bull of the University of the West of England, Bristol presented an evolutionary algorithm that was a hybrid of the genetic algorithm for global search combined with an evolutionary strategy for local search. Andy Fairley of the University of Liverpool introduced some new genetic operators for use on high level representations. Nick Radcliffe of Edinburgh University introduced a representation-independent formalism for using local search and demonstrated it on a travelling sales-rep problem. Colin Reeves of Coventry University explored a perspective which views genetic algorithms as a generalisation of neighbourhood search methods.

There were two papers on multi-objective optimisation, one of which was in fact presented the day before to allow the speaker to attend two meetings scheduled at the same time, but I will mention them both here. Carlos Fonseca of the University of Sheffield discussed current evolutionary approaches to multi-objective optimisation, drawing attention to issues such as how they affect the fitness landscape, and the implications of that for the search process. Phil Husbands of the University of Sussex described his work on distributed coevolutionary genetic algorithms for multi-criteria and multi-constraint optimisation.

Finally, there were two papers in the relatively new area of genetic programming. Bill Buckles of Tulane University in the USA showed how genetic programming can be applied to an information retrieval system to improve Boolean query formulation. Howard Oakley of the Institute of Naval Medicine demonstrated how genetic programming can be used for the forecasting of chaotic series in the face of short data-sets with significant amount of noise.

The workshop was a good reflection of the state of research in evolutionary computation in the UK. It showed that there is a viable community of researchers in this country working on evolutionary computation and its application to significant scientific, commercial, and industrial problems. This was facilitated by the AISB, and

I would like to thank Ann Blandford of the MRC Applied Psychology Unit, Cambridge, and Charlie Brown of Leeds University for all the hard work they put in to make the workshop a success.

Terence C. Fogarty
University of the West of England, Bristol
16 August, 1994

Contents

Classifier Systems

Applications

Formal Memetic Algorithms

Nicholas J. Radcliffe and Patrick D. Surry

Edinburgh Parallel Computing Centre
King's Buildings, University of Edinburgh
Scotland, EH9 3JZ

Abstract. A formal, representation-independent form of a memetic algorithm—
a genetic algorithm incorporating local search—is introduced. A generalised form
of N-point crossover is defined together with representation-independent patching
and hill-climbing operators. The resulting formal algorithm is then constructed
and tested empirically on the travelling sales-rep problem. Whereas the genetic
algorithms tested were unable to make good progress on the problems studied,
the memetic algorithms performed very well.

1 Motivation

The rôle of local search in the context of genetic algorithms and the wider field of evolutionary computing has been much discussed. The traditional view, which can be traced
back to Holland (1975), has been that the primary search operator in evolutionary computing should be recombination. In its most extreme form, this view casts mutation and
other local operators as mere adjuncts to recombination, playing auxiliary (if important)
rôles such as keeping the gene pool well stocked and helping to tune final solutions.
There have, however, long been advocates of a greater rôle for mutation, hill-climbing
and local refinement. The arguments for serious consideration of operators other than recombination for primary search come in many forms and are inspired by widely differing
applications. For example, Davis (1991) advocates *hybridisation* of genetic algorithms
with domain-specific techniques for "real world" optimisation, by incorporating extra
move operators. He regularly uses sophisticated decoders that make use, for example,
of greedy algorithms and repair mechanisms. Ackley (1987) recommends *genetic hill-
climbing,* in which crossover plays a rather less dominant rôle. Muehlenbein (1992)
argues theoretically and Gorges-Schleuter (1989) provides empirical demonstrations
that local search can play a key rôle, and Muehlenbein (1989) incorporates it as a
fundamental component of his particular notion of a parallel genetic algorithm with
a structured population. Meanwhile, the *Evolution Strategies* community has always
placed more emphasis on mutation than crossover (Baeck *et al.,* 1991). Countless other
advocates of a greater emphasis on non-recombinative elements of evolutionary search
could be cited, especially from the ranks of those competing with domain-specific
techniques.

Moscato & Norman (1992) have introduced the term *memetic algorithm* to describe
evolutionary algorithms in which local search plays a significant part. This term is motivated by Richard Dawkins's notion of a *meme* as a unit of information that reproduces
itself as people exchange ideas (Dawkins, 1976). A key difference exists between genes

and memes: before a meme is passed on, it is typically adapted by the person who transmits it as that person thinks, understands and processes the meme, whereas genes get passed on whole. Moscato and Norman liken this thinking to local refinement, and therefore promote the term "memetic algorithm" to describe genetic algorithms that use local search heavily.

The purpose of this paper is three-fold. The first aim is to formalise Norman and Moscato's memetic algorithms and to provide a unified framework for considering both memetic and genetic algorithms. The second aim is to use forma analysis (Radcliffe, 1991; 1994a) to devise further representation-independent operators to augment those previously developed in Radcliffe (1991, 1994a and 1994b). In particular, a generalised form of N-point crossover (GNX) will be defined, as will a general hill-climbing operator. This will allow the construction of a representation-independent (formal) memetic algorithm. The third aim is to investigate the application of the ideas developed to the travelling sales-rep problem (TSP) to test their efficacy.

2   Memetic Algorithms

The first task in this work is to provide a homogeneous formal framework for considering memetic and genetic algorithms. Informally, the idea exploited to achieve this is that if a (true) local optimiser is added to a genetic algorithm, and applied to every child before it is inserted into the population (including the initial population) then a memetic algorithm can be thought of simply as a special kind of "genetic" search over the subspace of local optima (figure 1). Recombination and mutation will usually produce solutions that are outside this space of local optima (and can thus be regarded as "damaged") but a local optimiser can then "repair" such solutions to produce final children that lie within this subspace, yielding a memetic algorithm. Section 2.1 formalises these notions and section 2.2 discusses when such memetic search might be more appropriate than genetic search.

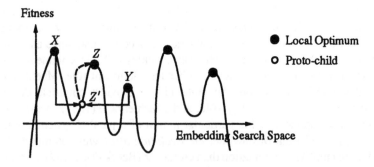

Fig. 1. Memetic algorithms search over the subspace of local optima within the embedding search space of all solutions. After recombination, the proto-child typically lies outside this subspace and a local optimiser is used to "repair" the child so that it lies at a local optimum. Here parents X and Y produce the proto-child Z', which is then optimised to produce the final child Z.

2.1 Formal Memetic Algorithms

Consider a search space S (of *phenotypes*) and a representation space C (of *genotypes*). Let

$$\rho : S \longrightarrow C \qquad (1)$$

be the representation function which, given any solution in S, returns the chromosome in C that represents it. It will be assumed throughout this paper that ρ is injective (so that every solution $s \in S$ has a well-defined, unique chromosome $\rho(s) \in C$ to represent it) but not that it is surjective, (so there may be chromosomes in C that do not correspond to any solution in S). Let f be the fitness function, which it will be convenient to regard as a mapping

$$f : C \longrightarrow \mathbf{R}^+. \qquad (2)$$

It will be assumed that the aim is to maximise fitness, and the set of global optima will be denoted $C^* \subset C$.

Let Q be a stochastic unary move operator over C. It will be convenient for the moment to accommodate the stochastic element of such an operator through a *control set*, $\mathcal{K}_Q$, from which a *control parameter* will be drawn to determine which of the (typically many) possible moves actually occurs. For example, in the case of mutation of binary strings, a binary mask might be used as the control parameter with the presence of a 1 at position i indicating that the ith bit should be mutated. The functional form for Q will then be

$$Q : S \times \mathcal{K}_Q \longrightarrow S. \qquad (3)$$

A chromosome $x \in C$ will be said to be *locally optimal with respect to Q*, or *Q-opt*, if no chromosome of higher fitness than x can be generated from it by a single application of Q, i.e. if and only if

$$\forall \kappa \in \mathcal{K}_Q : f(Q(x, \kappa)) \leq f(x). \qquad (4)$$

Let $C_Q \subset C$ be the set of Q-opt chromosomes in C, i.e.

$$C_Q \triangleq \{ x \in C \mid x \text{ is } Q\text{-opt} \} . \qquad (5)$$

A genetic algorithm applied to the task of optimising f over C has some goal such as finding some or all optima in C^* or making rapid improvements towards fitter chromosomes. It is clear that for any move operator Q, all chromosomes in C^* are Q-opt, and thus $C^* \subset C_Q$. It would be perfectly satisfactory, therefore, to formulate the search instead over C_Q.

Given a representation space C, a move operator Q, and the subspace C_Q of local optima as above, define a *hill-climber* to be any stochastic, parameterised operator that, given a chromosome $x \in C$, returns a local optimum in C_Q. Thus a hill-climber $\mathcal{H}$ with control set $\mathcal{K}_\mathcal{H}$ is any function

$$\mathcal{H} : C \times \mathcal{K}_\mathcal{H} \longrightarrow C_Q. \qquad (6)$$

Notice that there is no requirement that the solution returned be in any sense "near" the starting solution, though of course this will often be the case in practice.

Typical genetic algorithms produce new chromosomes by recombination of two parents followed by some small level of mutation, so that if

$$\mathcal{X} : \mathcal{C} \times \mathcal{C} \times \mathcal{K_X} \longrightarrow \mathcal{C} \qquad (7)$$

is the recombination operator (with control set $\mathcal{K_X}$), and

$$\mathcal{M} : \mathcal{C} \times \mathcal{K_M} \longrightarrow \mathcal{C} \qquad (8)$$

is the mutation operator (with control set $\mathcal{K_M}$), the combined genetic reproductive function $\mathcal{R}_g$ would typically be given by the composition of mutation and recombination, $\mathcal{R}_g = \mathcal{M} \circ \mathcal{X}$, yielding

$$\mathcal{R}_g : \mathcal{C} \times \mathcal{C} \times \mathcal{K_M} \times \mathcal{K_X} \longrightarrow \mathcal{C}, \qquad (9)$$

defined by

$$\mathcal{R}_g(x, y, \kappa_\mathcal{M}, \kappa_\mathcal{X}) \triangleq \mathcal{M}(\mathcal{X}(x, y, \kappa_\mathcal{X}), \kappa_\mathcal{M}). \qquad (10)$$

If, however, $\mathcal{R}_g$ is further composed with a hill-climber $\mathcal{H}$ (with respect to some unary move operator $\mathcal{Q}$), and restricted to $\mathcal{C}_\mathcal{Q}$, a memetic reproduction function $\mathcal{R}_m \triangleq \mathcal{H} \circ \mathcal{M} \circ \mathcal{X}$ results:

$$\mathcal{R}_m : \mathcal{C}_\mathcal{Q} \times \mathcal{C}_\mathcal{Q} \times \mathcal{K_H} \times \mathcal{K_M} \times \mathcal{K_X} \longrightarrow \mathcal{C}_\mathcal{Q}, \qquad (11)$$

defined by

$$\mathcal{R}_m(x, y, \kappa_\mathcal{H}, \kappa_\mathcal{M}, \kappa_\mathcal{X}) \triangleq \mathcal{H}(\mathcal{M}(\mathcal{X}(x, y, \kappa_\mathcal{X}), \kappa_\mathcal{M}), \kappa_\mathcal{H}). \qquad (12)$$

2.2 Decomposable Fitness Functions

While the general question of when it might be appropriate to use a memetic algorithm in preference to a genetic algorithm is beyond the scope of this paper, one special situation can be considered that seems likely to be relatively favourable to the memetic variety. This arises when the fitness function is *decomposable*, in the sense that computing the fitness of a solution given the fitness of another solution that is "close" to it (in the sense, informally, of having much genetic material in common with it) is significantly less computationally expensive than computing the fitness of a solution "from scratch". In the TSP, for example, computing the length of a tour that shares most of its edges with another tour whose length is already known is very much cheaper than computing the length of a general tour, so the fitness function is in that case decomposable. Contrariwise, when solving a system of non-linear equations, for example to compute the flow of gas through a pipe network, a small change in the chromosome can often have global effects and therefore computing the fitness of a chromosome is made no easier by knowing that of another similar chromosome. Given that in most real-world optimisation problems calculation of fitness accounts for almost all the time spent in a genetic algorithm, it seems likely that memetic algorithms will be at an advantage when the fitness function is decomposable, provided that the moves it makes while hill-climbing are "small".

3 Representation-Independent Operators

The principal complication that arises in defining representation-independent operators is that some combinations of gene values may be incompatible. While this is most obviously a problem for recombination operators, it is also a serious consideration for other move operators. As forma analysis (Radcliffe, 1991) has been developed, significant efforts have been made to define representation-independent operators and to understand and classify the kinds of representations that can arise in evolutionary search. In particular, one representation-independent mutation operator—binomial minimal mutation (BMM; Radcliffe, 1994b)—and three representation-independent recombination operators—random respectful recombination (R³; Radcliffe, 1991), random transmitting recombination (RTR; Radcliffe, 1992), and random assorting recombination (RAR; Radcliffe, 1994a)—have been developed. It is not necessary to revisit all of these for the purposes of this paper, but it is necessary to define some recombination operator and some mutation operator. BMM and RAR will therefore be reviewed briefly in sections 3.2 and 3.3, after which, in section 3.4, a further representation-independent recombination operator will be introduced—generalised N-point crossover (GNX). Attention will then be turned to generalised memetic operators, with a consideration of representation-independent *patching* operators in section 3.5 and representation-independent *hill-climbing* in section 3.6. Before any of this can be achieved, however, it is first necessary to introduce a distinction between two kinds of formal representations—genetic and allelic.

3.1 Genetic Representations and Allelic Representations

The notions of genes and alleles are very familiar, but need to be defined rather carefully for present purposes. A distinction will be drawn between *genetic* representations and *allelic* representations. A formal genetic representation is precisely a formal version of the familiar string composed of genes, and should cause little confusion. It will be assumed that a genetic representation consists of a string of n *genes*, numbered 1 to n, and that each gene takes on values from some (typically but not necessarily finite) set $\mathcal{A}_i$. Thus in the case of a genetic representation, the representation space will be assumed to have the form

$$\mathcal{C} = \mathcal{A}_1 \times \mathcal{A}_2 \times \cdots \times \mathcal{A}_n, \tag{13}$$

so that a chromosome is formally a vector of gene values. The only complication with respect to the typical case is that, as before, it will not be assumed that all members of $\mathcal{C}$ correspond to solutions in the search space $\mathcal{S}$, so some combinations of gene values may be "illegal".

A formal *allele* in the context of a genetic representation will be considered to be an ordered pair consisting of a gene and one of its possible values, so that a chromosome $x = (x_1, x_2, \ldots, x_n)$ has alleles $(1, x_1), (2, x_2), \ldots, (n, x_n)$. This formulation of alleles, so far from being new, was suggested in Holland (1975), albeit with different motivation.

There are situations in which a suitable genetic representation of the form described above is not straightforwardly available. In such situations, it may be appropriate to drop the requirement that genes be defined, working instead with an *allelic represent-ation* (Radcliffe, 1994b). In such an allelic representation, instead of being a vector, a chromosome is a *set* whose elements are drawn from some universal set $\mathcal{A}$. In order to qualify as a formal allelic representation, all that is necessary is that the representation function ρ of equation 1 be injective, as required previously, and that C be a subset of $P(\mathcal{A})$, where $P(\mathcal{A})$ denotes the *power set* (set of all subsets) of $\mathcal{A}$. Again, there is no requirement that all members of C represent solutions in S.

More concretely, consider representations of the TSP based on edges (city-to-city links). If these edges are considered to be directed, then a genetic representation is arrived at simply by letting the ith gene take the value of the city visited after city i, so that $(4, 3, 1, 2)$ represents the tour that goes from city 1 to city 4, to city 2, to city 3, to city 1 (sic). If, however, the edges are considered to be undirected (so that the 3–2 edge and the 2–3 edge are equivalent) it is no longer straightforward to identify genes, because each city is connected to two others. In this case, one approach is simply to let $\mathcal{A}$ be the set of all possible edges and represent a tour by the set of (undirected) edges it contains. The tour represented by $(4, 3, 1, 2)$ in the directed-edge representation is then represented by { 1–4, 2–4, 2–3, 1–3 } in the undirected edge representation, where the edges have all been written with the lower-numbered city first the emphasize their directionless nature.

It is obviously trivial to construct an allelic representation from a genetic represent-ation by taking $\mathcal{A}$ to be the set of all alleles, so that (referring to equation 13)

$$A = \bigcup_{i=1}^{n} A_i. \tag{14}$$

Under this scheme a solution is represented simply by its set of (formal) alleles, so that $(4, 3, 1, 2)$ in the directed edge representation gives rise to $\{(1, 4), (2, 3), (3, 1), (4, 2)\}$ in the allelic representation. This motivates the term "allelic representation", and the members of $\mathcal{A}$ will henceforth be referred to as alleles whether they are alleles in the sense of ordered pairs of gene values from a genetic representation or simply members of a given set $\mathcal{A}$ used directly to construct an allelic representation.

It is only slightly less obvious that given an allelic representation it is also easy to construct a genetic representation from it by creating for each member of $\mathcal{A}$ a binary gene that takes the value 1 if the (allelic) chromosome contains that allele and 0 if it does not. It should be noted, however, that such an induced genetic representation is very different from the initial allelic representation, so much so that if an allelic representation is then constructed from the (induced) genetic representation it will be quite different, in general, from the original allelic representation. For this reason, most of the operators introduced below are defined with respect to allelic representations, which allows them to be used for (natural) allelic representations or genetic representations without complication.

To reduce possible confusion, genetic chromosomes will be denoted with lower case letters x, y, z and allelic chromosomes will take upper case letters X, Y, Z.

3.2 Binomial Minimal Mutation (BMM)

In a general representation it will often not be possible to use standard gene-wise mutation because the new allele chosen may well be incompatible with other alleles in the chromosome. For convenience it will here be assumed that chromosomes have a fixed number n of alleles, though it is simple to relax this restriction. Allelic representations will be considered, so that a chromosome will be taken to be a set of exactly n alleles from $\mathcal{A}$. A distance measure, D, between two chromosomes can then be introduced, and will be taken to be the number of alleles present in one chromosome but not the other:

$$D(X, Y) \triangleq n - |X \cap Y|. \tag{15}$$

Y will be said to be a *minimal mutation* of X if and only if there is no other chromosome in $\mathcal{C}$ closer than Y to X with respect to D. Thus the set $M_D(X)$ of minimal mutations of X is given by

$$M_D(X) = \{Y \in \mathcal{C} \mid \forall Z \in \mathcal{C} \setminus \{X\} : D(X, Y) \leq D(X, Z)\}, \tag{16}$$

where $\setminus$ denotes set subtraction. For example, in the undirected edge representation for the TSP, any tour that can be constructed from another by reversing some section of it is one of its minimal mutations, because reversing a section involves breaking only two edges and there is no pair of tours that differ by only one edge (figure 2).

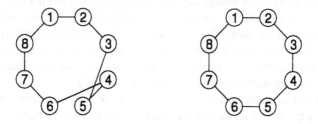

Fig. 2. In the undirected edge representation, the left-hand tour is { 1–2, 2–3, 3–5, 4–5, 4–6, 6–7, 7–8, 1–8 } and the right-hand tour is represented by { 1–2, 2–3, 3–4, 4–5, 5–6, 6–7, 7–8, 1–8 }. In this representation, these two tours are minimal mutations of each other, because they differ by exactly two edges and no pair of distinct tours, in this representation, differ by fewer than two edges.

The *binomial minimal mutation* operator (BMM) takes a single parameter p_m, which specifies the probability of performing each possible minimal mutation. A number k of mutations to perform is then selected from the binomial distribution $B(n, p_m)$, where n, as above, is the number of alleles in each chromosome. This choice ensures that in the case of orthogonal representations BMM's behaviour mimics that of conventional gene-wise mutation. A sequence of k chromosomes is generated, each of which is a minimal mutation of the previous, the first in the sequence being the chromosome to be mutated and the last being the resultant chromosome. Thus if

$$\hat{\mathcal{M}} : \mathcal{C} \longrightarrow \mathcal{C} \tag{17}$$

is a stochastic operator that returns a randomly (uniformly) chosen member of $M_D(X)$, BMM is its kth iterate:

$$\text{BMM}(X, p_m) \triangleq \hat{\mathcal{M}}^k(X), \tag{18}$$

where

$$k \sim B(n, p_m). \tag{19}$$

Note that this operator does not exclude the possibility that subsequent mutations reverse earlier ones, but in practice the likelihood of this is low for small values of p_m. (In the case of orthogonal representations, this is the only difference between BMM and conventional gene-wise mutation.) Note also that this operator is really only appropriate provided that any (legal) point in the representation space can be reached by a finite sequence of minimal mutations from any other point, so that BMM satisfies the requirements of ergodicity (Radcliffe, 1991).

3.3 Random Assorting Recombination (RAR)

Random assorting recombination (RAR_w) may be viewed as a generalisation of uniform crossover, though this was not its genesis. Informally, it proceeds to choose alleles from those of the parents, inserting them in the child when it can, and discarding them otherwise. If the parents' alleles become exhausted before the child is fully specified, its remaining alleles are set either at random (from among the legal combinations) or by some form of *patching*. As with uniform crossover, locus has no effect on the likelihood that a group of alleles will be inherited, and—neglecting the fact that alleles from one parent are known to be compatible, whereas those from different parents may not be— the number of alleles taken from each parent is binomially distributed. Indeed, in the limit of orthogonal genetic representations (those in which all allele patterns are legal) RAR_w reduces to uniform crossover (with parameter half).

RAR_w takes a parameter w that specifies a relative weighting between alleles common to the parents and those that are present only in one. $RAR_w(X, Y)$ begins by assigning to each allele $a \in X \cup Y$ a weight $W(a)$ given by

$$W(a) = \begin{cases} w, & \text{if } a \in X \cap Y, \\ 1, & \text{otherwise.} \end{cases} \tag{20}$$

It then initialises an empty child $Z_0 = \emptyset$ and selects an allele a_0 from $\mathcal{G}_0 \triangleq X \cup Y$, with probability proportional to its weight. This allele is added to the proto-child to form Z_1. The following process is then repeated for steps indexed by i:

Repeat until $\mathcal{G}_i = \emptyset$:

1. Let $\mathcal{G}_i \triangleq \mathcal{G}_{i-1} \setminus a_i$.
2. Choose a new allele a_i from $\mathcal{G}_i$ with probabilities proportional to the weights of the alleles in $\mathcal{G}_i$.
3. Let $Z_i \triangleq \begin{cases} Z_{i-1} \cup \{a_i\}, & \text{if } a_i \text{ is compatible with those in } Z_{i-1}, \\ Z_{i-1}, & \text{otherwise.} \end{cases}$
4. $i \leftarrow i + 1$.

In step 3 above, "compatible" means that there exists a solution in $\mathcal{S}$ whose representative in $\mathcal{C}$ has all the alleles in Z_{i-1} and also a_i.

At this stage it is possible that the child will be completely specified, but in general this will not be the case. If it is not, a patching algorithm must be used to complete the child. The most general way to achieve this is to select randomly (uniformly) from the chromosomes that include all the alleles in the proto-child constructed thus far. Section 3.5 introduces more sophisticated memetic patching operators.

3.4 Generalised N-point Crossover (GNX)

In constructing a generalised form of N-point crossover, it is convenient to consider only genetic representations. The difficulty in applying conventional crossover operators is that not all combinations of gene values are legal. Let $\mathcal{L} = \{\ell_1, \ell_2, \ldots, \ell_N\}$ be a set of cross points, with $0 < \ell_1 < \ell_2 < \cdots < \ell_N < n$. This breaks a parent (genetic) chromosome x into $N + 1$ segments

$$(x_1, x_2, \ldots, x_{\ell_1-1}), \ (x_{\ell_1}, x_{\ell_1+1}, \ldots, x_{\ell_2-1}), \ \ldots, \ (x_{\ell_N}, x_{\ell_N+1}, \ldots, x_n), \qquad (21)$$

and breaks up the second parent y into corresponding segments.

GNX proceeds by picking a random order to visit the $N + 1$ segments from alternate parents, and within each segment "tests" each allele in a random order. An allele is "tested" by seeing whether it can be placed in the child—whether it is compatible with those alleles that have already been placed in it. If compatible, the new allele is inserted, otherwise it is discarded. Because in general after this process has terminated the child will still be incomplete, the process is then repeated with the alternating untested segments from the parents, again visiting these in a random order and testing the alleles within them in random sequence. If the child is still incomplete after this, the child is completed at random or by patching in the same way as for RAR. The general pattern of progress of GNX is shown in figure 3.

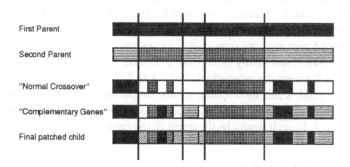

First Parent

Second Parent

"Normal Crossover"

"Complementary Genes"

Final patched child

Fig. 3. GNX first copies gene value from alternating segments of the parent chromosomes, visiting the segments and testing the genes within these segments in a random order. Gene values are copied to the child only if they are compatible with those already present. For genes not able to be assigned by this process, alleles from the unused (complementary) segments of the parent genomes are then tested, again in random sequence, for inclusion. Genes still not assigned after this process are assigned either at random, from the set of legal combinations, or by some heuristic or other patching procedure.

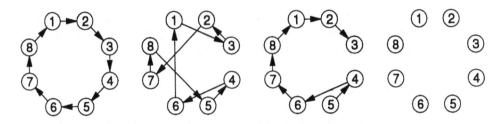

Fig. 4. Two parent tours, one possible partial child they can produce under GNX, and an empty grid for the reader to use while following through the example.

An example using the TSP may help to clarify this. In order to follow this example, the reader may find it helpful to try connecting up the "empty" tour in figure 4 to understand the edge acceptances and rejections. Consider the directed edge representation for the TSP and G2X with cross points 3 and 6 with parents as shown in figure 4 and given by

$$x = (2,3,4 \mid 5,6,7 \mid 8,1),$$
$$y = (3,7,2 \mid 6,4,1 \mid 8,5). \tag{22}$$

(Recall that in this representation the ith gene represents a directed edge from city i to city x_i.)

Suppose the permutation of the segments chosen is $(3, 2, 1)$. Then the third segment of x (visited first) will be inserted whole, giving edges $7 \rightarrow 8$ and $8 \rightarrow 1$ (or the proto-child $(\square,\square,\square,\square,\square,\square,8,1)$). Then alleles in segment 2 from y will be tested in a random order, say $5 \rightarrow 4, 6 \rightarrow 1, 4 \rightarrow 6$ and the first and third (in this case) will be accepted, giving the proto-child $(\square,\square,\square,6,4,\square,8,1)$. The first segment of x is then tested, and the edges $1 \rightarrow 2$ and $2 \rightarrow 3$ will be accepted giving $(2,3,\square,6,4,\square,8,1)$. This completes the first phase.

The untested segments are then visited in random order, say first $(5,6,7)$ from x, then $(3,7,2)$ from y, and finally $(8,5)$ from y. During this process only the edge $6 \rightarrow 7$ will be accepted, giving the proto-child $(2,3,\square,6,4,7,8,1)$.

Since this child is still incomplete, it must be patched. In this case however, only one legal chromosome (with directed edges) has the required allele pattern, namely $(2,3,5,6,4,7,8,1)$, so it would be the result of the cross.

3.5 Patching by Forma Completion

Both RAR and GNX produce (in general) partially-specified children that then need to be completed in some manner. In the case of GNX, it is reasonably natural to think of the partially completed child as a schema. In the case of RAR (which works with allelic representations) this is less natural, but a child is specified precisely by a set of alleles it should contain, and such a specification qualifies as a *forma*—a set of chromosomes sharing certain alleles. A schema may be viewed as a special case of a forma, applicable to the case of genetic representations. The question that arises for both RAR and GNX is thus how to choose a child from a given forma. The "default" method is to choose one randomly but two other methods of patching (or "completing") formae to produce children will be considered.

One option is to choose the best solution in the forma. In general, this would be prohibitively expensive, but if the number of unspecified alleles were small and the fitness function were decomposable this method can reasonably be considered. This method will be called *globally optimal forma completion*.

A more practical method in many circumstances is to find a local optimum within the forma. With *locally optimal forma completion*, this is achieved by completing the forma at random and then testing minimal mutations that remain within it in sequence, accepting any that are better. This process continues until there is no minimal mutation within the forma that is better than the current solution.

3.6 Allelic Hill-Climbing

In section 2.1, a hill-climber (with respect to a move operator Q) was defined to be any operator $\mathcal{H}$ having the functional form given in equation 6:

$$\mathcal{H} : \mathcal{C} \times \mathcal{K}_{\mathcal{H}} \longrightarrow \mathcal{C}_{Q}. \tag{6 bis}$$

It is easy to construct a hill-climber from Q by repeatedly applying Q to the chromosome to be optimised, cycling through all the control parameters from $\mathcal{K}_{Q}$. There is considerable freedom in exactly how such a hill-climber operates. In particular, there are many ways to decide when to accept a move and in which order to cycle through the control parameters. The hill-climber constructed here will be *greedy*, that is, it will accept any improvement generated by the operator immediately (and will never backtrack), as opposed, for example, to testing all control parameters and then accepting the move that generates the biggest improvement.

The order in which to test the control parameters in $\mathcal{K}_{Q}$ is more open. Any order will suffice provided that all parameters are tested (preferably without repetition) but a fixed order will afford the operator considerably less freedom than a random order. Testing the parameters in a totally random order (excluding only repetition) is by far the most appealing theoretically, and will almost certainly show the best performance because it minimises the correlations between applications of $\mathcal{H}$. In practice, however, this requires the generation of a very large number of (pseudo-) random numbers, and maintenance of a list of the moves that have been tested (or of those that remain to be tried). Nevertheless, this form of hill-climbing is sufficiently important to be named, and will be referred to as *ideal greedy hill-climbing*.

Many compromises between a totally random and a fixed order of sampling $\mathcal{K}_{Q}$ across applications of Q are possible. Two in particular will be considered. The first is to construct a random permutation of the control parameters to Q when the hill-climber is invoked, and to sample these in sequence. When an application of Q yields an improvement (and is therefore accepted), a new starting point within the permuted values is chosen, but the parameters are then sampled in the same order. This scheme will be called *rotated cyclic greedy hill-climbing*. A minor augmentation of this scheme involves also exchanging a randomly chosen pair of control parameters in the permutation when a move is accepted. This scheme will be called *rotated transposed cyclic greedy hill-climbing*.

For the purposes of the formal memetic algorithm that is the subject of this paper, the move operator defining local optimality will be minimal mutation ($\hat{\mathcal{M}}$).

4 Empirical Setting: Application to the TSP

Earlier sections having constructed the formal components required for a memetic algorithm, the present section seeks to construct an instantiation for comparison with its genetic counterpart and for simple experimentation with its coarser parameters. The problem tackled will be the travelling sales-rep problem (TSP) because this has a decomposable fitness function (and should therefore be favourable to memetic search), is well-known, and is relatively hard for evolutionary techniques. Most studies of large TSP instances have previously concluded that augmentation with local search is essential, (e.g. Verhoeven *et al.*, 1992; Gorges-Schleuter, 1989) and the aim here is not to achieve good performance as such but rather to understand how well an unembellished implementation of a formal memetic algorithm can work in this problem domain. For this reason, a modest but non-trivial TSP instance is used for these experiments.

Previous studies by Whitley *et al.* (1989) and Radcliffe (1994b) have provided evidence, both theoretical and empirical, that undirected edges are a relatively suitable basis for a representation of this problem, so the undirected edge representation discussed above will be used. This provides a small difficulty, however, in that the undirected edge representation is allelic, but it would be interesting to apply the GNX operator to this problem. This difficulty can be overcome as follows. In the context of alignment for crossover (only), the alleles (edges) making up a chromosome X will be arranged in the order they are visited in the tour, followed in a consistent direction, starting from city 1, to form a corresponding genetic chromosome x. This guarantees that every city occurs exactly once at the "start" of an edge, and gives appropriate linkage to adjacent edges. When brought together for crossover, the alleles in the second parent are then re-ordered to align with those in the first parent, and this allows GNX to proceed sensibly (figure 5). This borrows from the original view provided in Holland (1975) of crossover as a locus-independent operator in the context of re-linking operators such as inversion.

It may seem as if the effect of this procedure is to manipulate the edges as directed, but it is important to appreciate that this is not the case, for when an edge is tested for compatibility with those in the proto-child, no account is taken of its direction. Thus the direction of the tour affects only the order in which alleles from the parents are tested for inclusion in the child, not the direction in which they occur in the child.

5 Results and Discussion

Empirical studies were undertaken using the Reproductive Plan Language RPL2 (Surry & Radcliffe, 1994) running on super-scalar SPARC processors. The problem instance used was the 100-city Krolak 'C' problem from TSPLIB (Reinelt, 1990). All experiments used a panmictic population of size 100 with elitism, non-generational ("one-at-a-time") update, binary tournament selection with parameter 0.7, binary tournament replacement with parameter 0.7, recombination with probability 1.0 and mutation probability p_m of 0.02. The weight used for RAR was 2.0, and the number of cross points used

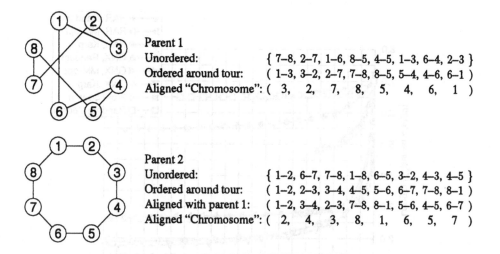

Parent 1
Unordered: { 7–8, 2–7, 1–6, 8–5, 4–5, 1–3, 6–4, 2–3 }
Ordered around tour: (1–3, 3–2, 2–7, 7–8, 8–5, 5–4, 4–6, 6–1)
Aligned "Chromosome": (3,   2,   7,   8,   5,   4,   6,   1)

Parent 2
Unordered: { 1–2, 6–7, 7–8, 1–8, 6–5, 3–2, 4–3, 4–5 }
Ordered around tour: (1–2, 2–3, 3–4, 4–5, 5–6, 6–7, 7–8, 8–1)
Aligned with parent 1:   (1–2, 3–4, 2–3, 7–8, 8–1, 5–6, 4–5, 6–7)
Aligned "Chromosome": (2,   4,   3,   8,   1,   6,   5,   7)

Fig. 5. In order to use GNX, which requires a *genetic* representation, with the undirected edge representation, which is *allelic*, a special form of alignment must be performed, creating a "pseudo-genetic representation". Each parent is re-ordered so the the order and sense of the edges are taken by following the tour around. The second parent is then further re-ordered to align it with the first. "Pseudo-genomes" that GNX can then manipulate are then created. Notice, however, that the sense of an edge may be reversed by GNX when it inserts it in the child tour.

for GNX was 2. Experiments were conducted using random and $\hat{\mathcal{M}}$-opt (i.e. minimal-mutation-based) patching, and also using Karp's heuristic (Lawler, 1985), which is specific to the TSP, for comparison. In the memetic experiments, rotated cyclic greedy hill-climbing was applied both to the initial population and before children were inserted into the population. Comparisons with random search and with repeated generation of 2-opt solutions are also shown. The results are shown in figures 6–8, and are on the basis of "wall-clock" time. In all cases, the length of the best tour in the population is plotted, normalised by the length of the optimum tour. Different algorithms are run for different numbers of updates to give broadly comparable total run times. It should, however, be noted that the implementations of the operators used are not tuned, and the implementation of RPL2 itself is still under beta test at the time of writing, so times should be taken as indicative rather than definitive.

As expected, given the decomposable nature of the evaluation function and the large number of possible alleles for the TSP, the memetic algorithms all significantly out-performed their genetic counterparts. Note particularly that with the exception of the algorithms using Karp stitching, all implementations are direct instantiations of the formal, representation-independent algorithms discussed above. The choice of patching algorithm has a large effect on the genetic algorithms, with the $\hat{\mathcal{M}}$-based and Karp stitching providing substantially higher performance, whereas the choice of recombin-ation operator has little effect. Conversely, for memetic search the choice of patching

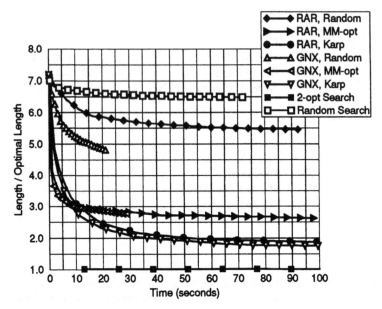

Fig. 6. The performance of genetic algorithms on the 100 city Krolak C TSP is shown. The length of the best solution in the population (relative to the length of the optimal tour) is shown as a function of wall-clock time. Error bars are not shown as they are smaller than the tick marks. For comparison, random search (the top line) is shown, as is the performance achieved by repeatedly generating 2-opt solutions (bottom line). Notice that the genetic algorithm is not remotely competitive with random search over 2-opt solutions. Ticks are placed every 5 generations (500 updates) except for the non-adaptive searches. The ticks for random search over 2-opt solutions occur every 100 updates and those for pure random search every 2,500 updates.

algorithm has relatively little influence over performance, but here G2X is significantly superior to RAR$_2$. Notice also that genetic algorithms fail by a large margin to match the performance achieved by repeatedly generating 2-opt solutions.

The patching results can be understood since in the memetic case the local search will be able to fix any poor patches, whereas this ability is not present in the genetic algorithm. The results for the two recombination operators seem to indicate that for this problem (in the context of the particular reproductive plans chosen) G2X is genuinely superior to RAR$_2$. The only cases in which this superiority is not exhibited are the genetic runs with good patching, but here the results suggest that it is the patching that is performing almost all the useful search, masking any distinction between the recombination operators' performances.

Clearly the problem instance chosen is rather easy for memetic search. (The other Krolak 100 city problems have also been tested, with very similar results.) To emphasize this point further, when a 3-opt-based hill-climber was tested, an initial population of 50 solutions was found to contain two copies of the optimum. The problem was, nevertheless, appropriate for this study given the level of difficulty it provided for genetic search. Extensive efforts will now be made to tackle larger problems with memetic search, using structured population models, parallelism and still more sophisticated operators; indeed, this work has already commenced.

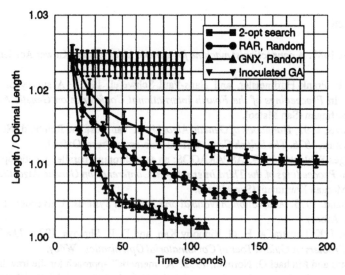

Fig. 7. The performance of memetic algorithms with random patching on the 100 city Krolak C TSP is shown. The top line shows the performance of the best genetic algorithm (i.e. using G2X and Karp stitching) inoculated with a starting population of randomly generated 2-opt solutions. The second highest line is the same as the bottom line of the previous graph, i.e. shows the performance of random search over 2-opt solutions, but notice the massively expanded scale on the y-axis. The bottom two lines show that G2X significantly out-performs RAR_2 on this problem. Tick marks are shown every generation (100 updates) except in the case of the inoculated genetic algorithm, where they are every five generations (500 updates), and error bars indicate standard errors.

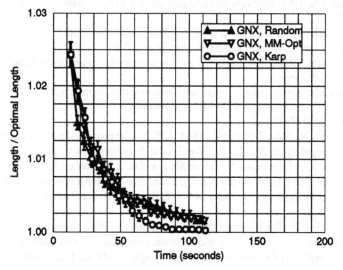

Fig. 8. The performance variation of memetic algorithms using G2X as a function of patching method is shown for the 100 city Krolak C TSP. Notice that the effect of the patching is rather small, in contrast to the large effect it has on the genetic algorithm (figure 6), though Karp stitching still performs best. A similar pattern emerges if RAR is used (not shown), but the performance for each patching method is worse than with G2X. Ticks are shown every generation (100 updates) and error bars indicate standard errors.

References

D. H. Ackley, 1987. *A connectionist machine for genetic hillclimbing.* Kluwer Academic Press, Boston.

Thomas Bäck, Frank Hoffmeister, and Hans-Paul Schwefel, 1991. A survey of evolution strategies. In *Proceedings of the Fourth International Conference on Genetic Algorithms.* Morgan Kaufmann (San Mateo).

Lawrence Davis, 1991. *Handbook of Genetic Algorithms.* Van Nostrand Reinhold (New York).

Richard Dawkins, 1976. *The Selfish Gene.* Oxford University Press (Oxford).

Martina Gorges-Schleuter, 1989. ASPARAGOS: an asynchronous parallel genetic optimization strategy. In *Proceedings of the Third International Conference on Genetic Algorithms*, pages 422–427. Morgan Kaufmann (San Mateo).

John H. Holland, 1975. *Adaptation in Natural and Artificial Systems.* University of Michigan Press (Ann Arbor).

E. L. Lawler, J. K. Lenstra, A. H. G. Rinnooy Kan, and D. B. Shmoys, 1985. *The Travelling Salesman Problem: A Guided Tour of Combinatorial Optimisation.* Wiley.

Pablo Moscato and Michael G. Norman, 1992. A "memetic" approach for the travelling salesman problem — implementation of a computational ecology for combinatorial optimisation on message-passing systems. In *Proceedings of the International Conference on Parallel Computing and Transputer Applications.* IOS Press (Amsterdam).

H. Mühlenbein, 1989. Parallel genetic algorithms, population genetics and combinatorial optimization. In *Proceedings of the Third International Conference on Genetic Algorithms*, pages 416–421. Morgan Kaufmann (San Mateo).

Heinz Mühlenbein, 1992. How genetic algorithms really work. part I: Mutation and hillclimbing. In R. Männer and B. Manderick, editors, *Parallel Problem Solving from Nature, 2.* Elsevier Science Publishers/North Holland (Amsterdam).

Nicholas J. Radcliffe, 1991. Equivalence class analysis of genetic algorithms. *Complex Systems*, 5(2):183–205.

Nicholas J. Radcliffe, 1992. Genetic set recombination. In Darrell Whitley, editor, *Foundations of Genetic Algorithms 2.* Morgan Kaufmann (San Mateo, CA).

Nicholas J. Radcliffe, 1994a. The algebra of genetic algorithms. *To appear in Annals of Maths and Artificial Intelligence.*

Nicholas J. Radcliffe, 1994b. Fitness variance of formae and performance prediction. Technical report, To appear in Foundations of Genetic Algorithms 3.

Gerhard Reinelt, 1990. TSPLIB.

Patrick D. Surry and Nicholas J. Radcliffe, 1994. RPL2: A language and parallel framework for evolutionary computing. In *Parallel Problem Solving from Nature III (to appear).*

M. G. A. Verhoeven, E. H. L. Aarts, E. van de Sluis, and R. J. M. Vaessens, 1992. Parallel local search and the travelling salesman problem. In R. Männer and B. Manderick, editors, *Parallel Problem Solving From Nature, 2*, pages 543–552. Elsevier Science Publishers/North Holland (Amsterdam).

Darrell Whitley, Timothy Starkweather, and D'Ann Fuquay, 1989. Scheduling problems and traveling salesmen: The genetic edge recombination operator. In *Proceedings of the Third International Conference on Genetic Algorithms.* Morgan Kaufmann (San Mateo).

A Statistical Mechanical Formulation of the Dynamics of Genetic Algorithms

Jonathan Shapiro[1], Adam Prügel-Bennett[1], and Magnus Rattray[1]

Department of Computer Science, Manchester University, Oxford Road, Manchester M13 9PL

Abstract. A new mathematical description of the dynamics of a simple genetic algorithm is presented. This formulation is based on ideas from statistical physics. Rather than trying to predict what happens to each individual member of the population, methods of statistical mechanics are used to describe the evolution of statistical properties of the population. We present equations which predict these properties at one generation in terms of those at the previous generation. The effect of the selection operator is shown to depend only on the distribution of fitnesses within the population, and is otherwise problem independent. We predict an optimal form of selection scaling and compare it with linear scaling. Crossover and mutation are problem-dependent, and are discussed in terms of a test problem – the search for the low energy states of a random spin chain. The theory is shown to be in good agreement with simulations.

1 Introduction

Although genetic algorithms have been used to good effect on a variety of problems, it is still not well-understood in what situations they are effective and why. For many problems, when a good representation is chosen, and if the parameters of the algorithm (e.g. population size, mutation and crossover rates, etc.) are set to appropriate values for the particular problem, the genetic algorithm can be very effective at searching for good solutions. When poor choices are made, the method can perform like random search. Yet, we are far from an understanding which would allow us to make these choices in a systematic way.

An understanding of the dynamics of GAs would be useful in understanding how to produce effective GAs. Clearly, if there was a theory which predicted the improvement in fitness as a function of the parameters of the algorithm, it could be used to find optimal values of those parameters. More generally, a theory of the evolution of genetic algorithms could be useful in producing insights as to how the algorithms behave and when they are likely to be effective. Unfortunately, the dynamics of genetic algorithms is very difficult to analyse. This is in part because the change to any member of the population depends upon the rest of the population. Thus, the stochastic parts of the dynamics are correlated with the system; this makes the system much more complicated than the usual stochastic dynamical systems in which the noise is independent and uncorrelated.

There have been numerous attempts to describe the dynamics of GAs. The simplest is the Schema picture of Holland [1, 2] (extended by Radcliffe [3]). Although this may provide a qualitative picture of how GAs work, it does not predict the dynamics of a particular genetic algorithm in detail. Alternatively, there are detailed mathematical formulations of genetic algorithms which do give a detailed description. The two most developed are the Markov Chain formulation of Vose and collaborators (for example, see [4]), and the Walsh-function analysis introduced by Bethke [5] and much used by Goldberg and collaborators [6, 7, 8]. The former is a beautiful and exact theory of the evolution of a GA. It has been useful in revealing the types of dynamics which can occur (e.g. punctuated equilibria). However, it is of little practical utility, as in order to make predictions, knowledge of transition probabilities between all strings is required. These are never known in real applications. A similar type of objection can be made for Walsh-function analysis – the Walsh expansion for a fitness function will not be known in general, so it is not a formalism which can be used to learn about specific problems. In addition, Walsh-function analysis has been criticized by Grefenstette [9] as not correctly describing essential aspects of genetic algorithm dynamics.

In this paper, we present a new formalism for studying the dynamics of a simple genetic algorithm consisting of selection, mutation, and crossover. The formalism is based on ideas from statistical physics. The purpose of using statistical mechanics is to attempt to describe the dynamics of the GA in terms of macroscopic and measurable quantities, rather than attempting to follow the evolution of each individual string over time. It is our belief that this approach will lead to a description of the evolution of genetic algorithms which is predictive, can be applied to realistic, high-dimensional optimization problems, and can take into account realistic, finite-population size effects.

2 Statistical Mechanics Formulation of Genetic Algorithms

Statistical mechanics is a branch of physics which deals with systems consisting of a huge number of interacting entities. It provides methods for calculating the gross or average properties of these systems by treating the small-scaled motions (which are probably unknown in detail) as random. It is usually applied to the computation of thermodynamic properties of particles (e.g. molecules, electrons) interacting with a heat bath, but has also been applied recently to a host of other problems in coding theory, neural networks and optimization (for examples, see [10]). As an example, statistical mechanics was found to be useful in setting the parameters for simulated annealing [11]. Statistical Mechanics has been previously applied to the study of genetic dynamics (see for example [12]). However, in these previous studies there was no notion of *fitness*; these were studies of genetic variability and convergence through genetic drift.

There may seem to be no relationship between the motions of atoms in a gas and the dynamics of genetic search, so one might wonder what the relevance of statistical mechanics might be. Intuitively, there is a connection – both involve motion which is both directed and random. Atoms in a gas would relax into a state of lowest energy, except that thermal motion causes states of high entropy to be more likely. The result is a trade-off between energy minimization and entropy maximization; the relative importance of each is controlled by the temperature. Likewise, in genetic search, selection tends to move the population towards states of high fitness, while the randomizing operators mutation and crossover explore states of high entropy. However, it is complicated to make this connection precise, because in statistical physics one is interested in equilibrium, whereas in GA dynamics we are interested in the transient behaviour (and there may not be an equilibrium), and because in statistical physics one often assumes an infinite size system, whereas in GA dynamics finite-size effects are very important.

We make this intuitive relationship precise by assuming that we know a few macroscopic quantities about the population. The fine details which we do not know we treat via a maximum entropy approach. This is precisely what one does in statistical mechanics; rather than following the detailed motion of each of the particles around, one assumes that the collective movements of the particles in the system are such that entropy is maximized subject to the constraint that the measured macroscopic quantities (e.g. the average energy) are known. For example, consider the physics of a glass of water. One could describe the states of this system by attempting to solve the equations of motion for the 10^{23} molecules in the glass. Clearly, this would tell you everything if you could solve it, although you would be overwhelmed with information. The alternative approach is to use statistical mechanics. Here you assume knowledge of a few macroscopic quantities only. The detailed motion is treated by assuming that the states which dominate are those with maximal entropy.

The analogy in GA dynamics to solving for the detailed motion, we believe, is to solve the Markov chain formulation of the population-to-population dynamics. This tells you the probability of arriving at any population given the starting population. However, it requires that you posess a very fine-grained level of detail about the system, and provides very fine-grained answers. The alternative which we are promoting is to assume only knowledge of coarse-grained quantities – cumulants of the fitness distribution and the average correlation within the population – and assume that the dominant states maximise entropy. How precisely this is done is described below. It is worth noting here that this will tell the average behaviour of the GA, i.e. the result of running the GA many times. This will be most useful when the fluctuations between runs is small.

We consider the evolution of the distribution of fitnesses within a population [13]. The goal is to predict the distribution of fitnesses at one timestep in terms of this distribution at the previous timestep. For example, suppose a GA is being used to find the minima of an objective function E. A typical evolution is shown in figure 1. Here the distribution of objective functions $\rho_t(E)$ of each individual in the population is shown at different times t in the genetic search. Initially, the distribution is nearly Gaussian; during the evolution of the GA the average fitness of the population increases (decreasing E), the variation in the population decreases, and the distribution loses the Gaussian shape (although this is not apparent from the figure). Throughout this paper, we shall consider the task of the GA to minimize an objective function E, and we will refer to the distribution of E rather than the distribution of fitnesses (so the fitness is $-E$ plus some constant to insure non-negativity). We will refer to E as the "energy" in analogy with statistical mechanics.

To study the evolution of the energy distribution, we calculate the effect of selection, crossover, and mutation on an arbitrary distribution. The full evolution of the GA is easily calculated by iterating

$$\rho_t(E) \xrightarrow{selection} \rho_t^s(E) \xrightarrow{crossover} \rho_t^c(E) \xrightarrow{mutation} \rho_t^m(E) = \rho_{t+1}(E), \tag{1}$$

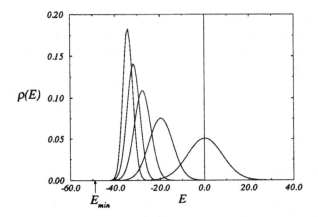

Fig. 1. A GA is used to find low energy states of a spin-glass. The distribution of energy in a population of 50 is shown after 0, 10, 20, 30 and 40 steps of the GA (the task is to *minimize* E, so earliest times are on the right; later generations are on the left). The search occurs in 63 dimensions. The curves show 1 000 samples averaged. Figure is from reference 12

starting from the initial population $\rho_0(E)$.

Of course, since the population size is finite, the actual distribution will consist of delta-functions. We will concentrate on the *statistical* properties of this distribution, the mean, variance, and higher moments averaged over samples. In particular, we consider a cumulant expansion for the distribution function. Let E^α denote the energy of a member of the population, $\alpha = 1, \ldots P$ where P is the size of the population. The cumulant expansion is defined by the generating function which is given by the log of the Fourier transform of the distribution function [14]

$$G(t) = \log(\langle e^{itE^\alpha} \rangle) \tag{2}$$

through the following equation for the nth cumulant κ_n

$$\kappa_n = (-i)^n \frac{\partial^n}{\partial t^n} G(t). \tag{3}$$

Here $< \cdots >$ denotes average over the distribution $\rho(E)$. (In statistical mechanics terminology, $G(t)$ is the log of the partition function, and the inverse temperature $\beta = -it$). The cumulants contain the same information as the distribution function. The first two cumulants are simply the mean and the variance. All cumulants except the first two are zero for a Gaussian distribution, so the higher cumulants measure the non-Gaussian nature of the function. The third cumulant is called the skewness – this measures the degree of asymmetry (about the mean) of the distribution. The fourth cumulant – called the excess – gives the first indication whether the function falls off more quickly or more slowly than a Gaussian.

Our description of the evolution of the GA will be in terms of the change in time of the cumulants due to the genetic operators. The cumulant expansion is convenient for a number of reasons. First, cumulants can readily be measured. Second, cumulants are a more stable measure than moments, because they tend to self-average. This means average values are the same as typical values, and the averages predicted from statistical mechanics will be a good approximation. Finally, the cumulants provide a interesting picture concerning how GAs work. For example, as we shall see, selection increases the skewness of the population. The low-fitness tail is long, whereas the high fitness tail is sharp. Thus, it is useful to study the effect of crossover and mutation on the skewness. The extend to which these operators decrease the skewness relative to the amount that they decrease the mean fitness is a measure of the effectiveness of these operators.

3 Selection

Selection is the mechanism which increases the mean fitness of the population by choosing more fit individuals with a higher probability than unfit ones. Many ways of performing selection have been proposed, the best-known being "roulette-wheel" selection. In this method, one chooses each individual with independent probabilities which are determined by the fitness. We use a Boltzmann weighting, in which the probability of selecting individual α is given by

$$p^\alpha = \frac{e^{-\beta E}}{Z}, \qquad Z = \sum_{\alpha=1}^{P} e^{-\beta E}. \tag{4}$$

Here β controls the amount of selection. For $\beta = 0$ each member of the population has the same weight, and the only change is due to genetic drift. When $\beta \rightarrow \infty$ only the most fit individual is selected. Small β corresponds to the usual proportional selection.

In order to compute the effect of selection on the cumulants, the generating function must be computed. This is given by

$$\langle \log(Z) \rangle_b = \left[\prod_{\alpha=1}^{P} \int_{-\infty}^{\infty} \rho(E_\alpha) \, dE_\alpha \right] \log(Z) \tag{5}$$

This is equivalent to a statistical mechanical model, the Random Energy Model, proposed by Derrida [15] as a model of spin-glasses, where fitness plays the role of energy. Using the methods developed for this model, the cumulants after selection can be computed from those before. Details will be given elsewhere [13, 16].

To leading order the first cumulant after selection becomes

$$\kappa_1^s = \kappa_1 - \beta \kappa_2 + \cdots. \tag{6}$$

Here we use the superscript s to denote the cumulants after selection, and we have expanded in powers of β. Note that the mean energy is shifted by an amount proportional to the selection parameter β times the variance, κ_2. The variance is changed by an amount

$$\kappa_2^s = (1 - 1/P)\kappa_2 - \beta \kappa_3 + \cdots. \tag{7}$$

In Fig. 2 we show the rate of convergence, κ_2^s/κ_2, versus the (scaled) selection parameter

$$\beta_s = \beta\sqrt{\kappa_2/2\log(P)},\qquad(8)$$

for populations of size 2^5, 2^{10} and 2^{20}, starting from a Gaussian distribution (i.e. $\kappa_n = 0$, $n > 2$). The curves have been calculated by numerical integration using Gaussian quadrature. The ordinate, to first order in β, is proportional to the shift in the mean energy of the distribution. We can see from Fig. 2 that even for $\beta = 0$ (arbitrary selection) there is an intrinsic convergence rate, which reduces the variance in the population by a factor $1 - 1/P$. This arises because, by chance, some members of the population will not be selected while other members will be selected more than once. This is the well-known phenomenon of 'genetic drift'.

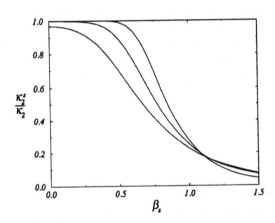

Fig. 2. The curves show the change in the variance of the energy distribution after selection starting from a Gaussian distribution for $P = 2^5$, 2^{10} and 2^{20} versus the scaled selection parameter $\beta_s = \beta(\kappa_2/2\log(P))^{1/2}$. These have been calculated by numerical integration.

The effects of selection on higher cumulants can also be computed [13]. Selection introduces a skewness into the distribution which slows down the shift in the average energy and increases the rate of convergence. The distribution is skewed in favor of the higher energy states, i.e. the tail of low-energy states is shorter that the higher-energy tail. Perhaps a role of crossover and mutation will be to replace some of the higher energy tail with lower than average states. Of course, these two operators will also increase the mean energy; this shows the trade-off which it will be important to optimize.

3.1 Optimal Selection Parameters

If we assume that the goal of selection is to provide the optimal improvement in the mean fitness of the population while having the least deleterious effect on the diversity, an optimal amount of selection can be chosen. As can be seen from figure 2, the curve is initially flat for small β. The improvement in the mean energy is proportional to β so it pays to increase β, since there is no extra loss of diversity in the flat region. However, passed an optimal value around the shoulder of the curves in the figure, the diversity in the population decreases very rapidly with selection and the GA will be unlikely to find a good solution. Thus we see there is an optimal choice of the selection parameter for Boltzmann selection and we can predict how it scales. Since the variance decreases as the population converges the degree of selection should be increased, as has already been observed [2]. At each generation, set β to be

$$\beta = \beta_o \sqrt{\frac{2 \ln P}{\kappa_2}} \tag{9}$$

where β_o is the location of the shoulder in the curves of figure 2 (approximately 0.2), and P is the population size. We can compare this with the usual linear scaling [2] where the fitnesses are rescaled according to

$$f \to c \frac{f \cdot f_{ave}}{f_{max} - f_{ave}} + b, \tag{10}$$

where c is taken to be around 0.2 and b is the shift (c and b are typically chosen so that the average fitness is 1 and the fitness of the best is the average times a constant factor). If the distribution of fitness was a Gaussian, linear scaling would reduce to

$$f \to c \frac{f \cdot f_{ave}}{\sqrt{\kappa_2 \ln P}} + b, \tag{11}$$

The similarities between the two equations are interesting. Both scale as $1/\sqrt{\kappa_2}$, the differences are in the scaling with population size and in constants. In addition, with Boltzmann scaling, the problem of negative fitnesses does not arise.

4 Crossover and Mutation

4.1 A Test Problem

Crossover and mutation are problem specific. In order to study these, we must consider specific problems. Here we report studies of a GA which searches for low lying states of a one dimensional spin chain with random nearest-neighbor couplings. The energy of a configuration of spins, $\mathbf{S} = (S_1, S_2, \ldots, S_N)$ is

$$E(\mathbf{S}) = -\sum_{i=1}^{N-1} J_i S_i S_{i+1}, \tag{12}$$

where each spin S_i takes the value $+1$ or -1 and the couplings J_i are drawn from a Gaussian distribution. The average ground state energy is $E_{min} = -\sum_i |J_i|$. Although this problem is trivial to solve it is nevertheless interesting in that is has an exponential number of local minima under single spin flip dynamics (typically $2^{N/3}$), and it is an example of search on a high-dimensional (N dimensions) space. Thus, it would be very difficult for a hill-climbing algorithm to solve. On the other hand, it is perhaps an easy type of problem for a GA, because the fitness function is very local. In the language of GAs, it would have zero epistatis, except for the interactions between neighbors. The thermodynamics of this model have been extensively studied [17, 18].

4.2 Mutation

Mutation is the operator which changes genes with a small random probability. Mutation changes the mean energy by a small amount proportional to κ_1/N. The $1/N$ is because mutation only changes a single spin, which can only affects two of the $N - 1$ neighbor pairs of the chain; the proportionality to the mean energy is because the more likely it is that a pair of neighbors is correctly aligned, the more likely it is that mutation will have a negative effect. Mutation increases the variance if it is less than the natural variance for the spin chain and decreases it if it is greater than that variance. The natural variance is that of a random population. This is obvious, because repeated application of mutation should produce a randomized population. The effects on higher cumulants are small.

In order to compute the effect of mutation on the distribution of energies, we consider the probability that the change in the energy of amount δE given that the string had energy E, $P(\delta E|E)$. To get the distribution of fitnesses after mutation, $P(\delta E|E)$ is averaged over the fitness distribution. The calculation of $P(\delta E|E)$ is fairly technical, so we refer the reader to the more detailed paper [16] for the calculation. We note here that $P(\delta E|E)$ is computed as in a microcanonical ensemble, by summing over all degrees of freedom which have the given energy E.

The results for the first two cumulants after mutation are given here.

$$\kappa_1^m = \kappa_1 \left[1 - 4p_m\left(1 - p_m\right)\right]. \tag{13}$$

Here the superscript m denotes the cumulant after mutation, and p_m is the mutation probability. Note that if p_m is 1 the average energy is unchanged, a consequence of the quadratic energy function (equation 12). Otherwise, mutation brings the average energy towards the random value. The second cumulant is changed to

$$\kappa_2^m = \kappa_2 \left[1 - 8p_m\left(1 - p_m\right)^2\right] + 8p_m\left(1 - p_m\right)^2 \sum_i J_i^2. \tag{14}$$

This result looks complicated, but it has a fixed point as $\kappa_2 = \sum_i J_i^2$ which is precisely the variance of a set of random strings. Thus, as expected mutation moves the cumulants towards a natural set of values. Higher cumulants have also been calculated and will be discussed elsewhere [16].

4.3 Crossover

Crosssover is a most important operator, because it is what distinguishes GAs from stochastic gradient search. It is also the most complicated to understand, because the effect depends upon the correlations between the strings within a population. This information is not contained in the distribution function. For example, two strings with very similar energies could be very similar or very different genetically. The energies produced by crossing these strings will obviously depend on how correlated they are. However, the degree of correlation between the strings will not be determined by their energies, but by their common ancestry. Thus, we must model the evolution of the correlation between the spins as well as the dynamics of the energy distribution.

Because of the correlations between the individuals in the population, the search is not over all states with the given cumulants. The search space is reduced by the correlation. Therefore, we cannot simply make maximal entropy assumptions, we must introduce additional assumptions about the nature of the correlations. We assume that some of the bits in the strings do not fluctuate within the population.

Like mutation, crossover is likely to affect the alignment of the neighbors at the crossover point. This will increase the energy in proportion to κ_1/N. The constant of proportionality is $(1 - q)$ where q is the average correlation (for example, if the two strings were identical, there would be no energy change). In an earlier paper [13], this was approximated as κ_2/N. Beyond the contribution of the interface energy change, crossover has no effect on the first two cumulants. It does have the effect of decreasing the higher cumulants, and bringing them towards the natural values of the random two energy distribution function.

The mathematical details of the crossover results are too technical for this paper. A simple theory of crossover was presented in [13]. We show here a comparison of the theory with simulation. Figure 3 shows a comparison of simulation (solid line) with the theory (dashed line) of a GA consisting of crossover and selection. The theory was derived by iterating the equations for the first six cumulants. The agreement with simulations is very good provided β and P are not too large. For larger values of these parameters the approximations used begin to break down; a more accurate treatment is being developed.

5 Future Work

In this paper, we have outline a formalism with which to study GA dynamics. We have not gone very far in using the methods to address questions of GA theory. Further simulations must be done to determine the range of applicability of these methods. In addition, a comprehensive study should be made to determine the applicability of the prediction concerning the optimal selection rate made in section 3.1.

It is not clear from the above work how to optimize the rates of mutation and crossover. It is clearly very important that this be done. The strong localness of the spin-chain is an untypical aspect of this test problem. This meant that mutation was clearly less effective than crossover. In more realistic problems, crossover could have a much larger effect on the mean fitness, and mutation a much smaller one.

The spin-chain was chosen as a first test problem because it is a trivial model to solve. It will be useful to extend this work to higher dimensional spin-glass, where mutation will have a much more important role to play in reducing the amount of convergence caused by selection.

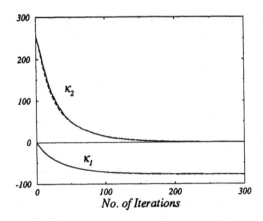

Fig. 3. Comparison of theory and simulations for selection and crossover using $\beta = 0.01$ with $P = 50$ and $N = 255$. The solid curves show the simulations averaged over 500 samples. The theory is shown by dashed curves, which are nearly obscured by the simulations. From reference [13].

References

1. Holland, J. H.: *Adaptation in Natural and Artificial Systems*, University of Michigan Press (Ann Arbor), 1975.
2. Goldberg, D. E.: *Genetic Algorithms in Search, Optimization and Machine Learning*, Addison-Wesley (Reading, Mass), 1989.
3. Radcliffe, N. J.: Equivalence Class Analysis of Genetic Algorithms, Complex Systems 5, 183 (1991).
4. Vose, M. D., Liepins, G. E.: Punctuated Equilibria in Genetic Search, Complex Systems 5, 31-44, 1991.
5. Bethke, A. D.: *Genetic Algorithms as Function Optimizers*, Doctoral Dissertation, University of Michigan, 1981.
6. Goldberg, D. E.: Genetic Algorithms and Walsh Functions: Part I, a Gentle Introduction, Complex Systems, 3, 129-152, 1989.
7. Goldberg, D. E.: Genetic Algorithms and Walsh Functions: Part II, Deception and its Analysis, Complex Systems, 3, 153-171, 1989.
8. Goldberg, D. E., Rudnick, M.: Genetic Algorithms and the Variance of Fitness, Complex Systems, 5, 265-178, 1991.
9. Grefenstette, J. J.: Deception Considered Harmful, in *Foundations of Genetic Algorithms 2*, Whitley, L. D. editor, Morgan Kaufmann Publishers, Inc. (San Mateo), 1993.
10. Mezard, M., Parisi, G., Virasoro, M. A.: *Spin Glass Theory and Beyond*, World Scientific (Singapore) 1987.
11. van Laarhoven, P. J. M., Aarts, E. H. L.: *Simulated Annealing: Theory and Applications*, Kluwer Academic Press (Dordrecht) 1987.
12. Tsallis, C.: Exactly Solvable Model for a Genetically Induced Geographical Distributions of a Population, Physica A 194, 502-518, 1993.
13. Prügel-Bennett, A., Shapiro, J. L.: An Analysis of Genetic Algorithms Using Statistic Mechanics, Phys. Rev. Lett., 72(9) p1305, 1994.

14. Abramowitz, M., Stegun, I. A.: *Handbook of Mathematical Functions*, Dover Press, 1964.
15. Derrida, D.: Random-energy Model: An exactly Solvable Model of Disordered Systems, Phys. Rev. B24, 2613 (1984).
16. Prügel-Bennett, A., Shapiro, J. L.: The Dynamics of Genetic Algorithms for the Ising Spin-Glass Chain, in preparation.
17. Li, T.: Structure of Metastable States in a Random Ising Chain, Phys. Rev. B **24**, 6579 (1981).
18. Chen, H. H., Ma, S. K.: Low-Temperature Behavior of a One-dimensional Random Ising Model, J. Stat. Phys. **29**, 717 (1982).

Evolutionary Stability in Simple Classifier Systems

D.F. Yates and A. Fairley[1]

The Bio-Computation Group
Dept of Computer Science
University of Liverpool
Liverpool, L69 3BX.

Abstract

In this paper, the relatively new branch of mathematics known as Evolutionary Game Theory is proposed as a potentially useful tool when seeking to resolve certain of the more global, unanswered questions related to classifier systems. In particular, it is proved that, under certain mild assumptions, the performance of a classifier systems plan will, if the Bucket Brigade Algorithm is adopted, conform to what is referred to as 'an evolutionary stable state'. A simple example is also provided to confirm the theoretical findings.

1.0 Introduction

To date, Classifier Systems have proved to be a most popular vehicle for the application of Genetic Algorithms (GAs) in the field of machine learning. However, despite such widespread popularity, the vast majority of research involving classifier systems has focussed either on specific applications (see, for example, [1] and [2]), or on means for improving the performance of certain of their constituent elements, see [3], [4] and [5]. Arguably the most notable exception to this is the work performed by Holland [6] which attempts the construction of a mathematical framework for learning in classifier systems. Notwithstanding, many of the more 'global' questions regarding classifier systems remain unanswered. One such question is: does a classifier system optimise the overall effectiveness of its plan, or the effectiveness of the individual rules which constitute the plan - at the possible expense of the plan itself? In this paper, Evolutionary Game Theory (which offers a means of formally analysing evolution in biological systems) is proposed as a potential facilitator in

---

[1]Current Address: Naval Studies Department, Defence Research Agency, Southwell, Portland, Dorset, DT5 2JS

respect of answering this and related questions. Specifically, in this paper it is proved that the behaviour of a simple classifier system might be explained in terms of evolutionary stability, and this is underpinned with the aid of a simple example.

2.0 Evolutionary Game Theory

The branch of mathematics known as Evolutionary Game Theory (EGT) [7] is a relatively recent development, the motivation for which was the need for a theory which would "model evolutionary processes in populations of interacting individuals and explain why certain states of a given population are - in the course of the selection process - stable against perturbations induced by mutations" [8]. Based upon pre-programmed behavioral policies called *strategies*, EGT differs from classical game theory in that the effectiveness of strategies are measured not in terms of *utility*, but *fitness*, and *natural selection*, rather than *rationality*, is used as a means for selecting strategy.

In EGT, a population consists of individuals possessing but a single inherited trait, a *pure strategy*, which is an in-built collection of behavioral characteristics selected from a finite set of such strategies. The environment in which the individuals interact takes the form of a set of contests, in each of which two randomly selected individuals from the population compete with one another, each playing its respective strategy. Each individual also has an associated 'fitness measure' which represents that individuals effectiveness in the environment.

Evolution in a population occurs as follows. After a contest between a pair of individuals has been concluded, the outcome results in the rewards (payoffs), P and P′, reflecting the success of the respective strategies adopted, being made to the two combatants. After a large number of such contests, an individuals average payoff corresponds to its utility (fitness), and therefore, the relative fitness of each individual can be calculated. Using natural selection to replicate the individuals in proportion to their fitness, the next generation of individuals is generated, and by repeating this sequence, the population evolves over time.

EGT provides a means of analysing evolving populations and can be used to determine whether there exists a state wherein the distribution of the various strategies amongst the population is in equilibrium. If such an equilibrium state does exist, the population is said to be *evolutionarily stable* and the collection of strategies is said to constitute an *Evolutionarily Stable Strategy* (ESS).

The concept of an ESS may be clarified by the following example, the *Hawk-Dove Game*, taken from Dawkins [9]. In this game, the objective of each contest between randomly selected pairs of individuals from a population, Σ, is to take control of territory. There are only two strategies that can be adopted by any one of the population of identical individuals. These two strategies are called *hawk* and *dove*. In a contest, any individual adopting the hawk strategy will, at the risk of injury, attack its opponent in order to gain territory. However, an individual adopting the dove strategy will not risk a confrontation, and so, in the face of an aggressive opponent (that is, one playing a hawk), will retreat and remain unscathed. If two hawks meet, they will fight and one will emerge the winner, whilst the other will sustain injury. If two doves meet, each will evade the other until ultimately, one wins by, almost accidentally, being in control of the territory.

After a confrontation, both contestants are given a 'reward' consistent with their performance. There is a reward of 50 points for gaining territory, 0 for ceding ground immediately but remaining uninjured, -10 for being involved in a protracted confrontation and -100 for losing a fight.

Given such a scenario, what strategy should an individual adopt? The answer to this question depends upon the strategy adopted by the rest of the population. For example, given a population consisting entirely of doves, the average payoff for each member is 15 points (since in each contest, one dove will gain 50 points for winning the territory but lose 10 for being involved in a protracted confrontation, while the other will lose 10 overall, and thus the average pay-off is (40+(-10))/2). Now, if just one of the population were to adopt the hawk strategy, it would be awarded 50 points after every confrontation, thereby increasing that individual's reward above the mean for the population. If then other members of the population learn from this example, the hawk strategy would spread through the population. However, as the number of hawks grows, their average fitness would decrease since, on some occasions, hawks would meet each other, and one of them would lose 100 points.

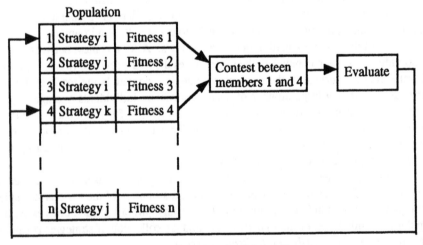

Update fitness of 1 and 4

Figure 1: Main Cycle of an evolutionary game

In order to find the ratio of hawks to doves in this population at which stability is reached, let x be the probability that a member of the population plays the hawk strategy and $(1-x)$ the probability that he plays a dove. At equilibrium, the average pay-off for playing hawk, P_{hawk}, equals the average pay-off for playing dove, P_{dove}, and

$$P_{hawk} = (x(-25) + (1-x)50)/2, \text{ and } P_{dove} = (x(0) + (1-x)15)/2.$$

Therefore, at equilibrium, $(50 - 75x) = (15 - 15x) \Rightarrow x=7/12$, and so stability is only reached when the population is divided in a ratio of 7/12 hawks to 5/12 doves. At this ratio, any mutant individual, that is one which changes its strategy from hawk to dove or vice versa, will receive a lower average pay-off than the population mean and so the population is evolutionarily stable and forms an ESS.

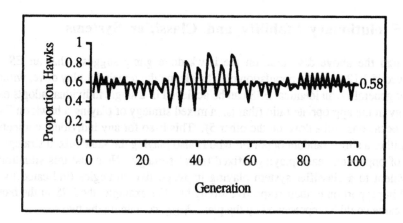

Figure 2: Results when simulating the Hawk-Dove Game

To investigate this result, a 'simple' experiment was performed. In the experiment, a population of 500 individuals was allowed to evolve over 100 generations. The initial population was divided equally into hawks and doves, and in every generation, each individual competed against all others five times (giving a total of over six hundred thousand contests per generation). Natural selection was mimicked using roulette wheel selection, and the 500 individuals selected were used to form the next generation. Figure 2 depicts the number of hawks in each of the 100 generations of the single simulation run. The results clearly demonstrate that the ratio of hawks to doves oscillates about the predicted theoretical equilibrium value, the oscillations being due, in the main, to the sampling errors in respect of the (small) population size (that is, genetic drift).

It is important to note that the above does not imply that an ESS is optimal in terms of the expected reward for the population. For example, at the ESS, the average reward for each of the population's members is 6.25 which is less than the 15 point reward for each member of a population consisting entirely of doves. Instead, the ESS is more like a pareto optimum where the success of one individual depends upon that of other individuals.

To be capable of evolving into such a stable state, a system must have some mechanism by which it can adapt its response to become more profitable. Such a mechanism in EGT is called a *Learning Rule*. A learning rule is a meta-rule which specifies the probabilities with which an individual will adopt one of a set of strategies as a function of the strategies it has adopted in the past and the results of adopting those strategies. An important question relating to learning rules is: is there any learning rule that will always cause an initially naive plan to adapt, in time, towards an ESS for the game which it is playing? Harley [10] proved that such a *rule for ESSs* is itself evolutionarily stable since a system adopting this could not be invaded by a mutant utilising an alternative learning rule. He called such a rule an *Evolutionarily Stable learning rule* or ES learning rule.

3.0 Evolutionary Stability and Classifier Systems

Although the above discourse on the hawk-dove game suggests that an ESS is achieved if 7/12ths of the population play hawk and the remainder play dove, without loss of generality, an identical ESS can be achieved if a single individual adopts each strategy in the appropriate ratio (that is, a mixed strategy of playing a hawk on 7 out of 12 occasions and a dove on the other 5). This is so for any individual competing with either a single random opponent who is also playing the same mixed strategy, or a set of opponents, each playing a fixed (pure) strategy. Note that this situation is equivalent to a classifier system playing its respective strategies (indicated by its rules) in proportion to their respective strengths. For example, the ESS in the hawk-dove game could be represented by the plan, P_{HDG}, containing the two rules:

P_{HDG}:   Rule 1: IF whatever THEN play_hawk   *Strength*=0.5833
            Rule 2: IF whatever THEN play_dove   *Strength*=0.4177

By representing strategies in this form, EGT can be applied to classifier systems to determine whether the strength of the rules (classifiers) approach evolutionary stability. Indeed, as Maynard Smith [7] points out, the idea of a strategy is the same as that of a behavioral phenotype and as such, "*can be applied equally well to any kind of phenotype; for example, a strategy could be the growth form of a plant, or the age at first reproduction, or the relative numbers of sons and daughters produced by a parent*".

However, for a classifier system's plan to change, there must be some mechanism by which the strengths of its rules can be adjusted - that is, there must be a learning rule. Moreover, if the system is to reach an ESS, then this rule must be an ES learning rule. In a classifier system, the mechanism by which rule strength is adjusted in order to adapt the system's response, is the Bucket Brigade Algorithm (BBA) and thus, if the system's rule set is to move towards evolutionary stability, the BBA must be an ES learning rule.

4.0 The Bucket Brigade Algorithm as an ES Learning Rule

As Harley [10] points out, "*it is unlikely that any animal has a learning rule which provides it with an unbeatable strategy for every game that it might encounter. Even an extremely simple frequency independent game such as the two-armed bandit requires a dynamic programming algorithm and a very large memory to solve for the best sequence of behaviours.*" Instead it is thought that some animals have a simple learning rule which allows them to do well in a wide range of situations. To determine whether such a learning rule is in fact an ES learning rule, Harley introduced five properties that it should possess. These properties are based on several assumptions about both the system and the game, of which the most important are:

(a)   each game considered actually possesses an ESS, which in principle is capable of being learnt;

(b)   although the reward (pay-off) may change in time, it does so sufficiently slowly that learning rules can establish stable frequencies of behaviours;

(c)   the pay-off, $P_i(t)$, that a behaviour B_i receives at time t is non-negative and evaluated in units of fitness. The pay-off at time t for any behaviour B_j not used at time t is zero;

(d)   the learning rule defines the probability of displaying each of the possible behaviours at each time step as a function of the previous pay-offs.

Given these assumptions, the five properties of an ES learning rule as specified by Harley are:

(1)   it must have the property that, after a long series of plays, the probability of selecting action A is equal to the total payoff for playing A divided by the total payoff so far. Harley called this the *relative payoff sum* or RPS property;

(2)   the ES learning rule will never completely abandon any action, K, nor will it ever fix the behaviour so that it always performs K. This is necessary because circumstances (and hence pay-off) may change;

(3)   the ES learning rule will attach greater significance to more recent pay-off information than less recent information. Again, this is necessary because circumstances may change;

(4)   the ES learning rule will be a function of prior expectation of pay-off;

(5)   the ES learning rule is also a function of actual pay-off.

By investigating the BBA in respect of these five properties, it is possible to determine whether the BBA can indeed be classed as an ES learning rule.

Property 1:

For the purpose of this proof, consider a classifier system, M, which employs the simplest form of BBA, namely:

$$S_i(t+1) = S_i(t) + P_i(t) + R_i(t) - B_i(t)$$

where $S_i(t)$, $P_i(t)$, $R_i(t)$ and $B_i(t)$ denote the strength of, pay-off to, payment received from other classifiers, and the bid made by rule i at time t respectively.

Given a sufficiently long time, the strength of a rule, X_n, which is used at the end of a sequence of actions, will be proportional to its payoff, P. As this rule pays a proportion of its strength to its immediate predecessor, X_{n-1}, in the sequence, then the strength of X_{n-1} will also be proportional to P. Repeating this argument for $i=(n-2),(n-3)...$, it can be seen that the strength of the first active rule in the chain, X_1, will also be proportional to P. In M the probability of selecting a rule is its relative strength divided by the sum of the strengths of all other candidate rules (that is, its relative payoff) and hence, the Relative Payoff Sum property is satisfied.

Property 2:

Assuming that pay-off is always non-negative, then a classifiers strength only decreases through taxation and bidding. As both of these quantities are only small percentages of a classifiers strength, then strength can only asymptotically approach zero, and so its probability will never be exactly zero. Correspondingly, as no classifier has zero strength, then no classifier can have a probability of 1.0 of being employed and hence fixation also can never occur.

Property 3:

The taxation element of the BBA causes a rule's strength to decay exponentially over time, and as such, causes greater significance to be placed on more recent pay-off information.

Property 4:

All classifiers are each assigned an initial strength for use with the BBA and therefore, any subsequent selection is a function of these initial values.

Property 5:

The fact that the actual pay-off a classifier receives is one of the parameters in function shows that the BBA is indeed a function of actual payoffs.

Thus, the BBA satisfies Harley's five properties for an ES learning rule. Of course, the BBA is not the only ES learning rule. Much more sophisticated and efficient rules are likely to exist for many problems. However, as the effectiveness of an ES learning rule is improved, its generality is reduced and its usefulness, in a general learning environment wherein a variety of different problems are encountered, would be limited.

5.0 HDCS: A Classifier System for the Hawk-Dove Game

If the BBA is, for a specific application, an example of an ES learning rule, then it should be possible to discuss a plan developed by a classifier system in terms of evolutionary stability. To determine whether a classifier system using the BBA does indeed exhibit such a characteristic, a simple classifier system for the hawk-dove game, HDCS (Figure 3), was developed.

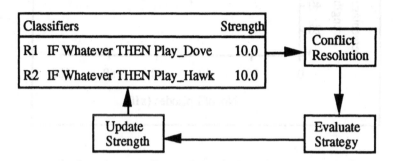

Figure 3: Outline of HDCS, a classifier system for the hawk-dove game

As a rule base, HDCS was invested with the plan P_{HDG} described earlier, and both rules were assigned an initial strength of 10.0. Since HDCS's environment would always be the same, and only a single action would ever be performed on any problem solving cycle, all the features which made use of internal and environmental messages, such as matching and message passing were superfluous and therefore omitted. This situation obtained also for the system's genetic algorithm.

HDCS was applied to 1000 instances of the hawk-dove game. At the start of each game instance, one of P_{HDG}'s two rules was selected stochastically in accordance with their respective strengths. When selected, a rule, R say, paid 10% of its strength (equivalent to its bid) to the environment. The action corresponding to R was then adopted in contests with 50 opponents whose respective strategies were selected in proportion to the strengths of the two strategies in the extant plan. For example, if the extant relative strength of rule 1 (play a hawk) were 0.6, then the system would adopt this rule on approximately 60% of occasions, and in the 50 consequent contests, the opponent would play a hawk in approximately 30 of them and dove in the remainder. The value of the pay-offs awarded at the end of each game were those detailed in section 2.0 but with each value being supplemented by 100 points in order to ensure that all pay-offs were non-negative (as demanded by Harley, see section 4.0). (It should be noted that only the pay-off to the 'system player' needs to be taken into account.) Games wherein both protagonists played either a hawk or a dove, were decided randomly.

The relative strength of rule 1 of P_{HDG} (that is, play hawk) was recorded at the conclusion of every tenth game instance. Figure 4 is a plot of the values derived against number of game instances. It can be seen from the plot that the strength of the two rules oscillates about the ESS equilibrium point. Correspondingly, given the finiteness of the population size and the number of game instances, it is not unreasonable to conclude that HDCS is playing the ESS for the problem.

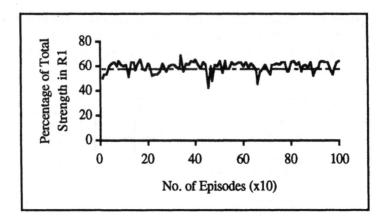

Figure 4: Results of Hawk-Dove Game using HDCS

6.0 Conclusions

In this paper, Evolutionary Game Theory has been proposed as a means for answering certain of the questions concerning the performance of plans invested in classifier systems. It has been proved that, provided the assumptions detailed in section 4.0 obtain, the Bucket Brigade Algorithm constitutes an ES learning rule, and that the plan of a classifier system which utilises the BBA will evolve in such a way as to constitute an ESS for the problem to which it is applied. It should be noted however, that this result is valid only for a 'static' plan, that is, one which is not changed as the result of applying a genetic algorithm. For, by applying a GA in a classifier system's regime, the evolutionary stability built up in the system's extant plan will likely be disrupted by the change in constitution of the plan that will almost certainly be the result of applying the genetic algorithm. Notwithstanding the characteristics of a plan, in respect of evolutionary stability, both in between consecutive applications of a GA, and as ultimately developed will, if the assumptions of 4.0 are valid, tend to conform to that of an ESS.

Acknowledgment

This research was performed under the CASE studentship scheme (SERC Award Ref. No. 9030996X) in co-operation with the Defence Research Agency, Portland.

References

[1]   Goldberg, D.E. *Computer Aider Gas Pipeline Operation Using Genetic Algorithms and Rule Learning*, Doctoral Dissertation, University of Michigan, Ann Arbor, 1983.

[2]   Wilson, S.W. *Classifier Systems and the Animat Problem*, Machine Learning, 2(3), 1987.

[3]   Yates, D.F. and Fairley, A. *An Investigation Into Possible Causes of, and Solutions to, Rule Strength Distortion Due to the Bucket Brigade Algorithm*, Proc. 5th Int. Conf. on Genetic Algorithms, Morgan Kaufmann, 1993.

[4]   Riolo, R.L. *The Emergence of Coupled Sequences of Classifiers*, Proc. 3rd Int. Conf. on Genetic Algorithms, Morgan Kaufmann, 1989.

[5]   Booker, L.B. *Triggered Rule Discovery in Classifier Systems*, Proc. 3rd Int. Conf. on Genetic Algorithms, Morgan Kaufmann, 1989.

[6]   Holland, J.H. *A Mathematical Framework for Studying Learning in Classifier Systems*, Physica 22D, 1986.

[7]   Maynard Smith, J. *Evolutionary Game Theory*, Physica 22D, 1986.

[8]   Bomze, I.M. and Potscher, B.M. *Game Theoretical Foundations of Evolutionary Stability*, in Lecture Notes in Economics and Mathematical Systems, Springer-Verlag, 1989.

[9]   Dawkins, R. *The Selfish Gene*, Oxford University Press, 1976.

[10]   Harley, C.B. *Learning the Evolutionarily Stable Strategy*, Journal of Theoretical Biology, 89, 1981.

Nonbinary Transforms
for Genetic Algorithm Problems

Paul Field
Department of Computer Science
Queen Mary and Westfield College
Mile End Road, London E1 4NS
email: paulf@dcs.qmw.ac.uk

Abstract

GA theory tends to be biased towards binary representations of problems. To try to redress the balance somewhat, this paper takes two important, related ideas, Walsh and partition coefficients, and generalises them to the nonbinary case. These coefficients provide an efficient way of calculating the fitnesses of low order schemata and of writing the conditions that characterise deceptive functions. Functions can be analysed for deception or created with varying degrees of deception by transforming between string fitnesses and coefficients or vice versa. In this paper, the matrix forms of the transforms are presented, the relationship between them examined and an efficient algorithm for performing the transforms is presented. Finally, an example of the coefficients' use is given.

1. Introduction

Problem representation is an important issue in genetic algorithm research because the way in which a problem is represented can affect the genetic algorithm's performance. At the heart of representation is the choice of an alphabet from which the strings that the GA manipulates are built. From the perspective of GAs as "schema processors", binary strings seem best because they provide the greatest number of schemata. But this is simplistic. We need, at least, to ask whether the extra schemata provided by a low cardinality representation are in some sense useful. Although it is sometimes used to support the binary case, Goldberg's (1991) principle of minimal alphabets states "the user should select the smallest alphabet that permits a natural representation of the problem". We should consider the implications of forcing a problem to be binary if this is "unnatural"; we could be doing damage that outweighs the supposed benefits of extra schemata. Unfortunately, the majority of GA theory centres on binary representations and provides no help with these important questions.

Perhaps the most important tool for examining the difficulty of binary represented problems for a GA is the Walsh transform. By looking at a problem through the eyes of Walsh coefficients rather than individual fitnesses we can get a feel for the significance of particular alleles and combinations of alleles to a problem. In particular, Walsh coefficients simplify the computation of low order schemata fitnesses. Since these schemata survive genetic operators better than high order ones, they and their fitnesses are important. It should be possible to use Walsh coefficients and their close relatives partition coefficients to build stronger arguments in the alphabet debate. However, until recently, such coefficients had only been computed for binary problem representations. Mason (1991) began to address this by generalising partition coefficients to the nonbinary case. This paper extends his work by presenting matrices to transform between partition coefficients and fitnesses. In addition, a generalisation of Walsh coefficients, called *incremental combination coefficients*, is examined. The efficient implementation (in terms of both time and space) of these transforms is also discussed. Finally, nonbinary coefficients are used to write the conditions that characterise a fully deceptive nonbinary function for a particular representation. However, to begin with we will discuss the intuitive meaning of the coefficients and how they are generalised to the nonbinary case. For an introduction to Walsh coefficients and their uses see Goldberg (1989). For a simpler and mainly non-mathematical description of the Walsh and partition transforms see Field (1994).

2. Generalising Binary Coefficients

Partition coefficients allow us to compute a string or schema's fitness from the contributions of its alleles. For example:

$$f(11) = \varepsilon_{\#\#} + \varepsilon_{1\#} + \varepsilon_{\#1} + \varepsilon_{11}$$

$\varepsilon_{\#\#}$ is the average fitness of all strings. $\varepsilon_{1\#}$ is the extra fitness that, on average, a string gets for having a 1 as its first bit and $\varepsilon_{\#1}$ is the average extra fitness that a 1 at the second locus brings. Either or both of these coefficients could be negative in which case the 1 is, on average, pulling the string's fitness down. ε_{11} is the extra fitness that a combination of ones brings over and above the individual contributions of the 1s. This accounts for the whole being more (or less) than the sum of the parts.

In the binary case there is a nice symmetry. If a string doesn't have a 1 at a particular locus then it must have a 0. This means that the benefit a string has from a 0 at, say, the first locus is the loss it suffers from not having a 1 there i.e. $-\varepsilon_{1\#}$. Here's an example[1]:

$$f(10\#) = \varepsilon_{\#\#\#} + \varepsilon_{1\#\#} - \varepsilon_{\#1\#} - \varepsilon_{11\#}$$

The difference of sign between Walsh and partition coefficients in the binary case can be interpreted as a difference in which allele is implied from the absence of which. The Walsh coefficient $w_{1\#}$ can be interpreted as the benefit that a 0 at the first locus brings to a string and so $-w_{1\#}$ is the benefit for a 1.

The nonbinary case lacks the symmetry of binary. We must explicitly decide which allele we want to imply from the absence of the others. For notational convenience we will assume that 0 is this *implied allele*. The next section briefly covers the implications of choosing a different implied allele. Using the same intuitive meanings for the coefficients we can write the equations relating partition coefficients and fitnesses for a problem represented by a single cardinality 4 gene:

$$f(0) = \varepsilon_\# - \varepsilon_1 - \varepsilon_2 - \varepsilon_3$$
$$f(1) = \varepsilon_\# + \varepsilon_1$$
$$f(2) = \varepsilon_\# + \varepsilon_2$$
$$f(3) = \varepsilon_\# + \varepsilon_3 \qquad\qquad (2.1)$$

Here, ε_1 is the benefit of a 1, ε_2 the benefit of a 2, ε_3 the benefit of a 3, and the benefit

---

1 Here we depart slightly from Mason's definition of partition coefficients. He allows redundant coefficients which, because we produce the coefficients from a change of basis, we must temporarily ignore. We disregard any coefficient whose label contains the implied allele (defined above). Using these redundant *implied coefficients* we could write the footnoted equation as : $f(10\#) = \varepsilon_{\#\#\#} + \varepsilon_{1\#\#} + \varepsilon_{\#0\#}^\dagger + \varepsilon_{10\#}^\dagger$. Higher order implied coefficients are more intuitive as they hide the puzzling (from the view of coefficients as "benefits of combinations of 1s") questions of why $\varepsilon_{10\#}^\dagger = -\varepsilon_{11\#}$ and $\varepsilon_{00\#}^\dagger = \varepsilon_{11\#}$.

of a 0 is the benefit of not having a 1, 2 or 3.

The generalisation of the Walsh transform presented in this paper uses similarly-generated incremental combination coefficients and is called the *incremental combination transform*. The equations relating fitnesses and incremental combination coefficients are the same as equations (2.1) except that additions and subtractions are exchanged.

The section above gave an informal view of partition and incremental combination coefficients. In later sections the coefficients will be described formally by constructing general matrices to convert them to and from fitnesses and we will see how they can be used to state the conditions required for deception in a nonbinary function. First, however, we must describe some of the background notation and assumptions behind the formal presentation.

3. Preliminaries

We assume that a GA processes strings of length l and that we have a *fitness function* f which maps every possible string to a positive real number. A value (or *allele*) at locus i on the string is drawn from a set of alleles or *gene* G_i. Although, in general, alleles could be any sort of object, their values are unimportant in this paper and so we simplify matters by assuming that a gene consists of contiguous nonnegative integers. The cardinality of each gene is $c_i = |G_i|$. Since the number of alleles in a gene is important but their values are not, we can specify the *representation* of a problem by a vector of gene cardinalities $\bar{c}$ and the fitness function. The details of the fitness function are unimportant in this paper and we will refer to problems as having "a representation of $\langle 3,2,2 \rangle$" (i.e. the first gene has 3 alleles, the second gene 2, and so on) on the understanding that a fitness function is associated with the representation. Strictly speaking, matrices should be subscripted with the problem representation that they apply to i.e. $\bar{f}_{(3,2,2)}$ or $\bar{f}_{\bar{c}}$ but we will sometimes drop subscripts when they would clutter an equation without adding clarity.

We will be examining nonbinary transforms using matrix notation (e.g. $\bar{f} = N\bar{v}$) and we need an ordering on strings and coefficients to determine their positions within their vectors. We order strings and coefficient labels so that $a < b \leftrightarrow \exists_j a_j < b_j \wedge \forall_{i<j} a_i = b_i$. We also require that $\# < \alpha_i$ where α_i is the implied allele at loci i and is less than any other allele. The ordering of non-implied alleles is unimportant. Other orderings can be accommodated with adjustments to the transform matrices, but this would unnecessarily complicate the presentation. For the purposes of this paper we pick 0 as the implied allele and so strings are ordered by the numeric ordering of the alleles with the rightmost being in the least significant position. As an example, strings of two trinary genes: $G_1 = G_2 = \{0,1,2\}$ are ordered 00, 01, 02, 10, 11, 12, 20, 21, 22. We order the coefficient labels in the same way, with # being smaller than any number. e.g.:

$$\bar{f}_{(2,3)} = \begin{bmatrix} f_{00} \\ f_{01} \\ f_{02} \\ f_{10} \\ f_{11} \\ f_{12} \end{bmatrix} \qquad \bar{v}_{(2,3)} = \begin{bmatrix} v_{\#\#} \\ v_{\#1} \\ v_{\#2} \\ v_{1\#} \\ v_{11} \\ v_{12} \end{bmatrix}$$

4. The Partition Transform

In this section we describe the matrix form of the partition transform. This matrix is used to convert partition coefficients into fitnesses. To travel the other way, fitnesses to coefficients, requires the inverse transform which is described in the next section.

Obviously, the particular matrix we need depends on the problem's representation. We denote the partition transform matrix as P and subscript it with the representation it applies to. The matrix form of the transform is simply a neat way of writing the equations relating fitnesses and partition coefficients. Taking equations (2.1) as an example:

$$\bar{f}_{(4)} = P_{(4)}\bar{\varepsilon}_{(4)}$$

$$\begin{bmatrix} f_0 \\ f_1 \\ f_2 \\ f_3 \end{bmatrix} = \begin{bmatrix} 1 & -1 & -1 & -1 \\ 1 & 1 & 0 & 0 \\ 1 & 0 & 1 & 0 \\ 1 & 0 & 0 & 1 \end{bmatrix} \begin{bmatrix} \varepsilon_\# \\ \varepsilon_1 \\ \varepsilon_2 \\ \varepsilon_3 \end{bmatrix}$$

Each row of a transformation matrix corresponds to the calculation of a fitness. Any single gene transform matrix $P_{(c)}$ will be a $c \times c$ matrix with 1s in the first column (since every calculation requires the "average fitness" coefficient), 1s along the main diagonal (for the single coefficient that must be added to $\varepsilon_\#$), and −1s from the second column onwards in the top row (subtracting the coefficients for the implied allele). Diagrammatically:

$$P_{(c)} = \begin{bmatrix} 1 & \begin{array}{|c} -1 \\ \hline I_{c-1} \end{array} \end{bmatrix}$$

where I_n is an $n \times n$ identity matrix.

Having dealt with the single gene case, we must move on to general representations. We will begin by looking at $P_{(3,2,2)}$ and, by examining its structure, we will develop the general formula for $P_{\bar{c}}$.

$$P_{(3,2,2)} = \begin{bmatrix}
1 & -1 & -1 & 1 & -1 & 1 & 1 & -1 & -1 & 1 & 1 & -1 \\
1 & 1 & -1 & -1 & -1 & -1 & 1 & 1 & -1 & -1 & 1 & 1 \\
1 & -1 & 1 & -1 & -1 & 1 & -1 & 1 & -1 & 1 & -1 & 1 \\
1 & 1 & 1 & 1 & -1 & -1 & -1 & -1 & -1 & -1 & -1 & -1 \\
1 & -1 & -1 & 1 & 1 & -1 & -1 & 1 & 0 & 0 & 0 & 0 \\
1 & 1 & -1 & -1 & 1 & 1 & -1 & -1 & 0 & 0 & 0 & 0 \\
1 & -1 & 1 & -1 & 1 & -1 & 1 & -1 & 0 & 0 & 0 & 0 \\
1 & 1 & 1 & 1 & 1 & 1 & 1 & 1 & 0 & 0 & 0 & 0 \\
1 & -1 & -1 & 1 & 0 & 0 & 0 & 0 & 1 & -1 & -1 & 1 \\
1 & 1 & -1 & -1 & 0 & 0 & 0 & 0 & 1 & 1 & -1 & -1 \\
1 & -1 & 1 & -1 & 0 & 0 & 0 & 0 & 1 & -1 & 1 & -1 \\
1 & 1 & 1 & 1 & 0 & 0 & 0 & 0 & 1 & 1 & 1 & 1
\end{bmatrix}$$

The matrix has been divided up and shaded to show its structure. At the "top level", the matrix consists of 3 types of identical "block": all 0s, the light grey block and the dark grey block. The dark grey block is simply the light grey block with all its elements negated. We could write:

$$P_{(3,2,2)} = \begin{bmatrix} B & -B & -B \\ B & B & 0 \\ B & 0 & B \end{bmatrix}$$

which, structurally, looks very similar to $P_{(3)}$. Examining the structure of a block B in the same way:

$$B = \begin{bmatrix} 1 & -1 & -1 & 1 \\ 1 & 1 & -1 & -1 \\ 1 & -1 & 1 & -1 \\ 1 & 1 & 1 & 1 \end{bmatrix} = \begin{bmatrix} B_2 & -B_2 \\ B_2 & B_2 \end{bmatrix}$$

which, structurally, looks like $P_{(2)}$, as does B_2 itself. What we have is the leftmost gene (cardinality 3) controlling the global structure of the matrix using $P_{(3)}$ as a "template" into which smaller structures are placed. The second gene controls "mid-distance" structure in the same way using $P_{(2)}$ as a template and the rightmost gene controls local structure using $P_{(2)}$.

There is a matrix operator called the *Kronecker product* (also known as the *direct* or *tensor product*) which builds matrices in exactly this way:

$$A \otimes B = \begin{bmatrix} a_{11}B & a_{12}B & \dots & a_{1n}B \\ a_{21}B & a_{22}B & \dots & a_{2n}B \\ \vdots & \vdots & & \vdots \\ a_{m1}B & a_{m2}B & \dots & a_{mn}B \end{bmatrix}$$

where A is an $n \times m$ matrix, B is $r \times s$ and $A \otimes B$ is $mr \times ns$.

We note the following properties of the Kronecker product (see Graham (1981)):

$$(A \otimes B)^{-1} = A^{-1} \otimes B^{-1} \tag{4.1}$$

$$(A \otimes B)(C \otimes D) = AC \otimes BD \tag{4.2}$$

Using this we can write:

$$P_{(3,2,2)} = P_{(3)} \otimes P_{(2)} \otimes P_{(2)}$$

and in general:

$$P_{\bar{c}} = P_{c_1} \otimes P_{c_2} \otimes \dots \otimes P_{c_l}$$

5. The Inverse Partition Transform

The Walsh transform matrix consists of orthogonal column vectors of equal length and so it is, bar scaling, its own inverse. Unfortunately, in general, neither the incremental combination transform nor the partition transform has orthogonal column vectors and so we need to find their inverses explicitly.

From equation (4.3), we can write $P_{\bar{c}}^{-1}$ as:

$$P_{\bar{c}}^{-1} = \left(P_{(c_1)} \otimes P_{(c_2)} \otimes \dots \otimes P_{(c_l)} \right)^{-1}$$

We can expand this using equation (4.1):

$$P_{\bar{c}}^{-1} = P_{(c_1)}^{-1} \otimes P_{(c_2)}^{-1} \otimes \dots \otimes P_{(c_l)}^{-1}$$

which simplifies our problem somewhat. We state without proof that:

$$P_{(c)}^{-1} = \begin{bmatrix} \frac{1}{c} & \frac{1}{c} & \frac{1}{c} & \cdots & \frac{1}{c} \\ -\frac{1}{c} & 1-\frac{1}{c} & -\frac{1}{c} & \cdots & -\frac{1}{c} \\ -\frac{1}{c} & -\frac{1}{c} & 1-\frac{1}{c} & \cdots & -\frac{1}{c} \\ \vdots & \vdots & \vdots & & \vdots \\ -\frac{1}{c} & -\frac{1}{c} & -\frac{1}{c} & \cdots & 1-\frac{1}{c} \end{bmatrix}$$

Informally, the first row sums all the fitness and divides by c, the number of fitnesses. This is what we would expect for $\varepsilon_\#$, the "average fitness" coefficient. All the remaining rows are the negation of row 1, except that they have a single fitness added. This is what we would expect from rearranging $f_i = \varepsilon_\# + \varepsilon_i$ to give $\varepsilon_i = f_i - \varepsilon_\#$.

6. The Incremental Combination Transform

In the single gene case, the incremental combination transform matrix $N_{(c)}$ will be $P_{(c)}$ with all elements except those in the first column negated. Diagrammatically:

$$N_{(c)} = \left[\begin{array}{c|c} & \overline{1} \\ \hline 1 & -I_{c-1} \end{array} \right]$$

This simply reflects the difference in signs between incremental combination and partition coefficients. We note that the inverse of this matrix is:

$$N_{(c)}^{-1} = \begin{bmatrix} \frac{1}{c} & \frac{1}{c} & \frac{1}{c} & \cdots & \frac{1}{c} \\ \frac{1}{c} & \frac{1}{c}-1 & \frac{1}{c} & \cdots & \frac{1}{c} \\ \frac{1}{c} & \frac{1}{c} & \frac{1}{c}-1 & \cdots & \frac{1}{c} \\ \vdots & \vdots & \vdots & & \vdots \\ \frac{1}{c} & \frac{1}{c} & \frac{1}{c} & \cdots & \frac{1}{c}-1 \end{bmatrix}$$

In the same way as the partition transform, we form general transform matrices using the Kronecker product:

$$N_{\bar{c}} = N_{(c_1)} \otimes N_{(c_2)} \otimes \ldots \otimes N_{(c_l)}$$

We can easily show that this is a generalisation of the Walsh transform. Homaifar, Qi and Fost (1991) give the Kronecker product formulation of the Walsh (Hadamard) Transform:

$$W_1 = \begin{bmatrix} 1 & 1 \\ 1 & -1 \end{bmatrix}$$

$$W_l = W_{l-1} \otimes W_1$$

since $W_1 = N_{(2)}$, both this and the incremental combination formulation reduce to the Kronecker product of $l\ N_{(2)}$ matrices.

Being a generalisation of the Walsh transform, the IC transform is particularly useful to GA researchers. The Walsh transform is limited to binary representations whereas the IC transform can be used for nonbinary representations as well. By using IC coefficients in preference to Walsh coefficients, theoretical results can apply to a wider range of representations.

7. The Relationship Between IC & Partition Coefficients

We have already stated that the difference between incremental combination and partition coefficients is simply one of sign. However, not all of the coefficients have different signs. We can investigate the relationship formally by finding the matrix T that transforms one set of coefficients into the others (for a particular representation z). Notice that T effects a change of basis:

$$T = NP^{-1}$$

To find out the structure of T, we expand the right hand side (using equation (4.2)):

$$T_{\bar{z}} = (N_{(c_1)} \otimes N_{(c_2)} \otimes \ldots \otimes N_{(c_l)})(P_{(c_1)}^{-1} \otimes P_{(c_2)}^{-1} \otimes \ldots \otimes P_{(c_l)}^{-1})$$

$$= N_{(c_1)}P_{(c_1)}^{-1} \otimes N_{(c_2)}P_{(c_2)}^{-1} \otimes \ldots \otimes N_{(c_l)}P_{(c_l)}^{-1}$$

$$= T_{(c_1)} \otimes T_{(c_2)} \otimes \ldots \otimes T_{(c_l)}$$

where:

$$T_{(c)} = N_{(c)}P_{(c)}^{-1}$$

$$= \begin{bmatrix} 1 & & & & 0 \\ & -1 & & & \\ & & -1 & & \\ 0 & & & \ddots & \\ & & & & -1 \end{bmatrix}$$

Notice that since $T_{(c)}T_{(c)} = I$, $T_{(c)} = T_{(c)}^{-1}$.

From T, we can generate a formula relating individual coefficients:

$$\varepsilon_i = (-1)^d \nu_i$$

where d is the number of defined (i.e. not #) loci in the coefficient label i.

8. Implied Coefficients

Although they are redundant, it can be useful to calculated the coefficients for the implied allele. Without them, it is difficult to see the effects of allele combinations involving the implied allele. They are also useful for calculating fitnesses by hand since they reduce the complexity of expressions and remove the "problem" of deciding whether to add or subtract higher-order coefficients (see footnote 1 for an example).

The matrix for computing these implied coefficients is the same for both partition and incremental combination coefficients so we just present the partition transform case. As before, we assume that the implied allele is 0 and, for clarity, we mark implied alleles with a + superscript. The implied coefficient labels consist of all possibilities of # and the implied allele, so there are 2^l implied coefficients. Notice that $\varepsilon_{\#\#\ldots\#}$ is counted as a implied coefficient since it makes the following presentation far more elegant. Its calculation is redundant and could easily be omitted by removing the first row of the final G matrix produced.

We wish to compute the implied coefficients from the coefficients that we already have:

$$\bar{\varepsilon}^+ = G\bar{\varepsilon} \tag{8.1}$$

As usual, we will use the Kronecker product to form the general matrix, so we start with the single gene case:

$$G_{(c)} = \begin{bmatrix} 1 & 0 & 0 & \dots & 0 \\ 0 & -1 & -1 & \dots & -1 \end{bmatrix}$$

This is a $2 \times c$ matrix. We can see from an example calculation with $c = 4$:

$$\begin{bmatrix} \varepsilon_\#^\dagger \\ \varepsilon_0^\dagger \end{bmatrix} = \begin{bmatrix} 1 & 0 & 0 & 0 \\ 0 & -1 & -1 & -1 \end{bmatrix} \begin{bmatrix} \varepsilon_\# \\ \varepsilon_1 \\ \varepsilon_2 \\ \varepsilon_3 \end{bmatrix}$$

that the first row of G simply "copies" $\varepsilon_\#$ and the second row computes $\varepsilon_0^\dagger$ from the remaining coefficients. We (implicitly) discussed this calculation of $\varepsilon_0^\dagger$ in section 2. To form the general matrix, we use the Kronecker product:

$$G_{\bar{z}} = G_{(c_1)} \otimes G_{(c_2)} \otimes \dots \otimes G_{(c_i)}$$

If we wish to compute the implied coefficients directly from the fitnesses then we substitute $\bar{\varepsilon} = P^{-1}\bar{f}$ into equation (8.1):

$$\bar{\varepsilon}^+ = GP^{-1}\bar{f}$$

expanding GP^{-1} gives:

$$G_{\bar{z}}P_{\bar{z}}^{-1} = G_{(c_1)}P_{(c_1)}^{-1} \otimes G_{(c_2)}P_{(c_2)}^{-1} \otimes \dots \otimes G_{(c_i)}P_{(c_i)}^{-1}$$

we note that:

$$G_{(c)}P_{(c)}^{-1} = \begin{bmatrix} \frac{1}{c} & \frac{1}{c} & \frac{1}{c} & \dots & \frac{1}{c} \\ 1 - \frac{1}{c} & -\frac{1}{c} & -\frac{1}{c} & \dots & -\frac{1}{c} \end{bmatrix}$$

We could insert the second row of this matrix into $P_{(c)}^{-1}$ (as its second row) and compute the implied coefficients at the same time as the others. This is more efficient than calculating the coefficients in a separate stage but this is really an implementation detail. Such a matrix can be efficiently multiplied using the fast multiplication technique described below but its implementation as a computer program is slightly more involved.

Implied incremental combination coefficients can be similarly calculated using:

$$\bar{v}^+ = GN^{-1}\bar{f}$$

expanded as before and noting that:

$$G_{(c)}N_{(c)}^{-1} = \begin{bmatrix} \frac{1}{c} & & \frac{1}{c} & \frac{1}{c} & \dots & \frac{1}{c} \\ \frac{1}{c} - 1 & & \frac{1}{c} & \frac{1}{c} & \dots & \frac{1}{c} \end{bmatrix}$$

9. Fast Transforms

One use of IC and partition coefficients is to specify the conditions that are required to make a problem deceptive and then generate or analyse functions using the coefficients. To do this in practice requires an efficient way of performing the transform.

The transforms are written as a matrix multiplication (e.g. $\bar{f} = N\bar{v}$) but naive matrix multiplication will take $O(N^2)$ operations (where N is the size of the transform matrix). Even assuming we can create the transform matrix "on-the-fly" with no time overheads we will still need $2N$ units of memory; N for the fitness vector, N for the coefficient vector. We could take advantage of 0s in the matrix but this only helps for high cardinality representations. However, there is an algorithm, analogous to the fast Walsh transform, that can perform the transforms (and their inverses) in only $O(N \log N)$ operations for the worst case (i.e. no zeros in the matrix) and can transform the data in-situ thus needing only N units of memory.

The algorithm takes advantage of identical calculations in the transform matrix. For example, we look at $P_{(2,2)}$:

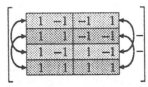

The arrows show identical calculations. Calculations in the right column have to be negated to become identical. Instead of $3 \times 4 = 12$ additions/subtractions, we only need 2 operations per column plus 4 to combine the columns = 8 operations.

These identical calculations are produced by the Kronecker product. A matrix generated by a Kronecker product is made up of identical blocks (bar scaling) and so only the calculations for one block in each column of blocks needs to be done. The incremental combination and partition transforms are, in general, created from many Kronecker products. We form matrices X_i that perform only the necessary calculations for each column of blocks at each "level" of structure in a matrix. The following construction of a fast algorithm is not specific to the incremental combination transform; it can be applied to any matrix formed by Kronecker products. This formulation has been developed independently but it is almost certainly not original; much work has been done on fast transforms. It does, however, have the virtue of being fairly simple and self-contained.

A matrix multiplication:

$$Mv = (B_1 \otimes B_2 \otimes \ldots \otimes B_n) v$$

where each B_i is a $b_i \times b_i$ matrix.

can be implemented as:

$$Mv = X_1(X_2(\ldots (X_n v) \ldots))$$

where

$$X_i = I_{\gamma(1, i-1)} \otimes B_i \otimes I_{\gamma(i+1, n)}$$

$$\gamma(l, u) = \begin{cases} \prod_{i=l}^{u} b_i & l \leqslant u \\ 1 & l > u \end{cases}$$

and I_n is the $n \times n$ identity matrix.

For brevity, the proof of this is omitted as is the derivation of the operation counts which we simply state:

$$\text{ops}(M) = \prod_{i=1}^{n} e_{B_i} - \prod_{i=1}^{n} r_{B_i}$$

$$\text{ops}\left(\prod X_i\right) = \left(\prod_{j=1}^{n} b_j\right) \sum_{i=1}^{n} \frac{e_{B_i} - r_{B_i}}{b_i}$$

where ops (A) is the number of additions/subtractions involved in multiplying Av

taking advantage of 0 elements, e_A is the number of non-zero elements in A and r_A is the number of non-zero rows in A.

10. The Fast Formulation Applied to the GA Transforms

The equations above were applicable to any Kronecker product formed matrix M. We now interpret them for our transforms:

General formulation	IC transform	Partition transform
M	$N_{\tilde{c}}$	$P_{\tilde{c}}$
v	v	ε
$B_1, B_2, \ldots B_n$	$N_{(c_1)}, N_{(c_2)}, \ldots N_{(c_l)}$	$P_{(c_1)}, P_{(c_2)}, \ldots P_{(c_l)}$
b_i	c_i	c_i

Remembering that the size of the transform matrix is $N = \prod_{i=1}^{l} c_i$ (the product of the gene cardinalities), and noting that $\hat{c} = \sqrt[l]{N}$ (i.e. the geometric mean of the cardinalities), we can calculate the number of operations for direct multiplication by the transform matrix as:

$$\text{ops}(P_{\tilde{c}}) = \text{ops}(N_{\tilde{c}}) = \prod_{i=1}^{l}(3c_i - 2) - \prod_{i=1}^{l} c_i$$

$$\approx O(3^{\log_{\hat{c}} N} N)$$

In contrast:

$$\text{ops}\left(\prod X_i\right) = 2\left(1 - \frac{1}{\tilde{c}}\right) N \log_{\tilde{c}} N$$

$$= O(N \log_{\tilde{c}} N)$$

where $\tilde{c}$ is the harmonic mean of the gene cardinalities: $\tilde{c} = l / (\frac{1}{c_1} + \frac{1}{c_2} + \cdots + \frac{1}{c_l})$.

We can obtain the same time complexities for the inverse transforms by noticing that:

$$P_{(c)}^{-1} = Q_{(c)} R_{(c)} = \begin{bmatrix} 1 & 0 & 0 & \ldots & 0 \\ -1 & 1 & 0 & \ldots & 0 \\ -1 & 0 & 1 & \ldots & 0 \\ \vdots & \vdots & \vdots & & \vdots \\ -1 & 0 & 0 & \ldots & 1 \end{bmatrix} \begin{bmatrix} \frac{1}{c} & \frac{1}{c} & \frac{1}{c} & \ldots & \frac{1}{c} \\ 0 & 1 & 0 & \ldots & 0 \\ 0 & 0 & 1 & \ldots & 0 \\ \vdots & \vdots & \vdots & & \vdots \\ 0 & 0 & 0 & \ldots & 1 \end{bmatrix}$$

and because:

$$e_Q = e_R = 2c - 1$$

$$r_Q = r_R = c$$

we can calculate that $\text{ops}(P_{(c)}^{-1}) = e_Q - r_Q + e_R - r_R = 2c - 2$. This is the same as for P (although we have a division as well) and so, informally, we can see that the inverse transform is as easy to compute as the transform. There is an equivalent breakdown for N^{-1}.

11. Implementing the Fast Transform

To implement the fast transform, we must implement multiplication by an X_i matrix. Although it might initially appear that we need to generate each X_i matrix, in fact they can be "hard coded". This is easy to see if we look at the structure of an X_i matrix. As an example, we'll look at X_2 for the partition transform with a problem representation $\langle 3,2,2 \rangle$. It's formula is:

$$X_2 = I_{\gamma(1,1)} \otimes P_{(2)} \otimes I_{\gamma(3,3)}$$

$$= I_3 \otimes P_{(2)} \otimes I_2$$

and the matrix itself is:

$$\begin{bmatrix}
1 & 0 & -1 & 0 & 0 & 0 & 0 & 0 & 0 & 0 & 0 & 0 \\
0 & 1 & 0 & -1 & 0 & 0 & 0 & 0 & 0 & 0 & 0 & 0 \\
1 & 0 & 1 & 0 & 0 & 0 & 0 & 0 & 0 & 0 & 0 & 0 \\
0 & 1 & 0 & 1 & 0 & 0 & 0 & 0 & 0 & 0 & 0 & 0 \\
0 & 0 & 0 & 0 & 1 & 0 & -1 & 0 & 0 & 0 & 0 & 0 \\
0 & 0 & 0 & 0 & 0 & 1 & 0 & -1 & 0 & 0 & 0 & 0 \\
0 & 0 & 0 & 0 & 1 & 0 & 1 & 0 & 0 & 0 & 0 & 0 \\
0 & 0 & 0 & 0 & 0 & 1 & 0 & 1 & 0 & 0 & 0 & 0 \\
0 & 0 & 0 & 0 & 0 & 0 & 0 & 0 & 1 & 0 & -1 & 0 \\
0 & 0 & 0 & 0 & 0 & 0 & 0 & 0 & 0 & 1 & 0 & -1 \\
0 & 0 & 0 & 0 & 0 & 0 & 0 & 0 & 1 & 0 & 1 & 0 \\
0 & 0 & 0 & 0 & 0 & 0 & 0 & 0 & 0 & 1 & 0 & 1
\end{bmatrix}$$

In the formula, the left hand identity matrix (I_3) controls the global structure and so we see three blocks along the main diagonal of the X_2 matrix (shaded very light grey). The right hand identity matrix controls the local structure, and a good way to look at this is as if the $P_{(2)}$ matrix has been scaled up or "exploded" and copies made of it along the main diagonal (exploded $P_{(2)}$ matrices are shown boxed with their elements shaded). From the formula for X_i we would expect to see $\gamma(1, i-1)$ blocks each containing $\gamma(i+1, n)$ exploded $P_{(c_i)}$ matrices. The distance between elements of each exploded matrix is $\gamma(i+1, n)$.

If we have a routines that can multiply matrices (i.e. $P_{(c)}$, $P_{(c)}^{-1}$, $N_{(c)}$, $N_{(c)}^{-1}$) taking account of a "scaling factor" and offset then the routine to multiply by X_i simply calls the appropriate routine repeatedly with the appropriate parameters. These parameters are easily generated "on-the-fly" so the X_i matrix does not have to be explicitly calculated or stored.

12. A Nonbinary Deceptive Function

In this section we use IC coefficients to write the conditions that characterise a fully deceptive nonbinary function with a representation of $\langle 3,2 \rangle$.

A deceptive function is one in which low order schemata "lead away" from the global optimum. For example, if 00 is the optimum of a 2 bit binary-represented function, then it contains deception if either $f(*0) < f(*1)$ or $f(0*) < f(1*)$ is true. IC and nonbinary partition coefficients provide a convenient and compact way of writing these conditions.

Using the terminology of Whitley (1991). in a *fully deceptive* problem all competitions between schemata of orders 1 to $l - 1$ (where l is the length of the GA's strings) lead towards a *deceptive attractor* i.e. a string which is not the optimum.

In the binary case, full deception implies that the deceptive attractor is the complement of the global winner (Whitley (1991)), but in the nonbinary case full deception implies only that, at every loci, the allele in the deceptive attractor is different from that in the global optimum. Of course, in the binary case there is only one possibility for a different allele: the complement. Since full nonbinary deception no longer implies a particular deceptive attractor, we will state the attractor being considered since each attractor has a slightly different set of deceptive constraints associated with it.

For a $\langle 3,2 \rangle$ problem to be fully deceptive with 00 as the optimum and 21 as the deceptive attractor, these conditions must hold:

Optimality constraints	*Deceptive constraints*
$f(00) > f(01)$	$f(*0) < f(*1)$
$f(00) > f(10)$, $\mathbf{f(00) > f(11)}$	$\mathbf{f(0*) < f(2*)}$
$f(00) > f(20)$, $\mathbf{f(00) > f(21)}$	$\mathbf{f(1*) < f(2*)}$

Only the conditions in bold are necessary; the other two optimality conditions are implied by them.

To use the nonbinary transforms, we must rewrite these constraints in terms of, for example, IC coefficients. To do this we need to calculate the fitnesses of schemata using the coefficients:

$$f(s) = \sum_{j \,:\, j \text{ subsumes } s} v_j N_{jx}$$

Where s is the schemata, j is a coefficient label, x is any member of s, j subsumes s iff all the defined (i.e. not * or #) loci of s and j agree and N_{jx} is the IC transform matrix element corresponding to f_x and v_j.

Calculating transform matrix elements individually can be performed with an "IC function", analogous to a Walsh function:

$$N_{jx} = N_j(x) = \prod_{i=1}^{l} m(x_i, j_i)$$

where

$$m(x_i, j_i) = \begin{cases} -1 & \text{if } x_i = j_i \\ 1 & \text{if } x_i = \alpha_i \lor j_i = \# \\ 0 & \text{otherwise} \end{cases}$$

Remember that α_i is the implied allele for loci i which, in this paper, is always 0. The constraints can now be written as:

Optimality constraints	*Deceptive constraints*
$2v_{\#1} + 2v_{1\#} + v_{2\#} + v_{21} > 0$	$v_{\#1} < 0$
$v_{1\#} + 2v_{2\#} + v_{11} + 2v_{21} > 0$	$v_{2\#} - v_{1\#} < 0$
$2v_{\#1} + v_{1\#} + 2v_{2\#} + v_{11} > 0$	$v_{1\#} + 2v_{2\#} < 0$

We will use these constraints to generate a fully deceptive function. We pick coefficients that satisfy the constraints (with $v_{\#\#}$ chosen to make all string fitnesses greater than 0) and then generate the fully deceptive function using the (fast) IC transform (i.e. $\bar{f} = N\bar{v}$):

Coefficient	Value	String	Fitness
$v_{\#\#}$	10	00	18
$v_{\#1}$	-1	01	0
$v_{1\#}$	0	10	4
v_{11}	5	11	16
$v_{2\#}$	-1	20	5
v_{21}	5	21	17

13. Conclusion

IC and nonbinary partition coefficients are an efficient way of calculating the fitnesses of low order schemata and of writing the conditions that characterise deception in both binary and nonbinary functions. They provide the foundations for generalising some binary based GA theory and, practically, they can be used in conjunction with the fast transforms to analyse and generate deceptive nonbinary functions. However, they are no more difficult to use than their binary counterparts despite their wider applicability.

In this paper, we have taken a detailed look at IC and nonbinary partition coefficients and their transforms and, in addition, the paper has produced a fast binary partition transform as a special case of the fast partition transform. This has not, to the author's knowledge, been previously published for binary problems. The C source code for the fast transforms is available from the author.

References

Belew, R.K. and Booker, L.B. (Eds) (1991), *Proceedings of the Forth International Conference on Genetic Algorithms*, Morgan Kaufmann

Field, P. (1994), Walsh and Partition Functions Made Easy, Presented at the AISB 1994 Workshop on Evolutionary Computing, University of Leeds, England.

Goldberg, D.E. (1989a), Genetic Algorithms and Walsh Functions: Part I, A Gentle Introduction, *Complex Systems* 3, pp.129-152

Goldberg, D.E. (1989b), Genetic Algorithms and Walsh Functions: Part II, Deception and Its Analysis, *Complex Systems* 3, pp.153-171

Graham, A. (1981), *Kronecker Products and Matrix Calculus with Applications*, Ellis Horwood.

Homaifar, A., Qi, X. and Fost, J. (1991), Analysis and Design of a General GA Deceptive Problem. In Belew, R.K. and Booker, L.B. (1991).

Mason, A.J. (1991), Partition Coefficients, Static Deception and Deceptive Problems for Non-Binary Alphabets. In Belew, R.K. and Booker, L.B. (1991).

Whitley, L.D. (1991), Fundamental Principles of Deception in Genetic Search. In Rawlins, G. (Ed.), *Foundations of Genetic Algorithms*, Morgan Kaufmann

Enhancing Evolutionary Computation using Analogues of Biological Mechanisms

Ray Paton
The Liverpool Biocomputation Group
Department of Computer Science
The University of Liverpool
Liverpool L69 3BX, UK

email: r.c.paton@csc.liv.ac.uk

Abstract

The biological sciences have provided the inspiration for the development of evolutionary computation. However, it is well known that the amount of biological ideas imported into evolutionary algorithms (EAs) is small. Beginning with a proposal of how certain biological details particularly, cycle, structure and ecology can be used to enhance EAs, this article considers a number of potentially useful source ideas. New mechanisms acting on strings, both as independent entities and within ecologies are considered and issues related to epigenetic and acquired inheritance systems are also discussed. The nature of hierarchical relations in gene ecologies is introduced with reference to the evolution of regulatory systems. Some comments about developmental and temporal systems are also made.

1 Introduction

Biological systems have provided the inspiration for the design of computational systems in a large number of ways. A very good example of how biology can inspire engineering solutions is the work of Professor O.H. Schmitt who introduced the term 'biomimetic' (emulating biology) into the US literature over a decade ago [32]. It is fascinating to see how, following his Ph.D. thesis on the simulation of nerve action, four well-known electronic devices emerged: Schmitt Trigger, Emitter-Follower, Differential Amplifier and Heat Pipe.

It has been argued elsewhere that in the transfer of biological sources to computational problems we can identify a number of key modelling emphases namely, architecture, mechanism, organisation and collective behaviour or whole system functionality ([29], [30]). The purpose of this article is to look at some ways of enriching Evolutionary Algorithms (EAs) using richer source ideas from biology, particularly biological mechanisms. As such it is far from a complete description not least because the biological sources continue to be better understood and evolutionary computation is an emerging discipline. The writer's underlying motivation has been to produce an overview of an approach to evolutionary computation that treats it as a sub-discipline of applied biology rather than its more common presentations namely, applied mathematics or A.I.

Biological adaptation describes a number of processes, some occurring within an organism's life-cycle and some between the life-cycles of successive generations. Morphological and physiological adaptation are within-life cycle process that describe phenotypic plasticity. Epigenetic adaptation is associated with the unfolding developmental process. Darwinian adaptation based on natural selection operates between generations. There are also heritable forms of non-Darwinian processes including the inheritance of acquired characteristics.

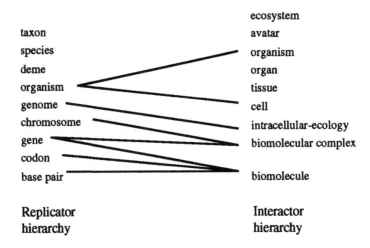

Figure 1 - Adaptor Hierarchies
(adapted and extended from Eldredge [16])

There is much debate about the nature of adaptation which cannot be discussed here. However, we note two kinds of adaptor in biosystems: replicators - entities which pass on their structure directly in replication and, interactors - entities that directly interact as cohesive wholes with their environment in such a way that replication is differential. Eldredge [16] attempts an analysis of hierarchical relations in biosystems by differentiating two kinds namely, the genealogical and the economic (ecological). These hierarchies are a valuable aid to the present discussion for they help to elucidate the nature of replicators and interactors and the relationships between the two classes and their members. Specifically, the genealogical applies to replicators and the economic to interactors.

One and the same entity may be a replicator and an interactor. Although gene is placed in the replicator hierarchy of Figure 1 it is also an interactor. Furthermore, strict adherence to a rigid hierarchy reifies the notion that there is no interrelationship between levels within and across hierarchies. This is far from the case. As we shall see in the next section, some interactors may have a replicator role - for example, the non-chromosomal components of chromatin. This leads to an important theme which pervades this paper namely, the need for a thorough understanding of how components interact. We shall call this understanding the ecology of the system.

An appreciation of the ecology of a system, whether biological or computational, can provide insights on how it is to be interpreted and how it makes interpretations (for a discussion of the interpretation problem see [28]). This in turn is related to issues associated with encoding and decoding representations and the nature of the various interactions between genotype/genotype, genotype/phenotype and phenotype/phenotype within the system. Although it is an extremely important issue, it will not be possible to pursue it further in this paper as the main focus is on mechanisms and models for heritable change. The reader will have to suspend the representational problem associated with epistasis and deception. Further discussion of issues associated with representation, and particularly the relations between genotype and phenotype can be found in [12] and [31].

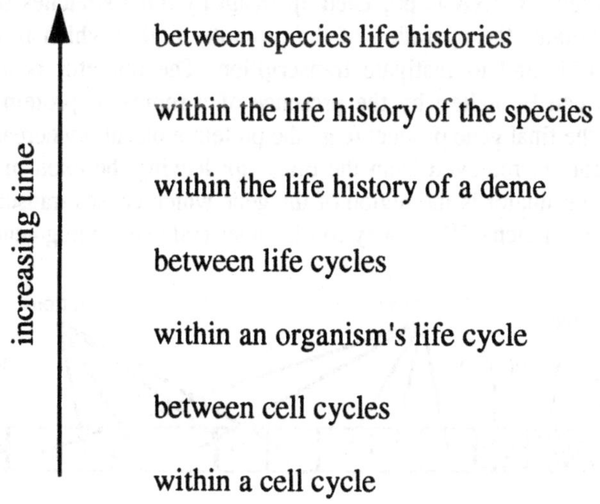

between species life histories

within the life history of the species

within the life history of a deme

increasing time

between life cycles

within an organism's life cycle

between cell cycles

within a cell cycle

Figure 2 - Self-similarity in a Temporal Hierarchy

In concluding this section we shall reflect on similarities between adaptational processes over different time periods. Figure 2 combines some of the features of the adaptor hierarchy in Figure 1 to reveal a number (not complete) of self-similar temporal epochs. Self-similarity is demonstrated in a number of ways in this hierarchy which can help in the development of EA systems - especially those in which both selection and learning take place. We shall see how an appreciation of the interactions within a cycle can influence adaptation over time. Clearly, events within a cycle recur across time. Furthermore, cycles are organised (in this case the adaptor is evolving and self-organising). We now consider events within subcellular and population cycles.

2 Exploiting Biological Mechanisms in EAs

The anatomy of an EA is quite simple and yet the genes, chromosomes and genomes, which are the biological source structures are highly organised biological systems [3]. We may envisage DNA, genes and the genome acting as parallel processing systems in a variety of ways for which a number of automata-based models can be described [27]. The computational capacities of a genome ecology are considerable. From the viewpoint of Figure 2, we find that they are evolutionary in the sense of time change in a given space, and connectionist in the sense of spatial differentiation within a given time. The ecological involvement of promoter, operator and terminator includes both replicator and interactor roles. In order to explore this further, we shall look at the structure of a gene and gradually incorporate thinking about the genome as a whole.

Figure 3 provides a simplified view of the anatomy of a structural gene that is, one which codes for a protein or RNA. That part of the gene which ultimately codes for protein or RNA is preceded upstream by three stretches of code. The enhancer facilitates the operation of the promoter region which is where RNA polymerase is bound to instigate transcription. The operator is a site where transcription can be halted by the presence of a repressor protein. Exons are expressed in the final gene product (e.g., the protein molecule) whereas introns are transcribed but are removed from the transcript leaving the exon material to be spliced. The terminator is the region of the gene which causes transcription to be terminated. One stretch of DNA may consist of several overlapping genes.

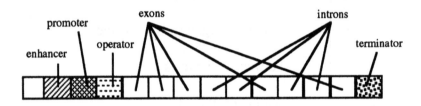

Figure 3 - Simplified Diagram of a Structural Gene

For example, the introns in one gene may be the exons in another. This shows that a gene is an ordered sequence (rather than a set of independent units) and that changes in some locations (e.g., promoter, operator or terminator) may have more dramatic effects than others. The meanings we associate with the gene as a biocomputational source may have to take account of its structuring if we are seeking to apply greater biological detail to EAs.

2.1 Mechanisms Acting on Strings

Evolutionary algorithms apply two basic types of operation to change the nature of strings - mutation and crossover. Figure 4 summarises the kinds of changes that can affect genes, chromosomes and genomes.

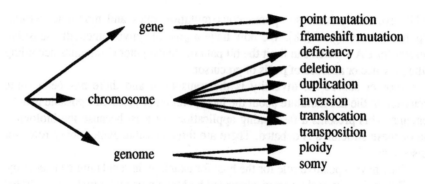

gene → point mutation
gene → frameshift mutation
chromosome → deficiency
chromosome → deletion
chromosome → duplication
chromosome → inversion
chromosome → translocation
chromosome → transposition
genome → ploidy
genome → somy

Figure 4 - Options for Change

We shall not pursue the application of mutational or crossover operators to EAs (for an excellent overview see Bäck [2]). For the purposes of the present discussion we shall go back to the biological structure from which the crossover/recombination idea is gleaned. For example, Head ([19]) provides an analysis of DNA splicing systems based on formal language theory. This model is based on the action of restriction and ligation enzymes on linear and circular strings. As Head notes, there are potential applications of this method to EAs in that constraints can be placed on where crossover can occur which could help refine the crossover strategies in GAs and ESs. In order to obtain a greater appreciation of crossover the structural organisation of a string must be considered. This may seem strange given that a string is a linear array so maybe the representation needs to change to more than one dimension.

Simple mutations can have complex origins and quite a few spontaneous mutations arise from an association of a number of discrete events. For example, base-pair substitutions and frameshift mutations may arise from complex events associated with incorporating sections of nearby sequence [13]. It could be argued that the spontaneity of a mutational event is modulated by the micro-ecology of the nucleus. Incorporation of such analogue mechanisms into an EA could make it more sensitive in a search.

Hall [17] discusses a number of possible strategies of how new metabolic functions may evolve through mutation in bacterial populations. Mutations in regulatory genes may allow a gene to be expressed under unusual conditions thus providing new metabolic (phenotypic) capabilities. Structural gene mutations may alter the catalytic capabilities of existing or even novel substrates. In most cases, both regulatory and structural mutations would be necessary for effective emergence of new functions. Hall further argues that the probability of certain advantageous mutations occurring when modulated by selective conditions can be higher than random. In terms of EA applications, this is related to independent co-evolving code.

Genes remain stable over generations because, if damaged, they can be repaired effectively. One of the factors facilitating effective repair is the adequate presence of DNA precursors specifically deoxyribonucleoside triphosphate (dNTP). Changes in dNTP levels may increase spontaneous mutation rates and modulate cellular responses to mutagens [28]. We now have a possible environmentally-sensitive operator for EA s which can limit the bit patterns in daughter chromatids according to the presence or absence of pools of precursor.

Another source for diversity is the transposon and there has been some discussion in the EA literature about the incorporation of these transposable genetic elements. This presents a fascinating application not least because the biological role of these structures is debated. There are three possible evolutionary roles for transposons:

1   They have a positive role for the host for example, in regulating gene activity. This is the classical position advocated by McClintock who first demonstrated the occurrence of transposable genetic elements in maize.
2   Transposition brings about genomic changes which can accelerate adaptation through selection processes (see flax example below).
3   Transposons are selfish DNA [6].

Each of these postulated roles could be applied to EA designs. In the first case, we may contemplate cross-generational control. The second provides a mechanism for directing adaptation and the third would help in exploring the coevolution of parasites. The challenge is to be able to decide on what kinds of problems would be appropriate for these techniques. In concluding this section, it is possible to say that there are a number of ways we could imagine EAs adopting some of the operational capacities of certain biosystems. So far, we have focussed on string operations as if the strings were a set. We now turn to hierarchical and ecological relations in a genome (see Figure 5). One challenge for EA design is to go from strings in sets to strings in ecologies.

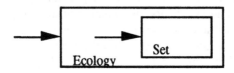

Figure 5 - Operators can be Applied in a Number of Ways

2.2 String Hierarchies and Ecologies

Hierarchical relations in Evolutionary Algorithms can be found for example, with the idea of a meta-level GA which optimises the parameter settings of a GA [2] and Voigt et al, [34] apply population-level hierarchical structuring in distributed GAs. Hierarchical relations in biological systems can be found when we explore the ecology of the genome. For example, in the expression of phenotype, some genes are regulated by the action of other genes. We shall consider two particular forms of regulation namely, operon systems (as summarised in simplistic terms in Figure 6) and regulon systems (see Figure 7). An operon system, which is located on one chromosome, consists of a regulator gene and a number of contiguous structural genes which share the same promoter and terminator and code for enzymes which are involved in specific metabolic pathways (the classical example is the *Lac* operon).

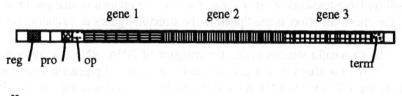

Key
reg - regulator gene
pro - promoter
op - operator
term - terminator

Figure 6 - Visualisation of an Operon

A regulon system consists of a regulatory gene which controls a number of other genes scattered at different locations throughout the genome. In order to illustrate these systems we shall briefly consider bacterial responses to reactive forms of oxygen (based on Demple [10]). Bacteria are able to respond to these physiologically and biochemically stressful environments by activating coregulated groups of genes in global responses which affect diverse cellular functions. For example, part of the cellular response to hydrogen peroxide stress involves the activation of at least eight genes which code for particular redox enzymes. These genes are under the control of a single regulatory gene, the oxyR-regulon.

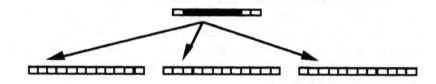

Figure 7 - Visualisation of Part of a Regulon Ecology

We find that a regulon ecology is just the basic stage in a hierarchy of regulatory systems. Most, if not all, the operons of a bacterial cell belong to higher level regulatory organisations. A regulon system tends to be conceived as one which is controls a number of operons by a single regulatory protein [26]. More general systems involve modulons. These are operon networks concerned with multiple pathways in which member operons may be governed by individual regulatory proteins and additionally, are under the control of a common pleiotropic regulatory protein. In a stimulon system, all the operons responding together to an environmental stimulus, no matter how many regulons and modulons are involved. There are higher order systems than this although they are not so well documented.

How can regulons be used as sources in evolutionary computation ? Firstly, we see that some genes are "smarter" than others. Structural genes code for enzymes or RNA whereas regulon systems have a gene which co-ordinates the activity of (possibly) large numbers of other genes. Clearly, if a regulon is changed in some way the knock-on effect is multiplied to the structural genes it regulates. We see that there is a hierarchical relationship.

The chromatin structure (i.e., the complex of DNA, RNA and protein) is inherited in reproduction, not just the DNA sequence. Epigenetic inheritance systems are additional to DNA replication systems for example, the DNA binding proteins which segregate with the chromosomes at meiosis may serve as heritable blueprints [20]. Epigenetic information is not totally erased in the germ line. This is especially noticeable in plants where there is a degree of indistinction between soma and germ plasm [6]. An associated though far from identical idea concerns the inheritance of acquired characteristics. Landman [24] describes a number of experimental systems which show a capacity to inherit acquired characteristics. These acquired inheritance systems comply with a fixed experimental protocol. Organisms or cells in culture are kept in conditions of little or no growth and exposed to an environmental stressor such as a chemical substance. (Note that lack of growth prevents the selection of mutants - because there is no reproduction). Following exposure the system under stress is returned to its original environment. Upon being returned, all or a large proportion of cells/organisms exhibit new characteristics that are passed on heritably to succeeding generations. A number of mechanisms have been described to account for phenomena of this kind some examples which Landman (ibid) includes are:

- Cortical inheritance - e.g., cell structure modifications in ciliates
- Inherited modification of DNA - e.g., methylation patterns
- Acquisition of foreign DNA - through plasmids and viruses

Acquired inheritance systems provide a further dimension to our collection of biological sources for adaptable systems. They seem to be able to change a replicators profile rapidly and although it would be unwise to speculate on their generality or applicability in biosystems it would be appropriate to consider their adoption in certain EA designs.

2.3 Directed Adaptation

Bäck [1] discusses results from experiments with genetic algorithms in which mutation rate was changed from a global external parameter into an internal item which changed during the search process (as in an Evolution Strategy). He shows how preliminary findings confirm the value of appropriate settings of environment-dependent self-adaptation in genetic algorithms and Davidor [8], makes use of a Lamarckian operator to improve machine learning strategies based on genetic algorithms. This suggests an addition to a basic selectionist scheme in which the environment tests but does not set a genome [25]. If there was a sufficiently robust model of the environment and adaptor-environment interaction it would be possible to specify adaptive strategies. It would then be possible to go further and consider contexts for the occurrence of such interactions and mechanisms which could bring it about.

Hall [18], in describing prokaryotic mutations, notes how some occur more frequently when they are advantageous to the cell than when they confer no advantage. From the point of view of the Evolutionary Synthesis, the probability that a particular mutation will occur is a constant property of the organism and is unresponsive to the environment or to any benefit the mutation may confer. This is arguably a central premise in the neo-Darwinian scheme. However, there are several experimental results which do not fit into this model. Following work on *Escherichia coli*, Cairns notes that the probability that a particular mutation will occur depends both on the external environment and the likelihood that the mutation will be useful ([4], [5]). From this standpoint, Weissman's barrier is NOT inviolate and information flow between germ and soma is in some ways bidirectional. A number of hypotheses have been proposed of mechanisms for accomplishing genomic instability which can be switched on and off depending on stress. Hall [18] describes three models for explaining *directed mutation*, the Cairns model based on instruction rather than selection, the Stahl model in which nutritionally depleted (i.e., stressed) cells carry out mutation repair more slowly and the Hall model in which stress gives rise to hypermutable states.

This kind of thinking indicates "talkback" between the environment and the genome. A simple EA application would be the environment-sensitive analogue of an enzyme-based DNA repair system. It is far from the case that selectionism is not the appropriate model. In the ecology of the genome interaction and the accumulation of data within and across states (i.e., the cycles of Figure 2) also plays a key role. Cullis [7] discusses environment-system interactions focussing on transposable elements as sources of variation. He hypothesises that a shift in the environment to which the plant has become adapted should increase the rate at which underlying mechanisms generate genotypic and phenotypic variation. Experimental evidence from investigations with flax shows transposon-mediated mechanisms for achieving such variation in which the rates of transposition are affected by the external environment and the stage of development. Transposon-mediated genomic restructuring can range from small changes of a few nucleotides to major modifications of large sections of a chromosome. Transposable elements do provide a mechanism by which the mutation rate and thereby the range of genetic diversity can increase during periods of stress.

We see further ecological issues raised in an insightful article about the notion of directed mutation by Keller [22]. She argues that there is a range of phenomena which are not easily susceptible to a neo-Darwinian interpretation. Her argument for providing an explanatory model of directed mutation is by analogy, and can be summarised as follows. Bacteria show a chemotactic effect to a substrate gradient despite the lack of sensors. It is achieved because their random motion increases with increasing substrate concentration. At the population level, this is seen as net flux into increased substrate concentrations and is an example of what ALife proponents call "Swarm Intelligence". In this case direction is at the population level and in Keller's evolutionary analogy, mutation is compared to the motion of a flagellum and net bacterial flux is like adaptation. Thus we see the emergence of self-organising behaviour (see also, Kauffman [21]). It is also possible to argue that cognitive properties are demonstrated at the level of population (see [27]). We see in this example how the behaviour of independent units, whether genes or flagellae, can collectively exhibit complex, directional behaviours. However, in the expression of phenotypes, genes are members of an ecology, interacting with each other through intracellular communication systems. Lamarckian and orthogenetic models of adaptation require greater interaction between environment and adaptor than in selectionist systems. That the debate of directed mutation and non-chromosomal inheritance goes on in biology is not our central concern here, our aim is to design better EAs. However, the problems even for EA design are considerable for example, what is stress and how will it be perceived in an EA ?

2.4 Developmental and Temporal Organisation in EAs

Spatial effects in population structure have been investigated in a variety of ways (see Sumida and Hamilton [33] and Davidor, [9] for two interesting examples). Another dimension to EA design concerns temporal effects. A scheme was presented in the first section of this paper which demonstrated some self-similar effects in biological cycles and the way in which some adaptors are both replicators and adaptors. Ebeling [14] applies developmental strategies to EA design.

In some of our experiments we have sought to structure populations based on temporal rather than spatial distribution using a life cycle as the source. Figure 8 summarises a three stage life cycle in which the 'f' values represent fecundity (reproducibility) of the stage and the 'm' values mortality.

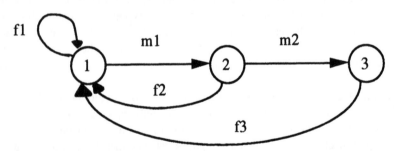

Figure 8 - A Three Stage Life Cycle

This digraph is represented in Leslie Matrix form as:

$$
\begin{pmatrix}
f1 & f2 & f3 \\
m1 & 0 & 0 \\
0 & m2 & 0
\end{pmatrix}
$$

The fecundity and mortality values are usually constrained by the algorithms regulating the GA such that exponential increases or decreases in the population are prevented.

The idea behind this approach is that a life cycle introduces a play-off between memory for old strings/recombinants and novelty with generating new ones. For example, low m values will reduce the memory effect and enhance novel recruitment. A kind of equivalent to inter-demic migration is the transition from one stage to another. So far, it has been applied to classifier GAs with some encouraging results. Some initial experiments have indicated that the introduction of the life history effect can enhance standard GA performance although a lot more work is needed to support any general statements about this approach.

This source idea will continue to be explored with respect to the temporal hierarchy of Figure 2. One area of ongoing research concerns the application of selectionist and connectionist ideas in the cell cycle and life cycles. We are also looking at a range of comparative life histories to indicate the relations between learning and selection. The case of Neural Darwinism and the way somatic selection can operate in the developmental process is a very good example of a biological theory of selection and development in cell populations [15]. Two sources for biocomputing are clonal and neuronal group selection. Indeed, these selectionist systems are also connectionist.

3 Concluding Remarks

For the purposes of the current discussion, the biological sources for gene ecologies have been related to two interactor levels namely, cell and ecosystem. However, both have been described with respect to them having an ecology so that it is possible to think of similarity effects. Figure 9 summarises the nesting of relations associated with biological sources at these two levels as they have been developed in previous sections.

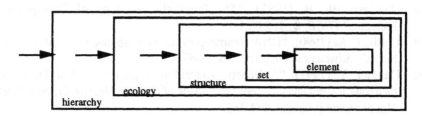

Figure 9 - Nesting of Relations

Some general statements can now be presented about the direction of this work:

- Genes are structures and not just sets that exist in ecologies and populations. They are not the only biomolecular replicators.
- Given that a greater understanding of the biology can help refine/extend the EA operators, strings which encode problem solutions are but one representational form. An appreciation of structure and ecology could extend

this. Some nonselectionist mechanisms could be valuable to EA design, *talkback* is an example.

- Hierarchical relations in genetic systems occur in the regulatory effects of certain genes and their products. This is best appreciated from the point of view of ecologies.
- The self-similarity of the temporal hierarchy, and the interplay between structure and ecology is a very rich source of ideas about interactions between evolutionary and connectionist systems.

Acknowledgements

Thanks to Barry Hall for some valuable details about talkback, Brian Sumida for information about regulons and their application, Geof Parker for insights into the working of animal systems and Peter Miller for clarifying some details about bacterial genetics. My thanks also to a number of past and present students who have carried out experimental work, John Coyne, Paul Devine, Tracey Fitzgerald and John Killoran.

References

[1] Bäck, T. (1992), "The Interaction of Mutation Rate, Selection and Self-Adaptation within a Genetic Algorithm", in Manner, R. & Manderick, B. (eds), *Parallel Problem Solving from Nature, 2*, Elsevier: Amsterdam, 85-94.

[2] Bäck, T. (1994), *Evolutionary Algorithms in Theory and Practice*, Ph.D. Thesis, University of Dortmund.

[3] Bernardi, G. (1989), "The Isochore Organisation of the Human Genome", *Annual Review of Genetics*, **23**, 637-661.

[4] Cairns, J., Overbaugh, J. & Miller, S. (1988), "The Origin of Mutants", *Nature*, **335**, 142-145.

[5] Cairns, J. & Foster, P.L. (1991), "Adaptive Reversion of Frameshift Mutation in *Escherichia coli*", *Genetics*, **128**, 695-701.

[6] Charlesworth, B. (1987), "The Population Biology of Transposable Elements",*TREE*, **2**, 1, 21-23.

[7] Cullis, C.A. (1990), "DNA Rearrangements in Response to Environmental Stress", *Advances in Genetics*, **28**, 73-97.

[8] Davidor, Y. (1991), "A Genetic Algorithm Applied to Robot Trajectory Generation", in Davis, L. (ed), *Handbook of Genetic Algorithms*, New York: Van Nostrand Reinhold, 144-165.

[9] Davidor, Y. (1994), "Free the Spirit of Evolutionary Computing: the Ecological Genetic Algorithm Approach", in Paton, R.C. (ed) *Computing with Biological Metaphors*, London: Chapman and Hall.

[10] Demple, B. (1991), "Regulation of Bacterial Oxidative Stress Genes", *Annual Review of Genetics*, **25**, 315-337.

[12] Desalles, J.L. (1992) "Biomimetic uses of Genetic Algorithms", in Manner, R. & Manderick, B. (eds), *Parallel Problem Solving from Nature, 2*, Elsevier: Amsterdam, 127-135.

[13] Drake, J.W. (1991), "Spontaneous Mutation", *Annual Review of Genetics*, **25**, 125-146.

[14] Ebeling, W. (1992), "The Optimisation of a Class of Functionals based on Developmental Strategies", in Manner, R. & Manderick, B. (eds), *Parallel Problem Solving from Nature, 2*, Elsevier: Amsterdam, 463-468.

[15] Edelman, G.M. (1988), *Topobiology*, New York: Basic Books.

[16] Eldredge, N. (1986), "Information, Economics, and Evolution", *Ann. Rev. Ecol. Syst.* **17**, 531-369.

[17] Hall, B.G. (1989), "Selection Adaptation and Bacterial Operons", *Genome*, **31**, 265-271.

[18] Hall, B.G. (1990), "Spontaneous Mutations that occur more often when Advantageous than when Neutral", *Genetics*, **126**, 5-16.

[19] Head, T. (1992), "Splicing Schemes and DNA", *Nanobiology*, **1**, 335-342.

[20] Jablonka, E., Lachmann, M. & Lamb, M.J. (1992), "Evidence, Mechanisms and Models for the Inheritance of Acquired Characteristics", *J. theor. Biol.*, **158**, 245-268.

[21] Kauffman, S.A. (1993) The Origins of Order, New York: Oxford University Press.

[22] Keller, E.F. (1992), "Between Language and Science: The Question of Directed Mutation in Molecular Genetics", *Perspectives in Biology and Medicine*, **35**, 2, 292-306.

[23] Kunz, B.A. & Kohlami, S.E. (1991), "Modulation of Mutagenesis by Deoxyribucleotide Levels", *Annual Review of Genetics*, **25**, 339-359.

[24] Landman, O.E. (1991), "The Inheritance of Acquired Characteristics", *Annual Review of Genetics*, **25**, 1-20.

[25] Manderick, B. (1994), "The Importance of Selectionist Systems for Cognition", in Paton, R.C. (ed) *Computing with Biological Metaphors*, London: Chapman and Hall.

[26] Neidhart, F.C., Ingraham, J.L. & Schaechter, M. (1990), *Physiology of the Bacterial Cell*, Sinauer Associates: Sunderland, MA.

[27] Paton, R.C. (1993) "Some Computational Models at the Cellular Level", *BioSystems*, **29**, 63-75.

[28] Paton, R.C. (1994), "Metaphors, Models and Bioinformation", Conference on Foundations of Information Science, Toledo '94, Madrid.

[29] Paton, R.C., Nwana, H.S., Shave, M.J.R., Bench-Capon, T.J.M. & Hughes, S. (1991), "Transfer of Natural Metaphors to Parallel Problem Solving Applications", in Schwefel, H-P. & Maenner, R. (eds), *Parallel Problem Solving from Nature, Lecture Notes in Computer Science*, Springer: Berlin. 363-372.

[30] Paton, R.C., Nwana, H.S., Shave, M.J.R. & Bench-Capon, T.J.M. (1994), "An Examination of Some Metaphorical Contexts for Biologically Motivated Computing", *British Journal for the Philosophy of Science*, in press.

[31] Schull, J. (1991), "The View from the Adaptive Landscape", in Schwefel, H-P. & Maenner, R. (eds), *Parallel Problem Solving from Nature, Lecture Notes in Computer Science*, Springer, Berlin, 415-427.

[32] Schmitt, O.H. (1993) personal communication.

[33] Sumida, B.H. & Hamilton, W.D. (1994), "Both Wrightian and 'Parasite' Peak Shifts Enhance Genetic Algorithm Performance in the Travelling Salesman Problem", in Paton, R.C. (ed) *Computing with Biological Metaphors,* Chapman and Hall.

[34] Voigt, H-M., Santibanez-Koref, I., Born, J. (1992), "Hierarchically Structured Distributed GAs", in Manner, R. & Manderick, B. (eds), *Parallel Problem Solving from Nature, 2,* Elsevier: Amsterdam, 155-164.

Exploiting mate choice in evolutionary computation: Sexual selection as a process of search, optimization, and diversification

Geoffrey F. Miller
Cognitive and Computing Sciences
University of Sussex
Falmer, Brighton BN1 9QH, England
geoffm@cogs.susx.ac.uk

Abstract

Sexual selection through mate choice is a powerful evolutionary process that has been important in the success of sexually-reproducing animals and flowering plants. Over the short term, mate preferences evolve because they improve the outcome of sexual recombination. Over the long term, assortative mate preferences can help maintain genetic diversity, promote speciation, and facilitate evolutionary search through optimal outbreeding; selective mate preferences can reinforce the speed, accuracy, and efficiency of natural selection, can foster the discovery and propagation of evolutionary innovations, and can function as aesthetic selection criteria. These strengths of sexual selection complement those of natural selection, so using both together may prove particularly fruitful in evolutionary computation. This paper reviews the biological theory of sexual selection and some possible applications of sexual selection in evolutionary search, optimization, and diversification. Simulation results are used to illustrate some key points.

1 Introduction: The Evolutionary Importance of Sexual Selection

The overwhelming biological success of sexually-reproducing animals and flowering plants has often been attributed to the raw power of sexual recombination. Yet the diagnostic feature shared by these two groups is not just sexual recombination per se (which bacteria and non-flowering plants also use), but rather, sexual selection through mate choice. Typically, animals are sexually-selected by opposite-sex conspecifics (Darwin, 1871) and flowering plants are sexually selected by pollinators such as insects and hummingbirds (Darwin, 1862). This suggests that the evolution of phenotypic complexity and diversity may be driven not simply by natural-selective adaptation to econiches, but by a complementary interplay of natural selection and sexual selection. If mate choice has been instrumental in the evolutionary success of higher animals and plants, perhaps it has been under-estimated as a process of search, optimization, and diversification.

Mate preferences evolve because they improve the outcome of sexual recombination for the organisms that use them. Selective or assortative mating based on mate preferences tends to produce offspring with higher viability, fertility, or attractiveness than random mating. Whereas the merits of recombination through random mating are so unclear that biologists still argue vehemently about why sex evolved (see Michod & Levin, 1988; Ridley, 1993), the merits of selective mate choice are now almost universally recognized (see Andersson, 1994; Cronin, 1992). Indeed, whenever a species uses sexual recombination, and has the sensory-motor capacity for mate choice, mate choice almost always evolves to guide who recombines with whom.

Biological interest in sexual selection has grown enormously in the last 15 years, but has not yet been integrated with the evolutionary computation view of evolution as a process of search and optimization. For example, genetic algorithms typically use recombination (crossover) without mate choice. Crossover can recombine useful schemata or building blocks from different parents, and this is the major advantage that genetic algorithms have over other stochastic, population-based, hill-climbing search algorithms (Eshelman & Schaffer, 1993). But recombination has two basic drawbacks under random mating: good schemata can be disrupted, and genetic diversity can be eroded. Mate choice, particularly assortative mating, is the major way that sexually-reproducing organisms lower these costs while preserving recombination's benefits.

This paper aims to promote interest in sexual selection theory, to suggest some practical ways that mate choice may improve evolutionary computation, and to inspire further work in this area. Space limitations preclude a complete discussion of relevant theory, methods, and results. For a longer review of sexual selection theory and its computational implications, see Miller and Todd (in press). For details of simulation methods and results see Todd and Miller (1991) on assortative mating and speciation, Miller and Todd (1993) on selective mating and stochastic runaway effects, and Todd and Miller (1993) on the evolution of parental sexual imprinting under sexual selection. For a discussion of the role of sexual selection in human evolution see Miller (1993; in press) and Ridley (1993); for a comprehensive, up-to-date biological review of sexual selection, see Andersson (1994).

2 Why Mate Preferences Evolve

Organisms that reproduce sexually should avoid mating randomly, because the genetic quality of one's mate will determine half the genetic quality of one's offspring (see Pomiankowski, 1988). Assortative mating with an organism similar to oneself can improve offspring fitness by providing them with co-adapted, strongly-linked genes that function well together, and selective mating with high-fitness organisms can improve offspring fitness by giving them genes that have clearly prospered in the current environment. The key to choosing mates adaptively is to evolve a mate choice mechanism that has 'internalized' the likely long-term fitness consequences of reproducing with different kinds of potential mates. The attractiveness of a potential mate should reflect the expected fitness of any offspring that one might have with it.

A major strength of evolutionary computation methods is that they can find solutions without using problem-specific knowledge, algorithms, or heuristics. However, if problem-specific methods could be evolved automatically, they would almost always be useful. Mate preferences can be viewed as evolved heuristics that improve the ways selection and recombination operate. The evolution of mate preferences then can be seen as a way of automatically incorporating knowledge about the current and past fitness landscape into evolutionary search. Just as evolutionary reinforcement learning (ERL) can 'internalize' utility functions from the environment into an organism's reinforcement learning system (Littman and Ackley, 1991), sexual selection can internalize fitness functions from the environment into an organism's mate choice system. At the level of an evolving species, mate preferences make a species more 'intelligent' by guiding future evolution based on past information about adaptive success and failure.

Evolvable mate preferences lead to complex evolutionary dynamics, because sexually-selected traits adapt to the current distribution of preferences in the population, but preferences also adapt to the current distribution and fitnesses of traits in the population. Over the short term, traits adapt to preferences and vice-versa; but over the long term, both traits and preferences can co-evolve in unpredictable directions (Miller & Todd, 1993). The following section describes some simple ways of incorporating mate choice into evolutionary computation methods.

3 Adding Mate Preferences to Genetic Algorithms

Traditional genetic algorithms use a fitness function to select parents and then pairs parents at random for sexual recombination. Sexual selection based on mate preferences requires an additional selection step after parents have been picked. The sexual selection process must somehow allow each parent to sample a number of potential mates and to 'choose' the most acceptable one based on its mate preferences, perhaps stochastically, and perhaps mutually. Although this section focuses on mate preferences for genetic algorithms, similar methods could be used in genetic programming, evolution strategies, classifier systems, and artificial life systems.

Mate preferences can be represented as probability-of-mating (POM) functions defined across an entire n-dimensional phenotype space (Todd & Miller, 1991; Miller & Todd, 1993). The POM function assigns to every possible phenotypic location, and thus to every possible mate, a certain per-encounter probability of mating, which can be represented as a height in the $n+1$st dimension. Peaks on each individual's POM function correspond to ideal, totally attractive mates; low points correspond to repulsive, totally unacceptable mates. Sexual selection is driven by the topography of evolving POM functions interacting with the evolving frequency distribution of available mates in phenotype space, and with the natural-selective fitness landscape.

Mate preferences can be genetically encoded by preference genes that specify the key parameters of POM functions. One way to focus evolution on an efficiently low number of POM parameters is to construct fairly standardized POMs with respect to a fairly limited class of "reference position" in phenotype space. Reference positions could be genetically specified such that they are set to one's own phenotype (as in assortative mating), one's parent's phenotype (as in parental imprinting), the current population average phenotype (as in human preferences for face shape), or an absolute position in genotype or phenotype space. Given the reference position, the remainder of the POM can be constructed around it, either as a radially symmetric, 'non-directional' function centered on the reference position (as in Todd & Miller, 1991), or as a vector-like 'directional' function that points away from the reference position in some direction, e.g. such that one prefers a mate much larger or more colorful than oneself (as in Miller & Todd, 1993). We have found that sexual selection works better if the total volume under POM functions is normalized to be equal across individuals; otherwise, many mating schemes will favor 'promiscuous' individuals who are attracted to everybody, thereby undermining the effects of sexual selection.

In nature, mate preferences must rely on observable phenotypic information. But simulation could allow individuals to choose mates based on a much richer database: complete and accurate information about their genotype, their phenotype, their past behavior, their performance on fitness tests, their ancestry, or even the expected performance of offspring produced with them. A preference based on direct genotypic information, for example, could be coded in a single gene specifying an ideal Hamming distance between one's own genotype and the genotype of a potential mate. The next section discusses the benefits of this type of assortative mating.

4 Assortative Mating I: Diversification Through Automatic Niching and Speciation

Genetic algorithms are good at hill-climbing and at overcoming some kinds of 'deception', but they have trouble escaping from local optima and finding multiple global optima in complex fitness landscapes. Short-term selection tends to reduce genetic variation through 'exploitation' of local optima, but the success of long-term evolution through 'exploration' depends on maintaining significant genetic variation. Mutation is not enough, because it produces only superficial variants tightly clustered around currently common genotypes, and sexual recombination with random mating is even worse, because given diverse parents it tends to produce lower-variance offspring with intermediate values on quantitative genetic traits. The problem is to preserve 'deep' genetic diversity as opposed to the superficial variation maintained by mutation and recombination. Two solutions have been developed for maintaining deep diversity: niching methods and spatially structured populations. They will be reviewed briefly, and then a new, complementary method for preserving deep diversity will be discussed: assortative mating.

Niching methods maintain deep diversity by distributing sub-populations across multiple fitness peaks. Normally this is difficult because genetic drift tends to disrupt the numerical balance between sub-populations, often leading populations to 'collapse' down to a single peak. Some methods fight genetic drift fairly passively, through replacing individuals with similar offspring somehow: in preselection (Cavicchio, 1970; see Mahfoud, 1992), offspring can only replace one of their parents; in the crowding method (De Jong, 1975), offspring replace the genotype most similar in

Hamming distance, selected from a randomly drawn subpopulation of a certain size. Other methods fight drift more actively through giving higher fitness to individuals in uncrowded regions of genotype or phenotype space. In the sharing scheme (Deb & Goldberg, 1989; Goldberg & Richardson, 1987), individuals within a certain genetic or phenotypic distance of each other must divide up the 'locally available' fitness. Sharing is like ecological competition: it tends to distribute individuals across fitness peaks in proportion to the heights of the peaks. Of these niching methods, the active ones work better than the passive ones, and of the active ones, phenotypic sharing seems to work better than genotypic sharing (Goldberg, Deb, & Horn, 1992; Goldberg & Richardson, 1987). However, these methods require the programmer to set various parameters rather carefully, based on assumptions about the number and distribution of fitness peaks.

Spatially structured populations maintain genetic diversity by maintaining geographic diversity, typically through local competition, local mating, local replacement, and fairly low migration rates. Such populations are naturally implemented on parallel computers, so the software method of spatial structuring has often been conflated with the hardware implementation of 'parallel GAs', through they are conceptually distinct (Gordon & Whitney, 1993). In 'coarse-grained' methods inspired by Wright's shifting balance theory (1932), the population is divided up into 'demes' with some degree of migration between them in each generation; demes can be spatially structured in relation to one another according to some topology and distance metric (as in 'stepping stone' models) or can simply exist as nodes equidistant from one another (as in 'island models'). In 'fine-grained' methods (a.k.a. 'diffusion', 'isolation-by-distance', and 'cellular' methods) individuals rather than demes are spatially arranged, typically as nodes on a 2-D grid. Spatial structuring has proven very useful in avoiding premature convergence, maintaining genetic diversity, allowing niche differentiation, finding global optima faster, finding multiple optima in the same run, and exploiting the power of parallel computers (e.g. Collins & Jefferson, 1991; Davidor et al., 1993; Gorges-Schleuter, 1989, 1992; Manderick & Spiessens, 1989; Mühlenbein, 1989; 1992). In spatial structuring, niching emerges somewhat passively, because different geographic areas may evolve towards different peaks due simply to different stochastic effects; nevertheless, spatial structuring reliably and efficiently explores multimodal fitness landscapes, and requires fewer assumptions than non-spatial niching methods.

For all the success of niching and spatial structuring in preserving deep genetic diversity, they overlook the main methods that nature uses in generating biodiversity: assortative mating and speciation. In assortative mating, animals pair up based on their similarity, such that "like mates with like". In speciation, a lineage splits apart into reproductively isolated populations (species) that can no longer interbreed. Speciation can be viewed as the extreme outcome of assortative mating. Speciation is important in evolutionary computation because it lets lineages specialize in exploring different peaks in phenotype space, without wasting effort on trying to interbreed across fitness valleys. In nature, speciation creates biodiversity to fill econiches; in evolutionary computation, speciation creates separate lineages doing parallel searches in fitness landscapes.

Speciation can be initiated by geographic isolation (Mayr, 1942) or ecological specialization (Dobzhansky, 1937), both of which tend to split populations apart. But for the initial split to be consolidated through reproductive isolation, mate preferences must reinforce the differentiation. Indeed, in Todd and Miller (1991) we showed that mate preferences can spontaneously differentiate in a population (through a combination of genetic drift, mutation pressure, and assortative mating), such that speciation can occur even without geographic isolation or ecological specialization. Our model of 'spontaneous sympatric speciation' through assortative mating is consistent with the fact that sexually-reproducing animals and flowering plants show vastly greater biodiversity and higher speciation rates than other kinds of organisms that do not use mate choice (Miller & Todd, in press). Of course, once speciation occurs, the well-known ecological 'exclusion principle' (two species cannot occupy the same niche in the same area at the same time) tends to push species into separate econiches or separate habitats. Thus, reproductive isolation through the spontaneous (stochastic) differentiation of mate preferences can initiate and consolidate other adaptive differences between populations, and can lead to ever-increasing biodiversity and niche differentiation.

Some genetic algorithms research has flirted with speciation. Booker (1985) explored the use of evolvable 'mating templates' that allow assortative and selective mating. Some sharing schemes use assortative 'mating restrictions' to minimize the disruptive effects of recombination across different fitness peaks; and it has been found that sharing plus assortative mating works better than sharing alone (Deb & Goldberg, 1989; Goldberg & Richardson, 1987). Still, a more systematic approach to maintaining genetic diversity through assortative mating might prove valuable. For example, mate preferences would be expected to register when a population has split apart on different fitness peaks, by evolving to discourage cross-peak mating (since preferences that encourage cross-peak mating get passed on to inviable hybrids and die out.) Evolvable preferences thus should be able to adaptively and spontaneously promote niching as a lineage evolves, without the programmer having to estimate the number and distribution of peaks ahead of time.

The main problem with maintaining speciation purely through assortative mating is that relative species numbers are subject to genetic drift if competition and replacement are handled as global operations: incipient species tend to disappear due to sampling error. We have developed a method for minimizing the effects of drift under assortative mating through allowing only one parent at a time to exert mate choice (Todd & Miller, 1991), but this only delays the inevitable extinction of species due to drift. Inman Harvey and I have recently been exploring the effects of assortative mating for adaptation in Kauffman's (1993) N-K model of fitness landscapes; so far, this problem of genetic drift seems to override any benefit of assortative mating, at least for a rough sample of K values between 2 and 5, N values of 10 to 100, population sizes of 50 to 500, and mutation rates from 0.3 to 3.0 per genome.

In nature, relative species numbers can be maintained because species diverge ecologically or geographically. In evolutionary computation then, it probably makes sense to combine ecological niching, spatial structuring, and assortative mating. GA methods could be developed to consolidate speciation by recognizing ecologically differentiated populations through phenotypic cluster analysis, and/or reproductively isolated populations through analyzing patterns of assortative mating, and then allowing each population to run autonomously as a separate GA 'lineage'. Such a method could be implemented on a parallel computer by starting with one species running on one processor, and assigning new species to other processors as they split off, up to the biodiversity limit imposed by the number of processors. Evolved mechanisms for assortative mating should prove critical to the success of such methods because it is the clearest way to allow a population to decide by itself when it 'wants' to speciate.

5 Assortative Mating II: Search Through Optimal Outbreeding

Aside from maintaining genetic diversity, another fundamental problem for GAs is maximizing the recombination of useful schemata and building blocks while minimizing the disruption of useful ones. In nature, this seems to be accomplished by 'optimal outbreeding' (see Bateson, 1983), a very common form of assortative mating where animals prefer the somewhat similar to the exactly similar. Optimal outbreeding strikes a balance between inbreeding, which tends to preserve useful schemata against disruption through crossover, and outbreeding, which allows the incorporation of useful new mutations and schemata.

Schaffer & Eshelman (1991) suggested that genetic operators such as crossover could be characterized by a "safety ratio", representing the ratio of the probability that an offspring will be better than its parents to the probability that it will be worse: P(better)/P(worse). Optimal outbreeding can be viewed as a way of improving the safety ratio of crossover. It allows individuals to avoid breeding with others who are so similar as to be make recombination useless, and to avoid breeding with others so different as to make recombination too risky.

Eshelman & Schaffer (1991) developed an 'incest avoidance' method of preventing premature convergence by allowing individuals to mate only if their Hamming distance is above a certain threshold. This threshold drops according to some schedule as the population converges. They found that incest-prevention significantly increases the speed and robustness of finding global optima on 11 of 13 test functions. Allowing optimal outbreeding preferences to evolve can be seen as a way of automating this incest avoidance system. In Todd and Miller (1991) we developed a single-parameter method for specifying degree of desired outbreeding, and we found that even with a moderate degree of incest avoidance, speciation can still occur through assortative mating. Thus, the benefits of incest avoidance at the local level need not undermine the benefits of assortative mating at the global level.

The advantage of allowing outbreeding parameters to evolve becomes clear through the following argument. The key problem in crossover is to match the genetic correlation between parents to the 'correlation length' of the fitness landscape,

(Manderick et al., 1991). Longer correlation lengths allow parents with larger Hamming distances between them to still produce reasonably fit offspring. For rugged fitness landscapes with short correlation lengths, crossover works best between similar parents, but for smoother landscapes with longer correlation lengths, crossover between more dissimilar parents explores the space more efficiently (Kauffman, 1993). Since we may not know the correlation length of a fitness landscape ahead of time, we should allow outbreeding parameters to evolve as mate preferences.

6 Selective Mating I: Optimization Through Fitness-Based Sexual Selection

In selective mating, animals prefer mates that display some desired trait, regardless of their similarity on that trait. For example, peahens prefer peacocks with large tails, regardless of their own tail size. Whereas assortative mating affects mainly the genetic linkages among alleles, selective mating affects allele frequencies themselves, because animals with popularly-sought traits have more offspring then those with rarely-sought traits. So, whereas assortative mating tends to preserve genetic diversity, selective mating usually imposes directional or stabilizing selection that reduces diversity.

Why would animals ever use selective mating if natural selection is already in force? Natural selection is a powerful force over the long term, but it can be horribly noisy, irregular, inaccurate, and inefficient within each generation. In the wild, small fitness differences are often too weak to drive evolution very efficiently, and large fitness differences are often due to chance (e.g. chance effects in predation, competition, and pathogen exposure). Both problems reduce heritable variation in reproductive success, which is what drives evolution. Evolutionary computation avoids the first problem by using perfectly accurate, deterministic fitness functions, and avoids the second problem through various forms of fitness scaling or rank-based reproduction. In nature, selective mating helps overcome both problems.

Noise can arise at any point in the mapping from genotype to fitness. Noisy development can perturb the mapping from genotype to phenotype, and noisy evaluations can perturb the mapping from phenotype to fitness, either because phenotypes face a limited sample of the full range of performance tests during their lifetime, or because the performance tests themselves are noisy. In the wild, natural selection has no incentive to reduce this noise or to provide an evolving species with accurate information about the local fitness landscape. By contrast, selective mate preferences evolve precisely in order to reduce this noise and to provide individuals with maximal information about the genetic quality of potential mates. Selective mating can thereby overcome noise in several ways. First, if natural selection is directional but noisy, mate preferences can internalize the current average selection 'vector', thereby smoothing out and speeding up the evolutionary trajectory of the population. For example, if a fitness function tends to select for a particular trait, but with high levels of noise, mate preferences should evolve to favor that trait, and can impose sexual selection with much less noise. Second, mate preferences could specify the sample size of performance evaluations to apply to each individual (just as female animals demand that male courtship displays be repeated a certain number of times before they choose a mate), so mate preferences could adjust the amount of noise across evaluations; this would be an automatic solution to the 'sample allocation problem' studied by Aizawa and Wah (1993) for noisy fitness functions. Third, mate preferences could also implement a form of adaptive sampling of the problem space by biasing search towards currently tricky problems (just as female animals prefer males who deliver difficult-to-find prey as nuptial gifts). Schultz (1991) found that this sort of difficulty-biased sampling leads to solutions more efficiently and robustly than uniform sampling.

Aside from the problem of noise, there is the problem of maintaining meaningful variation in reproductive success: fitness scores may be so similar that evolution happens very slowly. GA methods of fitness scaling, rank-based selection, and tournament competition have been developed to address this problem, but such methods require experimenters to set some scaling factor or bias term based on assumptions about the fitness landscape's structure. When fitness functions are noisy, such assumptions are especially tricky for rank-based and tournament methods (Fogarty, 1993). In nature, sexual selection is the fitness scaler: females convert slight differences in male quality into large differences in male reproductive success by mating disproportionately with higher-quality males. The scaling factor is determined by the degree of polygny, which in turn is determined by evolved mate preferences. Each individual has strong incentives for using mate preferences that register slight differences in the fitness of potential mates, so as to give their offspring a slight adaptive advantage in the next generation. These preferences can maintain substantial variance in reproductive success even when natural-selective differences in viability are minimal. Thus, selective mate preferences tend to impose an automatic, flexible form of fitness scaling that evolves to match the fitness landscape's structure.

A final advantage of selective mating is that 'directional preferences' can implement a form of 'evolutionary momentum' (Miller & Todd, 1993), which is analogous to 'momentum adaptation' developed by Ostermeier (1992) for evolution strategies, or the 'momentum term' used in neural network back-propagation learning. That is, directional preferences tend to keep a population moving in a particular direction in phenotype space. Mate preferences should automatically evolve to exploit and internalize current gradient information about the local natural-selection vector. Momentum is most useful in smooth fitness landscapes, and in non-stationary landscapes where fitness peaks are moving around in a somewhat predictable way (e.g. under co-evolution). In this latter case, directional mate preferences could evolve to point in the direction of consistent peak-movement, to anticipate selection pressures generation-by-generation (see Todd & Miller, 1993).

7   Selective Mating II: Search Through Neophilic Sexual Selection

Aside from preferring similarity and quality in mates, many animals prefer novelty. The 'rare-male' effect results from female choice in favor of rare phenotypes, and is a powerful form of negative frequency dependent selection. Darwin (1871) took neophilia (preference for novelty) seriously as a force in sexual selection, implying that it could often lead to evolutionary innovations unrelated to ecological demands. Neophilia can be implemented in GAs by allowing individuals to sample several individuals and to favor the most unusual one; neophilia can be combined with other forms of assortative and selective mating, either as a weighted factor applied to the same phenotypic traits, or as a force applied to independent traits.

Without neophilia, useful new mutations and schemata have trouble establishing a toehold in an evolving population: because they start out at such a low frequency, their disappearance through genetic drift is often more likely than their propagation through positive selection. With neophilic sexual selection, new mutations and schemata get an immediate boost that protects them from extinction and promotes their spread. From an individual animal's perspective, any novelty in a sexually-mature potential mate is at least promising, because it has survived at least one round of natural selection. Of course, deleterious novelties will be selected out eventually. Whereas the mutation rate scales the number of raw innovations introduced into a population per generation, neophilia scales the number of potentially useful innovations propagated towards higher frequency in the population (see Miller & Todd, in press, for more detailed discussion). Variable mutation rate schedules are useful in genetic search (Fogarty. 1989), and variable amounts of neophilia could approximate such schedules by adjusting the tendency of populations to propagate new mutations.

Neophilic selective mating has surprisingly important evolutionary implications. One central problem in evolution is how to start with a population already converged to a (locally optimal) fitness peak, and move somewhere else that proves better (Harvey, 1992). Sewall Wright (1932) suggested "The problem of evolution as I see it is that of a mechanism by which the species may continually find its way from lower to higher peaks ... In order that this may occur, there must be some trial and error mechanism on a grand scale", i.e. a scale over and above that of mutation plus selection. Wright's shifting balance theory, based on the geographic isolation of demes, was one way to promote divergence away from locally optimal peaks. Each deme can undergo a slightly different 'adaptive walk' in search of a better fitness peak (Kauffman, 1993). But neophilic sexual selection can provoke much more powerful, directional adaptive walks than Wright's model allows.

Directional mate preferences implement an indirect form of neophilia because the directional preferences usually favor regions of phenotype space that are sparsely populated. In our investigations of directional selection even without explicit neophilia (Miller & Todd, 1993), we found that directional preferences are powerful ways of escaping from local fitness peaks, to allow an entire population to explore the surrounding phenotype space in a coherent, directional way. Directional preferences tend to take populations on adaptive walks that have long 'foray lengths', capable of escaping from local optima in rugged fitness landscapes.

This sort of capricious exploration can be viewed as a form of 'runaway sexual selection', first postulated by Fisher (1930) and recently simulated by Collins and Jefferson (1992). Runaway is one of the most powerful evolutionary forces, accounting for some of the most important innovations seen in biology, such as the human brain (Miller, 1993). By contrast, mutation by itself simply creates 'mutational clouds' or 'quasi-species' whose density falls off exponentially with Hamming distance from fitness peaks (see Eigen, 1992); these mutational clouds are extremely weak at escaping from fitness peaks. 'Mutational search' relies on passive, diffuse, undirected replication errors to move populations away from local optima, whereas sexual selection relies on the active, focused, directed power of mate choice. The fact that most of the minor evolutionary innovations that distinguish species from one another are sexually-selected traits, suggests that most adaptive walks away from local optima occur under the power of sexual selection (see Miller & Todd, in press). In summary, neophilic sexual selection may be to macroevolution what mutation is to microevolution: the main source of potentially adaptive heritable variation.

8 Selective Mating III: Search Through Aesthetic Sexual Selection

Many design problems include both functional criteria and aesthetic criteria. The functional criteria are often easier to implement as an explicit fitness function. To evolve designs through aesthetic criteria, several researchers have fallen back on using human judges to award fitness scores to individuals. For example, Dawkins' (1986) Blind Watchmaker program allows human users to evolve 'biomorphs' on the screen through an evolutionary process. Caldwell and Johnson (1991) replaced the police sketch artist with a GA, by using human witnesses to judge the similarity of evolved faces to those of crime perpetrators. Sims (1992) has had spectacular success in evolving computer graphics images and dynamical systems based on human aesthetic judgments; Todd and Latham (1992) have also developed amazing computer graphics images through simulated evolution with human aesthetic input.

Humans can get tired of judging things during long GA runs, however. But how can aesthetic decisions be internalized into a GA selection system, when aesthetic criteria are so hard to formalize? Mate preferences offer one possibility. Consider a GA evolving two populations simultaneously: a population of designs subject to human aesthetic choice, and a population of 'preferences' (e.g. neural networks whose inputs are designs and whose outputs are aesthetic scores) that receive fitness scores for their ability to predict and match the human aesthetic choice. If the preferences are sufficiently rich and flexible, they should evolve to internalize the human user's aesthetic criteria; they could then be locked in and used as the fitness function that selects further designs, as a kind of aesthetic sexual selection. Such methods could prove useful in art, advertising, product design, and architecture; perhaps CAD systems of the future will learn the user's aesthetic preferences through the simulated evolution of mate preferences, which can then guide the evolutionary generation of new designs.

Aesthetic criteria are useful for more than just evolving computer artwork. Many complex design problems are made more tractable by reliance on aesthetic heuristics such as symmetry, simplicity, streamlining, smoothness, and modularity. If these criteria were used even for purely functional design problems, greater search efficiency might result. Such criteria could either be internalized in mate preferences by evolving preferences to imitate human aesthetic decisions, or could evolve autonomously because they improve the functional efficiency of the designs they favor. The reasoning is that mate preferences can pick out phenotypic traits that are different from those on which natural selection acts, but that are highly correlated with natural-selective fitness. For example, the travelling salesman problem favors smooth, non-overlapping tours, but the explicit fitness function ('minimize total distance travelled') cannot directly select for smoothness or lack of overlap; these 'aesthetic' but heuristically useful criteria could evolve as selective mate preferences that guide evolution.

9 Conclusions and Prospects

Kauffman (1993) and others have suggested that populations tend to tune their mutation rates, crossover rates, and developmental mechanisms (e.g. degrees of epistasis) to keep themselves perched on the 'edge of chaos' such that they optimize evolution's ability to search complex fitness landscapes. This paper has suggested that evolvable mate preferences offer an even more powerful, direct, and biologically well-established way for populations to guide their evolutionary dynamics. Indeed, mechanisms for assortative and selective mating evolve precisely because they improve they way in which sexual recombination operates. If evolutionary computation methods fail to incorporate mate choice mechanisms into the entities under selection, they are unlikely to be exploiting the full power of recombination.

Sexual selection through mate choice was overlooked in biology for over a hundred years following Darwin (1871). The evolutionary computation community need not repeat the same mistake. Sexual selection is a powerful way of doing several things that natural selection finds difficult, such as maintaining genetic diversity, guiding sexual recombination, optimizing adaptations under weak or noisy natural selection, discovering and propagating evolutionary innovations, and implementing selection based on human aesthetic criteria. Natural and sexual selection are complementary because natural selection typically favors convergent evolution onto a few (locally) optimal solutions, whereas sexual selection often results in unpredictable, divergent patterns of evolution, with lineages speciating spontaneously and exploring the space of phenotypic possibilities according to their evolved mate preferences. As a method of combining exploitation and exploration, natural plus sexual selection is much more powerful than selection plus mutation, because sexual selection delivers more complex, more integrated, more promising innovations than raw mutation does. In conclusion, the rich biological literature on sexual selection (see Andersson, 1994; Bateson, 1983; Cronin, 1991; Ridley, 1993) sets the stage for a sexual revolution in evolutionary computation.

Acknowledgments

This research was supported by NSF-NATO Post-Doctoral Research Fellowship RCD-9255323 and NSF Research Grant INT-9203229. Thanks to Peter Todd and Inman Harvey for collaborating on the research, and to Dave Cliff and Phil Husbands for useful feedback.

References

Aizawa, A. N., & Wah, B. W. (1993). Dynamic control of genetic algorithms in a noisy environment. In Forrest (Ed.), pp. 48-55.

Andersson, M. (1994). *Sexual selection.* Princeton U. Press.

Bateson, P. (Ed.). (1983). *Mate choice.* Cambridge U. Press.

Belew, R. K., & Booker, L. B. (Eds.). (1991). *Proc. of the Fourth Int'l Conf. on Genetic Algorithms.* Morgan Kaufmann.

Booker, L.B. (1985). Improving the performance of genetic algorithms in classifier systems. In *Proc. of an Int'l Conf. on Genetic Algorithms and their Applications,* pp. 80-92.

Caldwell, C., & V. S. Johnston (1991). Tracking a criminal suspect through "face-space" with a genetic algorithm. In Belew & Booker (Eds.), pp. 416-421.

Cavicchio, D. J. (1970). *Adaptive search using simulated evolution.* Ph.D. thesis, Univ. Michigan.

Collins, R. J., & Jefferson, D. R. (1991). Selection in massively parallel genetic algorithms. In Belew & Booker (Eds.), pp. 249-256.

Collins, R. J., & Jefferson, D. R. (1992). The evolution of sexual selection and female choice. In Varela & Bourgnine (Eds.), pp. 327-336.

Cronin, H. (1991). *The ant and the peacock: Altruism and sexual selection from Darwin to today.* Cambridge U. Press.

Darwin, C. (1862). *On the various contrivances by which orchids are fertilized by insects.* John Murray.

Darwin, C. (1871). *The descent of man, and selection in relation to sex (2 vols.)* John Murray.

Davidor, Y., Yamada, T., & Nakano, R. (1993). The ECOlogical framework II: Improving GA performance at virtually zero cost. In Forrest (Ed.), pp. 171-176.

Dawkins, R. (1986). *The blind watchmaker.* Norton.

De Jong, K.A. (1975) *An analysis of the behavior of a class of genetic adaptive systems.* Ph.D. thesis, Univ. Michigan.

Deb, K., & Goldberg, D.E. (1989) An investigation of niche and species formation in genetic function optimization. In Schaffer (Ed.), pp. 42-50.

Dobzhansky, T. (1937). *Genetics and the origin of species.* (Reprinted 1982). Columbia U. Press.

Eigen, M. (1992). *Steps towards life: A perspective on evolution.* Oxford U. Press.

Eshelman, L. J., & Schaffer, J. D. (1991). Preventing premature convergence in genetic algorithms by preventing incest. In Belew & Booker (Eds.), pp. 115-122.

Eshelman, L. J., & Schaffer, J. D. (1993). Crossover's niche. In Forrest (Ed.), pp. 9-14.

Fisher, R. A. (1930). *The genetical theory of natural selection.* Clarendon Press.

Fogarty, T. C. (1989). Varying the probability of mutation in the genetic algorithm. In Schaffer (Ed.), pp. 104-109.

Fogarty, T. C. (1993). Reproduction, ranking, replacement, and noisy evaluations: Experimental results. In Forrest (Ed.), p. 634.

Forrest, S. (Ed.). (1993). *Proc. of the Fifth Int'l Conf. on Genetic Algorithms,* Morgan Kaufmann.

Goldberg, D.E. (1989). *Genetic algorithms in search, optimization, and machine learning.* Addison-Wesley.

Goldberg, D.E., Deb, K., & Horn, J. (1992). Massive multimodality, deception, and genetic algorithms. In Männer & Manderick (Eds.), pp. 37-46.

Goldberg, D.E., & Richardson, J. (1987). Genetic algorithms with sharing for multimodal function optimization. In *Proc. of the Second Int'l Conf. on Genetic Algorithms,* pp. 41-49.

Gordon, V. S., & Whitney, D. (1993). Serial and parallel genetic algorithms as function optimizers. In Forrest (Ed.), pp. 177-183.

Gorges-Schleuter, M. (1989). ASPARAGOS: An asynchronous parallel genetic optimization strategy. In Schaffer (Ed.), pp. 422-427.

Gorges-Schleuter, M. (1992). Comparison of local mating strategies in massively parallel genetic algorithms. In Männer & Manderick (Eds.), pp. 553-562.

Harvey, I. (1992). Species adaptation genetic algorithms: The basis for a continuing SAGA. In Varela & Bourgnine (Eds.), pp. 346-354.

Hillis, W. D. (1992). Co-evolving parasites improve simulated evolution as an optimization procedure. In C. G. Langton et al. (Eds.), *Artificial Life II,* pp. 313-324. Addison-Wesley.

Kauffman, S. A. (1993). *The origins of order: Self-organization and selection in evolution.* Oxford U. Press.

Kirkpatrick, M. (1987). The evolutionary forces acting on female preferences in polygynous animals. In Bradbury & Andersson (Eds.), pp. 67-82.

Littman, M. J., & Ackley, D. H. (1991). Adaptation to constant utility non-stationary environments. In Belew & Booker (Eds.), pp. 136-142.

Mahfoud, S. W. (1992). Crowding and preselection revisited. In Männer & Manderick (Eds.), pp. 27-36.

Manderick, B., & Spiessens, P. (1989). Fine-grained parallel genetic algorithms. In Schaffer (Ed.), pp. 428-433.

Manderick, B., de Weger, M., & Spiessens, P. (1991). The genetic algorithm and the structure of the fitness landscape. In Belew & Booker (Eds.), pp. 143-150.

Männer, R., & Manderick, B. (Eds.). (1992). *Parallel problem solving from nature, 2.* North-Holland.

Mayr, E. (1942). *Systematics and the origin of species.* (Reprinted 1982). Columbia U. Press.

Michod, R. E., & Levin, B. R. (Eds.). (1988). *The evolution of sex: An examination of current ideas.* Sinauer.

Miller, G. F. (1993). *Evolution of the human brain through runaway sexual selection.* Ph.D. thesis, Stanford University Psychology Department. (To be published as a book by MIT Press/Bradford Books, 1995).

Miller, G. F. (in press). Sexual selection in human evolution: Review and prospects. For C. Crawford (Ed.), *Evolution and human behavior: Ideas, issues, and applications.* Lawrence Erlbaum.

Miller, G. F., & Todd, P. M. (1993). Evolutionary wanderlust: Sexual selection with directional mate preferences. In J.-A. Meyer, H. L. Roitblat, & S. W. Wilson (Eds.), *From Animals to Animats 2: Proc. Second Int'l Conf. on Simulation of Adaptive Behavior*, pp. 21-30. MIT Press.

Miller, G. F., & Todd, P. M. (in press). The role of mate choice in biocomputation: Sexual selection as a process of search, optimization, and diversification. For W. Banzaf & F. Eeckman (Eds.), *Proceedings of the 1992 Monterey Biocomputation Workshop.*

Mühlenbein, H. (1989) Parallel genetic algorithms, population genetics, and combinatorial optimization. In Schaffer (Ed.), pp. 416-421.

Mühlenbein, H. (1992). Darwin's continent cycle theory and its simulation by the Prisoner's Dilemma. In Varela & Bourgnine (Eds.), pp. 236-244.

Ostermeier, A. (1992). An evolution strategy with momentum adaptation of the random number distribution. In Männer & Manderick (Eds.), pp. 197-206.

Pomiankowski, A. (1988). The evolution of female mate preferences for male genetic quality. *Oxford Surveys in Evolutionary Biology, 5,* 136-184.

Ridley, M. (1993). *The red queen: Sex and the evolution of human nature.* Viking.

Schaffer, J. D. (Ed.) (1989). *Proc. of the Third Int'l Conf. on Genetic Algorithms.* Morgan Kauffman.

Schaffer, J. D., & Eshelman, L. J. (1991). On crossover as an evolutionarily viable strategy. In Belew & Booker (Eds.), pp. 61-68.

Schultz, A. C. (1991). Adapting the evaluation space to improve global learning. In Belew & Booker (Eds.), pp. 158-164.

Sims, K. (1992). Interactive evolution of dynamical systems. In Varela & Bourgnine (Eds.), pp. 171-178.

Todd, P. M., & Miller, G. F. (1991). On the sympatric origin of species: Mercurial mating in the Quicksilver Model. In Belew & Booker (Eds.), pp. 547-554.

Todd, P. M., & Miller, G. F. (1993). Parental guidance suggested: How parental imprinting evolves through sexual selection as an adaptive learning mechanism. *Adaptive Behavior, 2*(1): 5-47.

Todd, S., & Latham, W. (1992). Artificial life or surreal art? In Varela & Bourgnine (Eds.), pp. 504-513.

Varela, F. J., & Bourgnine, P. (Eds.) (1992). *Toward a practice of autonomous systems: Proceedings of the First European Conference on Artificial Life.* MIT Press.

Wright, S. (1932). The roles of mutation, inbreeding, crossbreeding, and selection in evolution. In *Proc. Sixth. Int'l Congr. Genetics,* 356-366.

An empirical comparison of selection methods in evolutionary algorithms

Peter J.B.Hancock

Department of Psychology
University of Stirling, Scotland, FK9 4LA

Abstract. Selection methods in Evolutionary Algorithms, including Genetic Algorithms, Evolution Strategies (ES) and Evolutionary Programming, (EP) are compared by observing the rate of convergence on three idealised problems. The first considers selection only, the second introduces mutation as a source of variation, the third also adds in evaluation noise. Fitness proportionate selection suffers from scaling problems: a number of techniques to reduce these are illustrated. The sampling errors caused by roulette wheel and tournament selection are demonstrated. The EP selection model is shown to be equivalent to an ES model in one form, and surprisingly similar to fitness proportionate selection in another. Generational models are shown to be remarkably immune to evaluation noise, models that retain parents much less so.

1 Introduction

Selection provides the driving force in an evolutionary algorithm (EA) and the selection pressure is a critical parameter. Too much, and the search will terminate prematurely, too little, and progress will be slower than necessary. There exist a variety of selection algorithms: this paper aims to shed further light on their relative merits. Goldberg and Deb [10] performed some analysis on some of the commoner algorithms used in Genetic Algorithms (GAs). This paper uses simulations to extend their work: a) by including selection schemes typical of Evolution Strategy (ES) and Evolutionary Programming (EP) approaches; b) by including stochastic effects they ignored; and c) by considering not only takeover times for the best string, but rate of elimination of weaker strings. This can be important, since weaker strings contribute to the overall genetic diversity, without which search may stagnate. The aim is to assist experimenters in making rational decisions about which selection method to employ.

Goldberg and Deb consider a simple model with just two fitness levels and observe the rate of convergence to the fitter string, under the action only of selection. Reproduction operators such as mutation and crossover are excluded. The first simulation used here is similar: $N = 100$ individuals are assigned random values in the range 0-1, except for one which is set to a value of 1. The effect of different selection schemes and parameters on the population are reported, based on an average of 100 runs.

Some of the selection methods, being stochastic, may lose the best value from the population. This is caused by errors in sampling, to be discussed further below. In tournament selection, for example, the best member of the population may simply not be picked for any contests. In this event it is replaced, arbitrarily overwriting the value of the first member of the population. If this were not done the graphs of takeover rate would be more affected by the particular number of runs that lost the best value than by real differences in the takeover rate in the absence of such loses. The number of occasions that reinstatement was necessary will be reported.

Some of the selection methods result in an exponential growth in the proportion of the fittest value. A second model therefore introduces mutation as a source of variation. The initial population is given random fitnesses in the range 0-0.1. When an individual is chosen for reproduction, it produces an offspring whose fitness differs by the addition of a Gaussian random variable, with mean zero and standard deviation 0.02. This allows observation of the different selection schemes' tendency to exploit such variations of fitness.

Many references will be made in this paper to *selection pressure*. As might be expected, increasing selection pressure will increase the rate of convergence. Unfortunately, it is difficult to define it other than circularly, as some measure of the rate of convergence for a given problem. Selection pressure as used here is therefore a relative term. A rough definition might be the expected number of offsping for the best member of the population. However, as will become clear, this by no means defines the convergence rate, which will depend on the allocation of offspring to all members of the population.

It is important to realise that fast convergence on these problems is not necessarily good. The tasks are kept extremely simple to highlight the effects of the selection schemes. The interest lies not in the ultimate speed of convergence, but in understanding the differences. Note also that many interesting variations on selection, such as crowding [5], niching [4] and mate selection (Miller, this volume), and the whole area of parallel EAs are not considered, these simple simulations being inappropriate.

2 Fitness proportionate selection (FPS)

The traditional GA model [12] selects strings in proportion to their fitness on the evaluation function, relative to the average of the whole population. This scheme has the merit of simplicity, but unfortunately suffers from well-known problems to do with scaling. Given two strings, with fitness 2 and 1 respectively, the first will get twice as many offspring as the second. If the underlying function is changed simply by adding 1 to all the values, the two strings will score 3 and 2, a ratio of only 1.5. It is unfortunate that mere translation of the target function will have a significant effect on the rate of progress.

Selection pressure is also affected by the scatter of fitness values in the population. At the beginning of a run, some individuals may be much better than most. These will be heavily selected, with a consequent tendency to premature convergence. Towards the end of a run, most will be of similar fitness, and the search stagnates. The latter problem has been addressed by techniques of windowing and scaling.

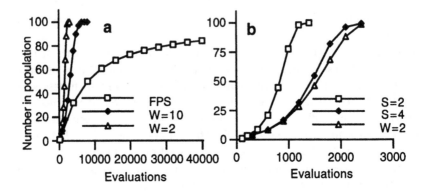

Fig. 1. a) Comparison of convergence to best value for proportional selection and for window sizes of 10 and 2 generations. b) Comparison between sigma scaling ($s=2$ and 4) and windowing ($w = 2$).

2.1 Windowing

Windowing introduces a moving baseline. The worst value observed in the w most recent generations is subtracted from all the fitness values, where w is known as the window size, and is typically of the order of 2-10.

The dramatic effect of this moving baseline is shown in Figure 1a, which shows the increase in the number of copies of the optimal value under selection only. The fitness proportional selection method initially converges rapidly, but then tails off as all of the population approaches a score of 1. Adding a baseline maintains much more constant selection pressure, stronger for smaller window size. Subtraction of the worst value also solves another problem with the direct use of fitness scores: what to do about negative values. A negative number of expected offspring is not meaningful (unless perhaps suggesting that the parent should be put down!). Simply declaring negative values to be zero is not sufficient, since with some evaluation functions the whole population might score zero.

2.2 Sigma scaling

As noted above, the selection pressure is related to the scatter of the fitness values in the population. Sigma scaling [9] exploits this observation, setting the baseline s standard deviations (sd) below the mean, where s is the scaling factor. Strings below this score are assigned a fitness of zero. This method helps to overcome a potential problem with particularly poor individuals ("lethals") which with windowing would put the baseline very low, thus reducing selection pressure. Sigma scaling keeps the baseline near the average.

It also provides a "knob" that may be twiddled to adjust the selection pressure, which is inversely related to the value of s. By definition, the average fitness of the scaled population will be $s \times sd$. Thus an individual that has an evaluation one standard deviation above the average will get $\frac{s+1}{s}$ expected offspring. Typical values of s are in the range 2-5. The effect on convergence to the best value is shown in Figure 1b, for s values of 2 and 4: takeover rate is somewhat greater than with a window of size 2.

These moving baseline techniques help to prevent the search stagnating, but may exacerbate the problem of premature convergence to a super-fit individual because they increase its advantage relative to the average. The sigma scaling method is slightly better, in that good individuals will increase the standard deviation, thereby reducing their selective advantage somewhat. However, a better method would be helpful.

2.3   Linear scaling

Linear scaling adjusts the fitness values of all the strings such that the best individual gets a fixed number of expected offspring. The other values are altered so as to ensure that the correct total number of new strings are produced: an average individual will still expect one offspring. Exceptionally fit individuals are thus prevented from reproducing too quickly.

The scaling factor s is a number, in the range 1 to 2, which specifies the number of offspring expected for the best string. It therefore gives direct control on the selection pressure. The expected number of offspring for a given string is given by $1 + \frac{(fitness - avg)}{(best - avg)} \times (s - 1)$.

It may be seen that this returns s for the best, and 1 for an average string. However, poor strings may end up with a negative number of offspring. This could be addressed by assigning them zero, but doing so would require that all the other fitness values be changed again to maintain the correct average. It also risks unpredictable loss of diversity. An alternative is to reduce the scaling factor so as to give just the worst individual a score of zero: $s = 1 + \frac{(top - avg)}{(avg - worst)}$.

The effects on convergence rate are shown in Figure 2a. As expected, increasing the scaling factor increases the convergence rate. With a linear scaling factor of 2, the convergence is between that obtained from a window size of 2, and a sigma scaling factor of 2. At low selection pressures, the convergence rate is proportional to s. Thus in this simulation, the best value takes over the population in 4000 evaluations for $s = 1.2$. With $s = 1.1$, it takes 8000 evaluations. This would suggest convergence in less than 1000 evaluations when $s = 2$, where in fact it takes 2000. The reason is the automatic reduction in selection pressure caused by the need to prevent negative fitness values. In this application the convergence produced with $s = 2$ is very similar to that produced with $s = 1.5$.

The growth rates in the presence of mutation for these scaling methods are shown in Figure 2b, All are quite similar, windowing and sigma scaling come out ahead precisely because they fail to limit particularly fit individuals. Fortuitous mutations are therefore able to reproduce rapidly.

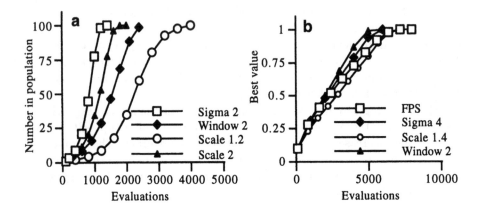

Fig. 2. a) Takeover rates for linear scaling with s=2 and 1.2, compared with window size of 2 and sigma scaling values of 2. b) Growth rate for proportionate selection and various scaling methods.

2.4   Selection and sampling

The various methods just described all deliver a value for the expected number of offspring for each individual. Thus with direct fitness measurements, a string with twice the average score should be chosen twice. That's easy enough, but what should be done about a string with a score of half the average, which should get half a reproductive opportunity? The best that can be done is to give it a 50% probability of being chosen in any one generation. Baker, who studied this problem in some detail [2], calls the first process selection, the second - actually picking the winners - *sampling*.

A simple, and lamentably still frequently used sampling method may be visualised as spinning a roulette wheel, the sectors of which are set equal to the relative fitness values of each string. The wheel is spun once for each string selected. The wheel is more likely to stop on bigger sectors, so fitter strings are more likely to be chosen on each occasion. Unfortunately this is not satisfactory. Because each parent is chosen separately, there is no guarantee that any particular string, not even the best in the population, will actually be chosen in any given generation. This sampling error can act as a significant source of noise. The results of Figures 1 and 2 were obtained using Baker's stochastic universal sampling (SUS) algorithm [2], an elegant solution to the problem marred only by the need to shuffle a sorted population, such as used for rank selection, prior to the sampling. Figure 3 shows the difference in results for the two methods with

FPS. The rate of takeover is reduced, a reflection of the fact that the roulette wheel simulation lost the best value from the population an average of 9.1 times per run. Conversely, the worst value current in the population increases more rapidly, because it is quite likely for poor strings to be missed by the random sampling. Both effects are likely to reduce performance.

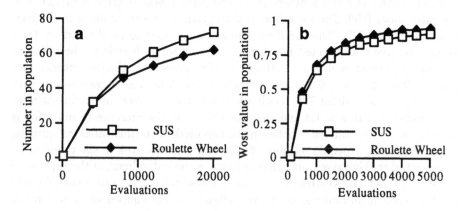

Fig. 3. Comparison of convergence to best value (a) and increase of worst value (b) for proportional selection, using roulette wheel selection and SUS algorithm.

3 Ranking

Baker [1] suggested rank selection in an attempt to overcome the scaling problems of the direct fitness based approach. The population is ordered according to the measured fitness values. A new fitness value is then ascribed, inversely proportional to each string's rank. Two methods are in common use.

3.1 Linear ranking

The best string is given a fitness s, between 1 and 2. The worst string is given a fitness of $2 - s$. Intermediate strings' fitness values are given by interpolation: $f(i) = s - \frac{2(i-1)(s-1)}{(N-1)}$ for $i = \{1..N\}$. Since this prescription automatically gives an average fitness of 1, the fitness values translate directly as the expected number of offspring. If s is set to 2, the worst string gets no chance of reproduction. In principle, s could be increased beyond 2 to achieve higher selection pressures, but then several of the worst strings would be given negative fitness values. These could be truncated to zero, but then the remaining fitness values would need rescaling to give the correct total number of offspring. A simpler method of achieving higher selection pressures, which also gives some chance to the worse members of the population, is to use a non-linear ranking, such as that described in the next section. The takeover rate produced by linear ranking is proportional to $s - 1$. Thus with $s = 1.1$, convergence takes about 10000 evaluations, with $s = 1.2$ it takes 5000, and with $s = 2$ it takes 1000.

3.2 Exponential ranking

The best string is given a fitness of 1. The second best is given a fitness of s, typically about 0.99. The third best is then given s^2 and so on down to the last, which receives s^{N-1}. The ascribed fitness values need to be divided by their average to give the expected number of offspring for each string. The selection pressure is proportional to $1 - s$, thus $s = 0.994$ gives twice the convergence rate of $s = 0.998$. With $s = 0.999$, convergence takes about 25000 evaluations, with $s = 0.968$, it takes about 700. With $s = 0.986$, the takeover time is identical to linear ranking with $s = 1.8$ and the growth rate in the presence of mutation is also the same. The difference between the two methods is illustrated in figure 4a, which shows the expected number of offspring for each rank in the population. Exponential ranking gives more chance to the worst individuals, at the expense of those above average. As a result, the rate of loss of the worst value is considerably less for exponential ranking, as shown in figure 4b. For equivalent growth rates, exponential ranking ought to give a more diverse population.

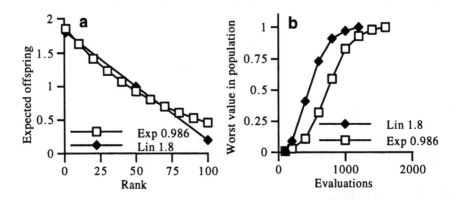

Fig. 4. a) Expected offspring for individual of given rank, linear ($s = 1.8$) and exponential ($s = 0.986$) ranking b) Elimination of worst values during takeover experiment.

4 Tournament selection

In tournament selection, n individuals are chosen at random from the population, with the best being selected for reproduction. A fresh tournament is held for each member of the new population. Goldberg and Deb show that the expected result for a tournament with $n = 2$ is exactly the same as linear ranking with $s = 2$. Tournament selection can be made to emulate linear ranking with $s < 2$ by making it only probable that the better string will win. The conversion between probability in tournament selection and s in linear ranking is to double the probability, thus a probability of 0.8 is equivalent to $s = 1.6$. The selection pressure generated by the tournaments may be increased by using $n > 2$. This produces non-linear ranking, that differs from exponential because the worst individual gets no chance. With $n = 3$ an average individual will expect to win a quarter of its three tournaments. If the tournaments are made stochastic, the result will be more similar to exponential ranking.

In practice, tournament selection differs from rank selection much as roulette wheel and SUS sampling differ in figure 3. Because each tournament is carried out individually, it suffers from exactly the same sampling errors and ranking, with Baker's selection procedure, should usually be used instead. The caveat is because tournament selection is particularly suited to parallel processing. Holding tournaments may also be the only sensible way to evaluate individuals, for example when evolving game-playing applications. However, those using it should be aware of the implied sampling errors.

5 Incremental models

One of the bigger arguments in the GA camp centres on whether to replace the whole population at a go (generational model), or some subset, one in the limit (incremental, also known as steady-state reproduction [14]). Whitley, with his Genitor system [17], is one of the major proponents of the incremental approach. Any of the above methods of selection could be used to pick the parents of the single offspring, but Whitley uses linear ranking [16]. Incremental reproduction inevitably carries the same kind of sampling errors as roulette wheel selection (see deJong and Sarma [6] for a further discussion).

Goldberg and Deb show that the Genitor model produces very high growth rates. Most of this comes from always replacing the worst member of the population. Changing the linear ranking scale factor has very little effect. A much softer selection is given by replacing the oldest member of the population (also known as FIFO deletion [6]). Figure 5a compares kill-oldest with kill-worst. In all the graphs, the x-axis units are evaluations, to allow direct comparison between the different population models. Kill-oldest is comparable with a generational model of the same selection pressure. However, it is faster at the start of run, and slower to finish off, probably the opposite of what is desirable. As might be expected, loss of the worst is more rapid as well (not shown). Slow finishing is the consequence of the sampling errors, like those of roulette wheel selection, that inevitably result from breeding one at once. With $s = 1.4$, kill-oldest lost the best value an average of 2.5 times per run.

One claim made for incremental reproduction is that it can benefit by exploiting good new individuals as soon as they are produced, rather than waiting a generation. DeJong and Sarma [6] refute this claim, on the grounds that the addition of good individuals has the effect of increasing the average fitness, and thus reducing the chance that any will be selected. However, their argument applies to takeover experiments: when reproduction operators are producing new best individuals there can be a difference. This appears to be the reason that kill-oldest converges more rapidly than a generational model with the same selection pressure in figure 5b.

There are alternative replacement strategies that may be used to reduce the selection pressure of kill-oldest. One possibility is to kill one of the n worst, another is to kill at random. Syswerda [15] shows that this is equivalent to a generational model. Syswerda uses kill by inverse rank, and shows that this is very similar in effect to kill-worst. However, he is using very high pressure (exponential ranking with $s = 0.9$). The effects at lower pressures are shown in figure 6, the reasons for them unfortunately too complex to discuss here. Kill-ranked incremental, with $s = 1.2$ from top for selection and from bottom for deletion, gives a growth rate almost identical to a linear ranked generational model with $s = 1.4$. This correspondence is not dependent on the particular conditions such as the mutation size and should provide a basis for comparisons in "live" EAs. Another correspondence at lower selection pressure is given by incremental with $s = 1.13$ and generational with $s = 1.2$ (not shown).

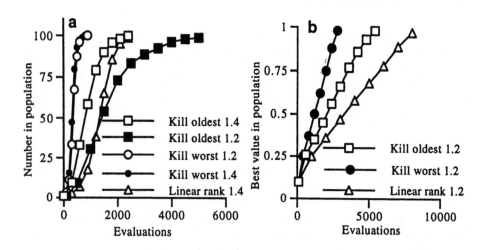

Fig. 5. a) Take over rates for kill-worst and kill-oldest incremental algorithms, compared with generational model b) growth rates

6   ES and EP methods

The selection methods used in ES and EP algorithms are rather similar, producing high selection pressures.

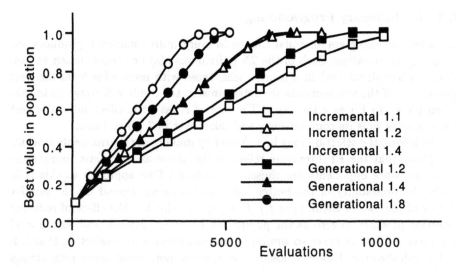

Fig. 6. Growth rates for kill-ranked incremental algorithm, compared with linear ranking generational model

6.1 Evolution Strategies

There are two main methods of selection used in ESs, known as $(\mu + \lambda)$ and (μ, λ), where μ is the number of parents and λ is the number of offspring [11]. The top μ individuals form the next generation, selection being from parents and children in the $(\mu + \lambda)$ case, children only for (μ, λ). Typically, λ is one or two times μ. With these schemes, takeover by the best value is exponential. Thus for a (100,200) ES, it takes $log_2(100) = 7$ generations, for (100+200), it takes $log_3(100)$ = 5 generations (1100 evaluations). Such figures are comparable to those given by the kill-worst incremental algorithm. Eshelman's CHC algorithm [7] uses an $(N + N)$ selection method.

Note a shift of strategy here. The traditional GA approach is reproduction according to fitness, the ES approach is more like survival of the fittest. In the (μ, λ) generational version λ must exceed μ if there is to be any selection. Allowing every member of the population to reproduce implies more evaluations per generation than the GA approach. Not necessarily better or worse, but different.

6.2 Top-n selection

Some workers select the n best individuals, and give them each N/n offspring [13]. This obviously has the potential for extremely rapid takeover - with $n = 10$, the best value will take over in two generations. It differs from the ES (μ, λ) approach only in what is called the population, thus Top-n with $n = 50$ and $N = 100$ is equivalent to a $(50, 100)$ ES.

6.3 Evolutionary Programming

EP selection also gives an equal chance to every individual: each produces one offspring by mutation. Each of the 2N individuals plays c others chosen at random (with replacement) in a tournament: those with most wins form the next generation. If the tournaments are deterministic, the result will converge to that of an $(N + N)$ ES as c increases. The size of c has little effect here: the best value always wins its competitions, and takes over in 7 generations.

As before, the selection may be softened by making the tournaments stochastic. (Note that the EP literature refers to the above as stochastic tournament selection, since opponents are chosen at random.) One approach to this is to make the probability of the better individual winning depend on the relative fitness of the two of them: $p_i = f_i/(fi + fj)$ [8]. This has the effect of reducing selection pressure to zero as the population becomes uniform, and with $c = 10$ produces a takeover curve so similar to simple proportionate selection that it is not worth showing. However, loss of the worst is more rapid, since poor strings initially get little chance to reproduce.

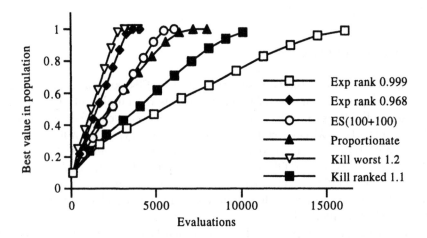

Fig. 7. Growth rates for for a variety of selection algorithms.

Figure 7 shows growth rates under mutation of a number of the selection methods described. Incremental kill-worst produces the highest growth rate, but note that exponential ranking can be made to converge similarly fast, or very slowly, by varying the selection pressure. Linear scaling and ranking can be tuned for similarly slow growth, as can an incremental model with inverse rank deletion, shown here for $s = 1.1$. Despite the very rapid takeover given by the ES methods, growth rate is less spectacular, and actually very similar to that provided by simple fitness proportionate selection. This grows relatively quickly because of the high number of offspring allocated to particularly fit individuals.

7 The effects of noise

Genetic Algorithms are traditionally held to be relatively immune to the effects of noise in the evaluation function. The susceptibility of the different selection methods was assessed by adding noise to the evaluation, and observing the effect on the growth in the presence of mutation. Another Gaussian random variable was added to the true value of each individual, and used to allocate its number of offspring. The true value was then passed to any offspring, subject to the small mutation as before. In order to have a significant effect on the rate of convergence, it was found to be necessary to add noise with a standard deviation of 0.2: 10 times that of the mutation, which gives some credence to the fabled noise immunity of GAs.

Figure 8a shows the effects of adding noise on two traditional GA methods, linear ranking and sigma scaling. Even with this level of noise, the convergence rate is less than halved. However, note that sigma scaling deteriorates rather less than the ranking method. The reason for this apparent anomaly is quite subtle, but similar to that responsible for the growth rates in Figure 2b. The effect of the noise is to reduce the accuracy of the selection procedure. In the limit, if noise swamps the signal entirely, all individuals would expect one offspring. Here, the best individual can expect somewhat more than one, but less than it should get in the absence of noise. With ranking and s=1.8, it will therefore get somewhere between 1 and 1.8 offspring. Measurements from the simulation indicate that it actually gets about 1.2 initially. Sigma scaling sets the baseline according to the spread of values in the population, which will be determined mostly by the noise. Lucky individuals can appear 2 or 3 standard deviations above the mean. Because there is not the upper limit imposed by ranking, the best individual averages higher, about 1.5 initially in this case. The faster growth rate is therefore effectively a case of premature convergence, but since the problem is so simple, there is only the correct solution for it to converge to.

Figure 8b show the effects on the ES selection methods. In the absence of noise, (100+100) converges rapidly. Allowing the parents to pass to the next generation ensures that nothing is lost, giving rapid gains on the simple task. A (100+200) ES converges more rapidly in terms of generations, but requires more evaluations in total. Comparison with the (100,200) model illustrates the advantage of conserving the best parents. In the presence of noise, however, the conservative approach fails. The (100,200) ES deteriorates by a similar amount to other generational techniques, (100+100) deteriorates dramatically. The deterministic version of EP selection performs very similarly to (100+100) ES, depending on the number of tournaments held (not shown).

Figure 9 shows the effects on incremental reproduction. Because Genitor kills the apparent worst in the population, lucky individuals that got a much better evaluation than they merited will linger in the population. The effects on convergence rate are disastrous. Killing the oldest performs much better, echoing the findings of Fogarty (unpublished results). Rank-based deletion is also relatively unaffected at low selection pressures - for higher values of s, it tends towards the behaviour of kill-worst.

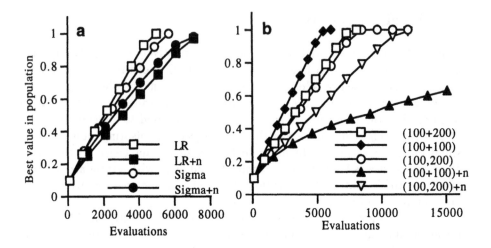

Fig. 8. a) Effects of adding noise (+n) on the growth rate of linear ranking (LR, s=1.8) and sigma scaled proportionate selection (s=4), b) effects on ES methods.

Re-evaluation is sometimes suggested as a means of reducing the noise sensitivity of incremental models. An individual is picked, either at random, or by (apparent) fitness, and the new evaluation averaged with the old. As expected, rank based selection for re-evaluation proved more effective than random choice in the simulations used here, but still managed only slightly less than double the evaluations required for convergence.

8 Conclusions

This work has employed some simple tasks to illustrate differences, and some surprising similarities between various selection methods. Fitness proportionate selection suffers from scaling problems, that are partially addressed by windowing and sigma scaling. However, these do not prevent premature convergence caused by particularly fit individuals. Linear scaling does address this problem, but the selection pressure achieved is still dependent on the spread of fitness values in the population. Ranking methods provide good control of selection pressure, but inevitably distort the relationship between fitness and reproductive success. Roulette wheel sampling suffers from errors and should not be used. It is perhaps less well recognised that tournament selection and incremental reproduction display similar errors. In addition, the latter suffers badly from noisy evaluation functions if kill-worst deletion is used. EP selection is similar to an (N+N) ES model if tournaments are deterministic, and remarkably similar to fitness proportionate selection with one form of stochastic tournament.

A common device that has not been illustrated is elitist selection [5], where the best member of the population is retained between generations. This is not a

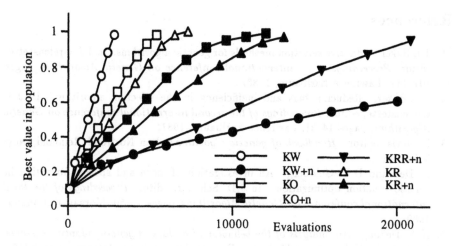

Fig. 9. Effects of noise (+n) on the growth rate of various incremental replacement strategies: kill-worst (KW), kill-oldest (KO) kill-ranked (KR) and kill-ranked with reevaluation (KRR), all with $s = 1.2$

hedge against inaccurate sampling, but to guard against disruption by crossover or mutation. It therefore has no effect on the takeover problem used here, and minimal effect on growth under mutation. In the presence of noise, it reduces convergence, for the same reason as incremental models.

At the end of their paper, Goldberg and Deb pose the question "selection: what should we be doing?". The author's preference is for exponential ranking, for the range of pressures achievable. Incremental methods are the choice of many expert practitioners (e.g. Davis [3]), but it remains unclear to what extent the apparent gains in convergence speed are due simply to their tendency to produce high selection pressure. Another crucial difference is that it is possible to run incremental models with the elimination of duplicates [17, 3], which, in the absence of evaluation noise, may have significant benefits not demonstrable with the simple tasks used here. Which works best in a real EA, with crossover, multimodalities, deception etc. is open to further experimentation. Certainly there will be an interaction, hence Eshelman's deliberate choice of a conservative, ES-style selection method to balance a radical recombination operator in his CHC algorithm [7].

9 Acknowledgements

This work was partly supported by grant no. GR/H93828 from the UK Science and Engineering Research Council.

References

1. J.E Baker. Adaptive selection methods for genetic algorithms. In J.J. Grefenstette, editor, *Proceedings of an international conference on Genetic Algorithms*, pages 101–111. Lawrence Earlbaum, 1985.
2. J.E Baker. Reducing bias and inefficiency in the selection algorithm. In J.J. Grefenstette, editor, *Proceedings of the second international conference on Genetic Algorithms*, pages 14–21. Lawrence Earlbaum, 1987.
3. L. Davis, editor. *Handbook of genetic algorithms*. Van Nostrand Reinhold, New York, 1991.
4. K. Deb and D.E. Goldberg. An investigation of niche and species formation in genetic function optimization. In J.D. Schaffer, editor, *Proceedings of the third international conference on Genetic Algorithms*, pages 42–50. Morgan Kaufmann, 1989.
5. K.A. DeJong. *An analysis of the behavior of a class of genetic adaptive systems*. PhD thesis, University of Michigan, Dissertation Abstracts International 36(10), 5140B, 1975.
6. K.A. DeJong and J. Sarma. Generation gaps revisited. In D. Whitley, editor, *Foundations of Genetic Algorithms 2*, pages 19–28. Morgan Kaufmann, 1993.
7. L.J. Eshelman. The CHC adaptive search algorithm: how to have safe search when engaging in nontraditional genetic recombination. In G.J.E. Rawlins, editor, *Foundations of Genetic Algorithms*, pages 265–283. Morgan Kaufmann, 1991.
8. D.B. Fogel. An evolutionary approach to the travelling salesman problem. *Biological Cybernetics*, 60:139–144, 1988.
9. S. Forrest. Documentation for prisoner's dilemma and norms programs that use the genetic algorithm. Technical report, Univ. of Michigan, 1985.
10. D.E Goldberg and K. Deb. A comparative analysis of selection schemes used in genetic algorithms. In G.J.E. Rawlins, editor, *Foundations of Genetic Algorithms*. Morgan Kaufmann, 1991.
11. F. Hoffmeister and T. Bäck. Genetic algorithms and evolution strategies: similarities and differences. Technical Report SYS-1/92, University of Dortmund, 1992.
12. J.H. Holland. *Adaptation in natural and artificial systems*. The University of Michigan Press, Ann Arbor, 1975.
13. S. Nolfi, J.L. Elman, and D. Parisi. Learning and evolution in neural networks. Technical Report CRL TR 9019, UCSD, July 1990.
14. G. Syswerda. Uniform crossover in genetic algorithms. In J.D. Schaffer, editor, *Proceedings of the third international conference on Genetic Algorithms*, pages 2–9. Morgan Kaufmann, 1989.
15. G. Syswerda. A study of reproduction in generational and steady-state genetic algorithms. In G.J.E. Rawlins, editor, *Foundations of Genetic Algorithms*, pages 94–101. Morgan Kaufmann, 1991.
16. D. Whitley. The genitor algorithm and selection pressure: why rank-based allocation of trials is best. In J.D. Schaffer, editor, *Proceedings of the third international conference on Genetic Algorithms*, pages 116–121. Morgan Kaufmann, 1989.
17. D. Whitley and J. Knuth. Genitor: a different genetic algorithm. In *Proceedings of the Rocky Mountain Conference on Artificial Intelligence*, pages 118–130. Denver Colorado, 1988.

AN EVOLUTION STRATEGY AND GENETIC ALGORITHM HYBRID: AN INITIAL IMPLEMENTATION AND FIRST RESULTS

Lawrence Bull
Faculty of Engineering
email l_bull@csd.uwe.ac.uk

Terence C Fogarty
Faculty of Computer Studies and Mathematics
email tc_fogar@csd.uwe.ac.uk

University of the West of England
Bristol, BS16 1QY, England.

Abstract

Evolution Strategies (ESs)[15] and Genetic Algorithms (GAs)[13] have both been used to optimise functions, using the natural process of evolution as inspiration for their search mechanisms. The ES uses gene mutation as it's main search operator whilst the GA mainly relies upon gene recombination. This paper describes how the addition of a second mutation operator, used in conjunction with the mutation and crossover operators of the normal GA, can improve the GA's performance on rugged fitness landscapes. We then show that by adding Lamarckian replacement the GA's performance on smooth landscapes can also be improved, further improving it's performance on rugged landscapes. We explain how the extra operators allow the GA to gain and exploit local information about the fitness landscape, and how this local random hill climbing can be seen to combine the search characteristics of the ES with those of the GA.

1 Introduction

A fundamental aspect of the basic genetic algorithm is it's exclusion of a phenotype in that a given genotype is directly evaluated with some fitness measure, which is subsequently used for a selective position during the simulated process of evolution; the processes of growth and learning are omitted. In 1896 J.M. Baldwin[3] proposed that learning can greatly influence evolution, citing the situation of when a learned trait becomes critical to the survival of individuals. Recent investigations (eg [11][1][4][10]) into the interaction between learning and evolution have shown that learning can speed up the search process of evolutionary approaches. We can imagine the scenario of a given genotype existing at the foot of an optimum within a search space. Although that genotype is very close to a good sub-space, because it represents a fixed point, the information is lost to the GA at that time unless it has generated another individual within the sub-space. Therefore if the genotype was able to do some local search, via some form of learning, it can be seen that it may be able to "move" partway up the optimum, thus obtaining a better fitness measure than it otherwise would. The genotype would then have an increased chance of selection and so guide the GA's search toward the sub-space.

The simplest form of learning, local random search, is easily used in conjunction with the GA by adding a second mutation operator. Here the select()-generate(crossover,mutate)-evaluate() sequence of the basic GA becomes select()-generate(crossover,mutate)-evaluate(mutate). Now instead of taking a given genotype and evaluating it directly, a copy of the genotype is made where there is some probability (p_{m+}) that any given bit is flipped during the copy process, just as in the generation of individuals from their parents. In this way the genotype at the foot of the optimum described above may have one or more bits flipped causing it's evaluation string (phenotype) to be within the good sub-space.

2 Random Learning

2.1 A Partially Deceptive Function

Mitchell, Forrest and Holland[14] presented a simple partially deceptive function who's fitness landscape features an isolated optimum, surrounded by a region of low fitness, all contained in an area of average fitness as follows:

$$F(x) = 5_{o**11}(x)-16_{o*111}+5_{o11**}(x)-16_{o111*}+31_{o1111}(x).$$

where average fitnesses $u(s)$ of the five schemas (with optimum point 1111) are:

$$u(**11) = 2$$
$$u(*111) = -1$$
$$u(11**) = 2$$
$$u(111*) = -1$$
$$u(1111) = 9.$$

The shape of this landscape allows for the scenario described above, ie a genotype in the "trench" around the optimum will be at the foot of a good sub-space. In the basic GA any genotype in the trench will be selected against, even though it represents a step in the right direction (hence the "partially deceptive"[9] label), whereas with the added mutation operator the genotype may be evaluated as either of average or optimum fitness and so will more than likely be selected during reproduction.

The following table shows the mean number of generations taken for an individual to reach the optimum for this function over fifty trials, using various population sizes

(*N*). In the GA with added mutation (GA+) the time taken for the genotype, not the phenotype, to reach the optimum is recorded - showing the true rate of progress of the prerequisite schema. The generational GA uses proportional selection, mutation (rate 0.005 per bit), single-point crossover and the extra mutation rate p_{m+} is set at *1/4* per bit (ie on average one mutation per genotype-phenotype copy).

	$N=2$	$N=4$	$N=8$	$N=10$
GA	73.5	28.9	20.9	18.0
GA+	5.5	3.9	2.9	2.5

Table 1.

It can be clearly seen that adding the extra mutation operator has greatly improved the GA's performance at the partially deceptive function (statistically significant at the 1% level), as expected, particulary with the smaller population sizes.

2.2 The "Max Integer" problem

This next function can be seen as a rugged version of the simple bit counting problem, "max ones", where each location *b* of the genotype string containing a set bit contributes 2^b to the string's overall fitness, rather than 1.

Table 2 shows the mean number of generations taken for an individual to reach the optimum for this function over fifty trials. Again in the GA with added mutation (GA+) the time taken for the genotype, not the phenotype, to reach the optimum is recorded. The generational GA again uses probabilistic selection, mutation (rate 0.005 per bit), single-point crossover, population size is 100, string length *l* is twelve bits (max int = 4095), and the extra mutation rate p_{m+} is set at *1/l* per bit.

	Mean	Std. Dev.
GA	24.1	27.5
GA+	15.3	16.1
Random	109.1	77.8

Table 2.

Analysis of variance on the results shown in table 2 finds the GA with added mutation statistically better than the basic GA at the 5% level of significance. We also see that the populations are less converged with the extra mutation operator. It can also be seen that both easily outperform a random search.

We will now look at the performance of the extra operator on the smoother landscape of the max ones problem to see if the added "noise" inhibits the GA's easy step-wise ascent to the optimum.

2.3 The "Max Ones" problem

Table 3 shows the mean number of generations taken for an individual to reach the optimum for this function over fifty trials. Again in the GA with added mutation (GA+) the time taken for the genotype, not the phenotype, to reach the optimum is recorded. The generational GA is the same as the one used above (max fitness now = 12).

	Mean	Std. Dev.
GA	7.3	2.8
GA+	7.2	3.1
Random	163.7	55.2

Table 3.

Analysis of variance on the results shown in table 3 finds no statistically significant difference between the two algorithms. From this we can conclude that the added mutation has had no real derogatory effect on the hyperplane sampling of the GA - which may have been initially suspected. No laudatory effect was anticipated because in this problem there are no deceptive regions or isolated optima for the operator to "smooth over" or find. It can also be seen that both easily outperform a random search.

3 Evolution Strategies

Evolution Strategies are another class of algorithms which use the process of evolution as inspiration for their search mechanisms. Developed at around the same time as the GA by Rechenberg[16] at the University of Berlin they, in their simplest form, use a mutation-selection scheme for their search, denoted as the $(1+1)$-ES. Here a population of one real-valued vector is mutated to produce a child. If the child's fitness is better than it's parent's it becomes the parent for the next generation, etc. Rechenberg successfully applied the ES to various optimisation problems, some highly non-linear, later extending it to the multi-membered $(\mu+1)$-ES in which two out of μ parents are selected to produce the child. Schwefel[17] developed this still further to the $(\mu+\lambda)$-ES (and (μ,λ)-ES) in which the μ parents produce λ children, the fittest of all these being reduced to the μ parents of the next generation, etc. Schwefel also viewed the mutation operator as an evolvable entity, storing it in the genotype and finding that it adapts during the search process (usually from a high rate to low with time).

The power of searching by mutation was first demonstrated by Fogel et. al.[7] and Davis[6] has recently shown it's ability on some of the classic GA benchmark problems. The GA of course uses mutation for searching, but at the genotypic level where it is traditionally viewed as the background operator[13], often to simply re-introduce lost schema.

3.1 An ES-GA Hybrid

Hoffmeister and Bäck[12] have compared the search characteristics of the ES to those of the GA, concluding that they are "diametrically opposed concerning their emphasis on global and local search ... ESs have a high bias towards local search". In the above functions we have seen that mutating a GA genotype slightly before evaluating it can give improved performance on rugged landscapes and we suggest this is because it allows the algorithm to gain some local knowledge - a form of random learning. On the smooth landscape of the max-ones problem we found no improvement though. If we use the "+" strategy of ESs in conjunction with the extra mutation operator we can turn the random learning into local hillclimbing. Here both the genotype and it's mutated offspring (phenotype) are evaluated, with the fittest of the two surviving to be a parent for the succeeding generation; if the mutated offspring is fitter than it's parent it replaces it in the population. Effectively we run a (1+1)-ES on each GA generated genotype. In GA terms this adds Lamarckian replacement to it's operators, something Braun[5] has shown to give improvements of 1-2 orders in magnitude.

Table 4 shows the averaged performances of this ES-GA hybrid on the functions of section 2, over fifty trials. All parameters remain the same as before except that we halve the population size N as each gene is now evaluated as both a genotype and a phenotype (using a population of the same size appears to make little difference here).

	Mean	Std. Dev.
Deceptive ($N=4$)	2.9	2.1
Max Integer	8.5	5.1
Max Ones	5.7	1.8

Table 4.

From Table 4 it can be seen that the hybrid's performance on the partially deceptive function is better than that of the basic GA (Table 1) but no better than that of the GA with just the extra mutation. This is perhaps as expected since the function was specifically chosen for it's shape to demonstrate the advantage possible with the extra mutation (random learning). It can be noted that this result is the same for all the values of N shown.

On the other two problems the hybrid performs better at the 1% level of statistical significance in comparison to both the GA and the GA with extra mutation (Tables 2&3). Therefore adding the ES/Lamarckian replacement strategy in the hybrid enables it to outperform the GA on both a smooth and a rugged landscape, as opposed to with the extra mutation operator on it's own which only performed better on the rugged case. Mülenbein has also shown (eg [15]) that the addition of hillclimbing can improve the efficiency of the (parallel) genetic algorithm.

4 Conclusion

Starting with the fact that the basic Genetic Algorithm omits any form of growth or learning we have shown that crudely approximating these processes using simple random mutation can improve the GA's performance on rugged landscapes. This can perhaps be seen as a form of the somatic mutation seen in the immune system[8]. We then showed that by further adding Lamarckian replacement, drawing from the field of Evolution Strategies, the GA's performance on smooth landscapes can also be improved, giving additional improvement on rugged landscapes. We have suggested that by adding these extra operators it allows the GA to maintain it's explorative global search whilst also being able to exploit some local search. The results indicate that the effect of these extra operators on the normal schema processing of the GA is complementary though Gruau and Whitley[10], for example, voice the concerns at the effects of these types of operators. In the first case of adding just mutation some disruption is inevitable, with it presenting noisy information about a given hyperplane, but we have seen that this can help the GA - whether it is indicating nearby optima and/or smoothing over lower areas of the search space. The second operators direct inheritance of phenotypic traits may "alter the statistical information about hyperplane subpartitions that is implicitly contained in the population"[10] (as in[9]). The effects of both operators does require closer analysis.

This work represents the initial implementation of an ES-GA hybrid. We have only touched on the possible crossover between the two fields and on the potential of growth and learning with GA generated genotypes. For example Bäck[2] has successfully shown how ES's storing of the mutation rate in the genotype can also be applied to the GA. We are now working with storing both p_m (as Bäck did) and p_{m+} in the genotype rather than having to fix them a priori, particulary p_{m+}. We are also investigating allowing a genotype to produce a number of children to gain more local information, where this too can be coded in the genotype. The effect of the operators with regards to population size and diversity is also being examined.

5 Acknowledgements

The authors acknowledge the support of HP labs through their European Equipment grants programme.

6 References

[1] Ackley D. H. & Littman M. (1991), "Interactions Between Learning and Evolution", Artificial Life II, Addison-Wesley.

[2] Bäck T. (1992), "Self-Adaption in Genetic Algorithms", Toward A Practice of Autonomous Systems, MIT Press.

[3] Baldwin J. M. (1896), "A New Factor in Evolution", American Naturalist, 30.

[4] Belew R. K., McInerney J. & Schraudolph N. N. (1991), "Evolving Networks: Using the Genetic Algorithm with Connectionist Learning", Artificial Life II, Addison-Wesley.

[5] Braun H. (1993), "Evolution - a Paradigm for Constructing Intelligent Agents", From Animals to Animats II, MIT Press.

[6] Davis L. (1991), "Bit-Climbing, Representational Bias, and Test Suite Design", International Conference on Genetic Algorithms IV, Morgan Kauffman.

[7] Fogel L. J., Owens A. J. & Walsh M. J. (1962), "Artificial Intelligence Through Simulated Evolution", John Wiley, N.Y.

[8] Forrest S., Smith R. E., Perelson A. S. & Javornik B. (1993), "Using Genetic Algorithms to Explore Pattern Recognition in the Immune System", Evolutionary Computation, 1, 3.

[9] Goldberg D. E. (1989), "Genetic Algorithms and Walsh Functions: Part II, deception and its analysis", Complex Systems, 3.

[10] Gruau F. & Whitley D. (1993), "Adding Learning to the Cellular Development of Neural Networks", Evolutionary Computation, 1, 3.

[11] Hinton G. E. & Nowlan S. J. (1987), "How Learning Can Guide Evolution", Complex Systems 1.

[12] Hoffmeister F. & Bäck T. (1990), "Genetic Algorithms and Evolution Strategies: Similarities and Differences", Parallel Problem Solving from Nature, Univ. of Dortmund.

[13] Holland J. H. (1975), "Adaption in Natural and Artificial Systems", Univ. of Michigan Press, Ann Arbor.

[14] Mitchell M., Forrest S, & Holland J. H. (1992), "The Royal Road for Genetic Algorithms: Fitness Landscapes and GA Performance", Toward A Practice of Autonomous Systems, MIT Press.

[15] Mülenbein H. (1991), "Evolution in Time and Space - the Parallel Genetic Algorithm", Foundations of Genetic Algorithms, Morgan Kaufman.

[16] Rechenberg I. (1973), "Evolutionsstrategie: Optimierung Technischer Systeme Nach Prinzipien der Biologischen Evolution", Frommann-Holzboog Verlag, Stuttgart.

[17] Schwefel H-P. (1977), "Numerische Optimierung von Computer-Modellen Mittels der Evolutionsstrategie", Interdisciplinary Systems Research; 26.Birkhäuser, Basel.

Genetic Algorithms and Directed Adaptation

John Coyne[†] and Ray Paton

The Liverpool Biocomputation Group, Department of Computer Science,
The University of Liverpool, Liverpool L69 3BX, U.K.

Abstract

This paper presents a case for the value of "talkback" in genetic algorithms. Talkback is described in terms of an environmentally imprinted adaptation occurring at loci on the genotype, not expressed phenotypically but acting as control parameters for evolution. Motivations are presented drawing on examples of directed adaptation in nature and the field of optimised search. A model is described based on directed adaptation and the results of experiments, acting at the level of both the population and the individual, are presented.

Keywords

genetic algorithms, directed adaptation, Lamarckianism, regulators, environmental talkback

1 Introduction

The genetic algorithm is a stochastic optimisation technique that draws on the evolutionary synthesis as its source metaphor and has traditionally adhered to a strict Darwinian interpretation of the theory. A central premise of the scheme is that of a selectionist system existing as genotype-phenotype independent of the environment, but whose contribution to the succeeding generation's gene pool is determined by a measure of fitness relative to that environment. Genetic operators function on the genotypic level while the selection mechanism operates on the phenotypic level. The role of the environment is to evaluate the phenotypic expression of the genotype and bias the system towards the selection of fitter variants.

In the GA the directed search through solution space is determined by the exogenous control parameters of crossover and mutation, corresponding respectively to exploitation (short term optimisation) and exploration (long term optimisation) of the environment. The parameterisation of these control settings, which determines

---

[†] Present address : Econostat Ltd., Hennerton House,
Wargrave, Berks. RG10 8PD *john@ecowar.demon.co.uk*

the effectiveness of the search strategy, has on the whole been carried out empirically, with *optimal* settings remaining constant throughout the evolutionary time frame [Shaffer 89].

What this paper sets out to determine is whether the information present in the environment, in the form of the state vector describing payoff values for the current population, can be harnessed not solely in determining the probability distribution function used for reproductive selection, but to strategically adjust the level of genetic instability. This may be achieved through adapting the level of mutation based on environmental feedback. Thus the mechanism can effect a rapid scaling of peaks in the exploration map and subsequently drive the search towards new regions by altering the copying fidelity at genotypic level.

2 Genetics of Genetic Algorithms

The genetic algorithm is the best known of a number of problem solving approaches used in computer science based on selectionist systems in nature. All are characterised by a system comprising a population of units containing a high degree of diversity where the difference between units produces a differential response in an environment. This discriminatory action enables better adapted units to propagate by means of a replicating mechanism. Diversity within the population is generated through specific techniques so as to prevent the system settling towards an attractor [Manderick 91]. In evolutionary systems this diversity is produced through mutation and is a measure of the copying fidelity of replication. In natural systems mutation can operate at the gene level where a gene undergoes point or frameshift mutation, but also at the level of the chromosome through inversion, transposition and amplification of strings of genetic code.

The guiding principles, if not the exact mechanisms, in natural selection are well understood and a simplified scheme, observing the phenotype-genotype distinction in biology was introduced by Holland [Holland 75]. The algorithm conforms to the predominant neo-Darwinian philosophy and has been successfully applied to a number of optimisation problems. A body of research within the field has focused on incorporating further aspects of evolutionary theory into the model [Sumida 94], but to date directed adaptation, the environmental setting of the genome, has been underexplored.

Without recourse to the much plied example of the blacksmith's arms the Lamarckian scheme [Dobzhansky 55] has now come to represent a concept of evolution in which the development of a species is driven by the efforts of organisms to adapt themselves to their environment. It thus stands in stark contrast to the random (i.e. non directed) genetic variation of Darwinian theory and replaces the arbitrary, yet passive, role of the environment with an adaptive feedback mechanism into the evolving agent. A modern orthogenetic interpretation might incorporate a feedback mechanism from the environment producing an adaptation on the genotypic level within life cycles, which is replicated between life cycles of successive generations.

Despite its obvious flaws Lamarckianism continues to surface in response to occurrences in biology that cannot be explained adequately in a neo-Darwinian framework. The *directed adaptation* can take the form of an amplification of genes or an increase in the level of mutation to hypermutatable states. Cullis proposes mutation at the chromosome level due to transposition in direct response to environmental shifts most notably in the amplification of genes observed in flax (Linum) when treated with high levels of fertiliser. Hall introduces a model in which stress gives rise to hypermutatable states, with the probability of a particular mutation occurring dependent both upon the external environment and the mutation being advantageous to the organism [see Paton 92 and this volume].

What this paper does not set out to advocate is an overturning of the central ideas of Darwinian evolution. Rather we seek in the above biological examples aspects of Lamarckian behaviour that may be incorporated into the general model of the genetic algorithm. It is through these features we seek to describe heritable forms of non-Darwinian processes.

3 Searching Solution Space

Heuristic optimisation techniques have been used in a wide range of problems where no *a priori* knowledge exists as to how a change will improve the performance of an agent in an environment. In common with all such methods the genetic algorithm describes a compromise between exploration and exploitation of the environment.

Exploitation may be viewed as the application of specific knowledge in guiding the search through solution space [Thurn 92], and without which the algorithm degenerates into a random walk. Conversely exploration is the mechanism by which sub-optimal basins of attraction in the search space are avoided and distinguishes this class of algorithm from their greedy or hill climbing counterparts.

In the genetic algorithm exploration and exploitation is achieved through the reproduction operators. Primary among these are crossover and mutation. Crossover both exploits and explores through recombining schemata drawn from the cluster of points in multi-dimensional bit space that make up the current population. The scope of the exploration is bounded by the information contained in the gene pool.

Mutation is described by Holland as a background operator [Holland 75] and extends the reach of the local mapping function described by reproduction to include those alleles lost to the population. The rate of mutation in the standard genetic algorithm is pre-determined with the setting fixed so as to maintain population diversity whilst minimising the disruptive effect of the operator on schemata building.

The parameterisation of control parameters runs counter to the intuitive idea that the probability of success of a mutation varies during a run of the genetic algorithm. This is a consequence of rising on-line fitness, and a corresponding reduction in the hamming distance to the optimal solution.

A variant on the standard genetic algorithm is to include a time dependent schedule for control parameters [Fogarty 89]. Scheduling is deterministic by definition and incapable of adapting to the evolving nature of the search.

In a recent paper Bäck [Bäck 92] demonstrates the dependence of mutation on quantities such as population size, representational length of bitstring and fitness function. Borrowing from evolutionary strategies Bäck introduces a self adaptable mutation rate encoded in the genotype and thus susceptible to the same environmental pressures as those elements of the genotype expressed phenotypically. Davis details a scheme for adapting the probabilities assigned to a host of genetic operators based on the phenotypic evaluation of individuals produced using the operator with respect to the environment [Davis 89]. While both variants on the genetic algorithm make use of environmental information in the adaptation of control parameters they cannot be said to operate outside of a Darwinian framework.

4 Directed Adaptation in Artificial Systems

Directed adaptation differs from the self-adapting models discussed above in that the environment tests but also talks back to the genotype.

Davidor has introduced a model based on a Lamarckian variant of the basic genetic algorithm to optimise robot trajectories by manipulating reproduction operators to bring about sub-goals. This takes the form of altering the probability distribution at crossover sites according to the error distribution, and by directing mutation at certain loci not co-adapted [Davidor 91]. Gruau has detailed a strategy exploiting (what Dawkins [Dawkins 82] calls the pseudo Lamarckian principle of) the Baldwin Effect. This is implemented by altering the fitness landscape to incorporate within life cycle acquired characteristics of the organism achieved through learned behaviour [Gruau 92].

One of the difficulties of using directed mutation in genetic algorithms is predicting the effect that a mutation on the genotypic level will have on the fitness value of the chromosome as a whole, and is dependent on the bit string representation and the objective function.

5 Analysis of the Models

Two classes of models are introduced here based on environment directed adaptation of the mutation rate. They may be classified according to the inclusion of exogenous or endogenous mutation, corresponding respectively to operators on the population and individual levels.

Exogenous adaptation [Coyne 92] is a strategy to exploit features in the fitness landscape by altering the level of mutation at the population level in response to changing off-line fitness. This introduces a short term population level memory with a rising off-line fitness depressing the mutation level within the community of organisms enabling interesting peaks in the exploration map to be scaled. It can thus be regarded as a breeding strategy rather than a genetic operator.

The system has two modes of operation. Once local features of the fitness landscape have been exhausted the mechanism is biased towards exploration. The search is driven towards new regions in the solution map through an increase in genomic instability effected through a higher rate of mutation. On rising off-line fitness the system is switched back into a stable state to exploit effectively the new schemata. The strategy has been implemented by using both a two state implementation, switching between hypermutatable and low mutational states, and alternatively by using a gradual step like, or creeping adjustment of mutation rate bounded by upper and lower limits.

The endogenous directed mutation scheme is based on a number of regulators contained within and operating at the level of the individual. Regulators have no phenotypic expression, instead they control the level of copying fidelity in an associated segment of code.

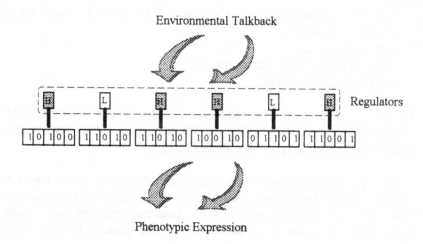

Environmental Talkback

Regulators

Phenotypic Expression

Figure 1. Environmental feedback on genotype

The genotype is subjected to differential mutation along its length. Stress in the external environment has the effect of switching on and off regulators introducing hypermutatable states at various loci on the binary string (see Figure 1).

A stressed organism is defined as one whose fitness is poor relative to the ecosystem (community of organisms). A fit organism will produce an aggregate low level of mutation being well adapted to its niche whereas a poorly adapted organism will have many regulators switched to an excitatory state. Switching on and off of regulators is determined probabilistically and is based on an aggregate evaluation of the organism.

6 Model Performance

The test problem used to benchmark all models was the knapsack problem. The objective function is to maximise the payoff for carrying a number of bars. Each bar is assigned a specific weight and value and the cargo is subject to a maximum loading constraint. This can be expressed as solving for

$$f(.) = \max\left(\sum_{i=1}^{N} \sigma_i v_i \right)$$

subject to

$$\sum_{i=1}^{N} \sigma_i \omega_i < \Phi$$

where
  $\sigma_i \in \{0, 1\}$ corresponds to a bar being included or absent from the knapsack,
  ω_i is the weight of bar i,
  υ_i the value of bar i,
  Φ the maximum weight constraint, and
  N the total number of bars

The teaser is a paradigm example of the NP hard class of problem for which no polynomial-time algorithms are known. The state space is both discrete and multi-modal with the set of all possible solutions being 2^N.

On-line and off-line experimental results are presented based on an aggregate result over ten runs of the program. Population size was defaulted to the length of the binary string. A crossover rate of 0.65, generation replacement, windowing of objective function, and a length of run of 200 generations was used in all experiments. Results were benchmarked against a standard genetic algorithm (SGA) based on a mutation rate of 0.012 without elitism.

6.1 Exogenous Adaptation

Adaptive exogenous mutation experiments were undertaken, using upper and lower bounds on the rate of mutation of 0.004 - 0.02 respectively, optimising over a 40 bit length binary string. Figure 2 which is based on the two state mutation strategy described above shows off-line fitness results compare with the SGA implementation. Performance degenerated considerably when the creeping mutation strategy was employed.

When combined with herd leader and elitist breeding strategies performance deteriorated for both creep and two-state adaptive exogenous mutation. This may suggest that the model is prone to cycling.

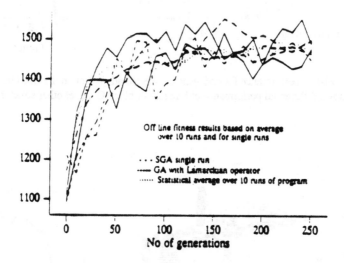

Figure 2. Off-line fitness performance for exogenous adapted and SGA

6.2 Endogenous Adaptation

All endogenous adaptation experiments are based on optimising over a 32 bit length binary string. Environmental talkback results in the switching in and out of hypermutatable and normal states corresponding to mutation rates of 0.02 and 0.004 respectively.

In order to determine the optimum number of regulators for the genome, experiments were undertaken by consecutively introducing added granularity and the results are presented below. The granularity was varied from the 2 regulator block

GA (each regulator controlling the mutation rate of 16 alleles), to a 32 regulator block (each regulator controlling a single allele). The graphic presented in figures 3 and 4 details on line and off-line fitness results. The performance characteristics support a 8 regulator block GA with differential performances (LHS of figures 3 and 4) highlighting the competitive advantage over other variants.

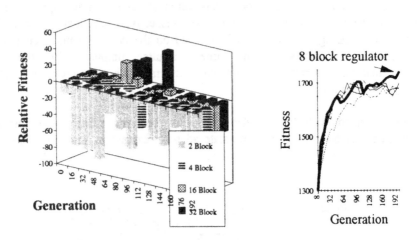

Figure 3. Off-line performance for endogenous directed adaptation in the GA. Graphic on left shows differential performance of 8 regulator block GA over other structures.

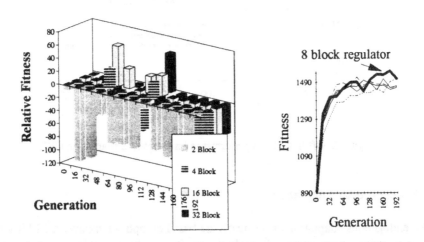

Figure 4. On-line performance for endogenous directed adaptation in the GA. Graphic on left shows differential performance of 8 regulator block GA over other structures.

The 8 regulator block directed adaptation GA (DAGA) was benchmarked against the SGA described above with the directed adaptation model demonstrating superior on-line and off-line characteristics (see Figures 5 and 6). However, these results were not mirrored with the introduction of an elitist breeding strategy into the models.

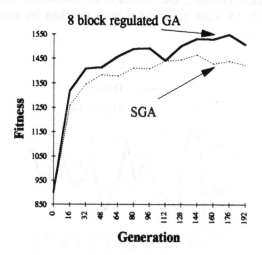

Figure 5. On-line fitness results for 8 regulator block and SGA.

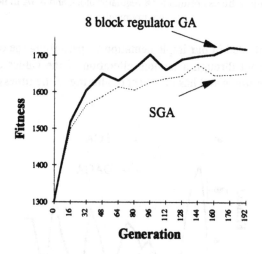

Figure 6. Off-line fitness results for 8 regulator block and SGA.

6.3 Non-stationary Environment

The complex interactions of organisms both at population and at species level together with the global chaotic effect of the atmospheric environment give rise to a system in nature in a continual state of flux. Niches are continually created and perpetual novelty results. The effect of a dramatic change to the system can dramatically change the nature of the environment through the elimination of a niche and the subsequent extinction of a species. Evolution in natural systems has accounted for the statistical properties of the environment through long term adaptation, anatomical structure, etc., robust to environmental change. This type of system is less likely to result in a global attractor than its static counterpart in function optimisation.

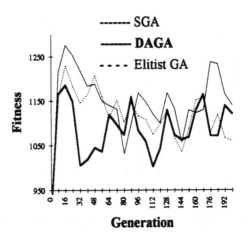

Figure 7. On-line fitness results for 8 regulator block and SGA in non-stationary environment.

The concept is simulated in our implementation by perturbing the objective function. This is brought about through a random alteration of the values υ assigned to the bars every 10 generations and has the effect of altering of the fitness landscape.

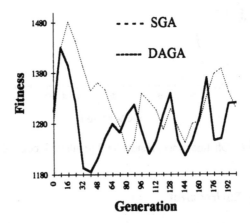

Figure 8. Off-line fitness results for 8 regulator block and SGA in non-stationary environment.

How would a system based on directed adaptation perform in such an environment? Poorly one would expect. A system that has adapted itself to a niche is less likely to contain sufficient generic properties to escape from what has now become a poor global solution.

These suspicions are borne out in our computer simulations. The DAGA underperforms both the SGA and the elitist GA in terms of on-line and off-line fitness measures (Figures 7 and 8).

7 Discussion

The results presented in this paper mark a divergence from the traditional approach used in the genetic algorithm. We have presented two alternative primitive classes of models of the genetic algorithm based on an exogenous and endogenous mutation control parameter respectively. For the latter the results detailed demonstrate how directed adaptation can give rise to improved on-line and off-line performance.

The environmental talkback technique outlined above relates to the aggregate properties of an organism. This information is diffused through regulators to effect a differential setting for reproductive control parameters along the genome. In this sense the genome can be regarded as hermetic since regulator setting is determined probabilistically dependent on the evaluation of the phenotype. This mechanism has been adopted in some part due to the information sparsity of the knapsack problem environment. The genotype is expressible and evaluation available only as a phenotypic whole.

In many optimisation problems information can be derived from the environment relating to the specific characteristics of the allele or grouping of alleles. In the two stage mechanism existing in nature model specification is defined through the genotypic-phenotypic transformation, projected on to discrete state space. Model evaluation is based on characteristics acquired within life cycle but parameterised by the genotype and defines a continuous solution space. In such an information rich environment a performance measure for differential phenotypic plasticity can be derived. This additional information may be incorporated into the model, again through varying the mutation rate, resulting in an environmentally driven regulator tightly coupled with physical position on the genome.

References

Bäck T. (1992), "The Interaction of Mutation Rate, Selection and Self-Adaptation Within a Genetic Algorithm". In Proceedings of the Second International Conference on Parallel Problem Solving for Nature, 2, Männer R., Manderick B. (Ed.), Elsevier, Amsterdam.

Coyne J. (1992), "The Genetic Algorithm and the Lamarckian Operator", Unpublished Technical Report, The Liverpool Biocomputational Group, University of Liverpool.

Davis L (1989), "Adapting Operator Probabilities in Genetic Algorithms". In Proceedings of the Third International Conference on Genetic Algorithms and their Applications, Shaffer J. (Ed.), Morgan Kaufmann, San Mateo.

Davidor Y. (1991), "A Genetic Algorithm Applied to Robot Trajectory". In Handbook of Genetic Algorithms, Davis L. (Ed.), Van Nostrand Reinhold, N.Y.

Dawkins R. (1982), The Extended Phenotype, Oxford University Press, pp 169.

Dobzhansky T. (1955), Evolution, Genetics and Man, John Wiley and Sons, N.Y.

Gruau F., Whitley D. (1993), "Adding Learning to the Cellular Development of Neural Networks : Evolution and the Baldwin Effect". In Evolutionary Computation 1 (3), pp. 213-233.

Fogarty T. (1989), Varying the Probability of Mutation in the Genetic Algorithm. In Proceedings of the Third International Conference on Genetic Algorithms and their Applications, Schaffer (Ed.), Morgan Kaufmann, San Mateo, pp. 104-109.

Holland J. (1975), Adaptations in Natural and Artificial Systems, The University of Michigan Press, Ann Arbor.

Manderick B. (1991), "Selectionist Systems as Cognitive Systems". In Proceedings of the First European Conference on Artificial Life, Paris, Dec. 11-13.

Paton R. (1992), "Adaptation and Environment", Unpublished Technical Report, The Liverpool Biocomputational Group, University of Liverpool.

Schaffer J., Caruana R., Eshelman L., Rajarshi D. (1989), "A Study of Control Parameters Affecting On-line Performance of Genetic Algorithms for Function Optimisation". In Proceedings of the Third International Conference on Genetic Algorithms and their Applications, Shaffer J. (Ed.), Morgan Kaufmann, San Mateo.

Sumida B., Hamilton W. (1994), "Both Wrightian Peak Shifts and Parasites Enhance Genetic Algorithm Performance in the Travelling Salesman Problem". In Paton (Ed.), Computing with Biological Metaphors, Chapman and Hall, London.

Thurn S, Moller K. (1992), "Active Exploration in Dynamic Environments". In Advances in Neural Information Processing Systems 4, Moody J. et al (Ed.), Morgan Kaufmann, San Mateo.

Genetic Algorithms and Neighbourhood Search

Colin R. Reeves

School of Mathematical and Information Sciences
Coventry University
UK
Email: CRReeves@uk.ac.cov.cck

Abstract. Genetic algorithms (GAs) have proved to be a versatile and effective approach for solving combinatorial optimization problems. Nevertheless, there are many situations in which the simple GA does not perform particularly well, and various methods of *hybridization* have been proposed. These often involve incorporating other methods such as *simulated annealing* or *local optimization* as an 'add-on' extra to the basic GA strategy of selection and reproduction.

Here, we explore an alternative perspective which views genetic algorithms as a generalization of *neighbourhood search* methods. It is not the intention to present a fully worked-out statement as to what sort of neighbourhood search a GA is. Rather, it is to investigate some of the parallels, and to suggest some areas for further research which may enhance our understanding of both neighbourhood search and genetic algorithms.

1 Introduction

Genetic algorithms (GAs) have frequently been applied to problems of combinatorial optimization, such as bin-packing [1, 2], machine-sequencing [3], vehicle routing [4], and of course the travelling salesman problem (TSP) [5, 6, 7]. These problems typically require some fairly major adjustments to the 'classical' GA paradigm as developed by Holland [8]. In what follows, we assume the reader has a basic understanding of the way in which GAs operate—if not, we refer them to Goldberg [9] or Reeves [10] for an introduction. A discussion of GAs in the particular context of combinatorial optimization can be found in [11].

One useful ploy is that reported by Jog *et al.* [5] (amongst others) in the context of the TSP. They suggest that following an application of crossover to produce a new chromosome, a neighbourhood search (or local optimization) should be carried out on this string to move it to a local optimum before injecting it into the population. In a similar vein, Goldberg [9] uses the idea of 'G-bit improvement' for binary strings. Other examples of this type of approach are in [12, 13, 14]. In these cases, the application of local optimization is kept distinct from the recombination operation.

The problem with this approach, in terms of the traditional account of how GAs work, is in interpreting what is a good 'building-block'. By interposing local optimization, the offspring become rather remote from their parents, and it is hard to see how the Schema Theorem for example has any relevance.[1] Radcliffe [15, 16] has pointed out the importance of ensuring that genetic operators can produce offspring that *respect* and/or *assort* their parents. This is clearly difficult to ensure when local optimization is included. There is also a practical problem in deciding whether it should be the 'original' offspring or its optimized version which should be used in the next generation. This is similar to the debate as to whether, in the context of constrained problems, to use an original but infeasible chromosome or the 'repaired' feasible one. Recently, Orvosh and Davis [17] reports some interesting experimental evidence which suggests that this decision should itself be a stochastic choice.

In this paper, rather than keeping the local optimization separate from recombination, we intend to explore the possibilities of integrating it directly into the standard genetic algorithm. This will entail a consideration of how GAs can be viewed from a neighbourhood search perspective. Before doing this, it is necessary to give a brief overview of traditional neighbourhood search methods.

2   Traditional Neighbourhood Search

Neighbourhood search (NS) is a widely used method in solving combinatorial optimization problems. A recent introductory treatment can be found in [10]. In neighbourhood search, we assume a solution is specified by a vector $\mathbf{x}$, where the set of all (feasible) solutions is denoted by $\mathbf{X}$, and the cost of solution $\mathbf{x}$ is denoted by $c(\mathbf{x})$—often called the objective function. Each solution $\mathbf{x} \in \mathbf{X}$ has an associated set of *neighbours*, $N(\mathbf{x}) \subset \mathbf{X}$, called the neighbourhood of $\mathbf{x}$. Each solution $\mathbf{x}' \in N(\mathbf{x})$ can be reached directly from $\mathbf{x}$ by an operation called a *move*, and $\mathbf{x}$ is said to move to $\mathbf{x}'$ when such an operation is performed. A description of the method is given below (adapted from Glover and Laguna [18]), where it is assumed that we are trying to *minimize* the cost.

Neighbourhood Search Method

1	(Initialization)
1.1	Select a starting solution $\mathbf{x}^{now} \in \mathbf{X}$.
1.2	Record the current best known solution by setting $\mathbf{x}^{best} = \mathbf{x}^{now}$ and define $best\_cost = c(\mathbf{x}^{best})$.
2	(Choice and termination)
2.1	Choose a solution $\mathbf{x}^{next} \in N(\mathbf{x}^{now})$. If the choice criteria used cannot be satisfied by any member of $N(\mathbf{x}^{now})$ (hence no solution qualifies to be $\mathbf{x}^{next}$), or if other termination criteria apply (such as a limit on the total number of iterations), then the method stops.
3	(Update)
3.1	Re-set $\mathbf{x}^{now} = \mathbf{x}^{next}$, and if $c(\mathbf{x}^{now}) < best\_cost$, perform Step 1.2. Then return to Step 2.

[1] It is fair to point out that some have argued positively for this process as an analogy of the natural 'maturation' process: offspring do alter in quite complex ways before they are mature enough to reproduce.

Here we assume the choice criteria for selecting moves, and termination criteria for ending the search, are given by some external set of prescriptions. By specifying these prescriptions in different ways, the method can easily be altered to yield a variety of procedures. Descent methods, for example, which only permit moves to neighbours that improve the current $c(\mathbf{x}^{now})$ value, and which end when no improving solutions can be found, can be expressed by the following provision in Step 2.

Descent Method

2	(Choice and termination)
2.1	Choose $\mathbf{x}^{next} \in N(\mathbf{x}^{now})$ such that $c(\mathbf{x}^{next}) < c(\mathbf{x}^{now})$ and terminate if no such $\mathbf{x}^{next}$ can be found.

The failing of NS is well-known: its propensity to deliver solutions which are only *local* optima. Randomized procedures, including the important technique of *simulated annealing*, offer one way of circumventing the local-optimum problem, and they can similarly be represented by adding a simple provision to Step 2.

Randomization Method

2	(Choice and termination)
2.1	Randomly select $\mathbf{x}^{next}$ from $N(\mathbf{x}^{now})$.
2.2	If $c(\mathbf{x}^{next}) \leq c(\mathbf{x}^{now})$ accept $\mathbf{x}^{next}$ (and proceed to Step 3).
2.3	If $c(\mathbf{x}^{next}) > c(\mathbf{x}^{now})$ accept $\mathbf{x}^{next}$ with a probability that decreases with increases in the difference $\Delta c = c(\mathbf{x}^{next}) - c(\mathbf{x}^{now})$. If $\mathbf{x}^{next}$ is not accepted on the current trial by this criterion, return to Step 2.1.
2.4	Terminate by a chosen cutoff rule.

Such methods continue to sample the search space until finally terminating by some form of iteration limit. Normally they use an exponential function to define probabilities—simulated annealing (SA), for example, uses $\exp(-\Delta c/T)$, where T is a 'temperature' parameter which diminishes monotonically toward zero as the number of iterations grows. Thus SA starts with a probability of accepting non-improving moves at Step 2.3 which is initially high and decreases over time.

These procedures can do better than finding a single local optimum since they effectively terminate only when the probability of accepting a non-improving move in Step 2.3 becomes so small that no such move is ever accepted (in the finite time allowed). Hence, they may wander in and out of various intermediate local optima prior to becoming lodged in a final local optimum, when the temperature becomes small.

Another approach to overcome the limitation of the descent method is simply to restart the method with different randomly selected initial solutions, and run the method multiple times. This is sometimes called *iterated descent*. Yet another possible method is to execute a few random moves for a period after reaching a local optimum, and then begin a new phase of neighbourhood search.

Tabu search is another recent addition to the family of NS methods. The central theme underlying tabu search is the idea of incorporating memory into the search process. How this is done is described in detail in Glover and Laguna [18], but we give a brief resumé here for the benefit of readers who may be unfamiliar with TS.

Tabu search maintains a selective history H of the states encountered during the search, and replaces $N(\mathbf{x}^{now})$ by a modified neighbourhood which may be denoted $N(H, \mathbf{x}^{now})$. History therefore determines which solutions may be reached by a move from the current solution, in that $\mathbf{x}^{next}$ is selected from $N(H, \mathbf{x}^{now})$. In TS strategies based on short term considerations, $N(H, \mathbf{x}^{now})$ is typically a subset of $N(\mathbf{x}^{now})$, and a *tabu classification* serves to identify elements of $N(\mathbf{x}^{now})$ excluded from $N(H, \mathbf{x}^{now})$.

TS can also use history to create a modified evaluation of currently accessible solutions. In effect, TS replaces the objective function $c(\mathbf{x})$ by a function $c(H, \mathbf{x})$, which has the purpose of evaluating the relative quality of currently accessible solutions. This modified function is relevant because TS uses aggressive choice criteria that seek a best $\mathbf{x}^{next}$, i.e. one that yields a best value of $c(H, \mathbf{x}^{next})$, over a candidate set drawn from $N(H, \mathbf{x}^{now})$.

It has been suggested previously [19, 20] that genetic algorithms, simulated annealing and tabu search can all be viewed as forms of a general search strategy. However, these accounts have been content with a formal statement of this equivalence, and in practice, GAs have usually been seen as forming a rather different approach from the others mentioned. In the next section we explore these claims further, by considering the form of the neighbourhoods that are being used in a genetic search.

3 GA Neighbourhoods

In what follows, we shall assume that $\mathbf{x}$ is binary vector of length n. The ith bit of $\mathbf{x}$ will be denoted by x_i. The set $\mathbf{X}$ is thus a subset of the n-dimensional hypercube $\{0, 1\}^n$. (We recognize that there is a continuing debate—see [21, 22] for example—as to the desirability or otherwise of using binary coding. However, binary encoding is still commonly used in applications of GAs, and it makes the exposition of the concepts we shall develop somewhat easier. Nevertheless, the ideas discussed below are clearly capable of being generalized to other situations.)

The neighbourhood search method defined above takes the standpoint of defining neighbours with respect to a *single* solution $\mathbf{x}$. For a binary vector $\mathbf{x}$, a simple neighbourhood might be the set of all single-bit changes to $\mathbf{x}$. For instance, if $\mathbf{x} = (0, 1, 1, 0)$, then

$$N_1(\mathbf{x}) = \{(1, 1, 1, 0), (0, 0, 1, 0), (0, 1, 0, 0), (0, 1, 1, 1)\}.$$

As this relates to changes of just one bit at a time in one solution, we might call this a *first-order 1-neighbourhood*. Similarly, we can define a *second-order 1-neighbourhood* consisting of all changes of two bits in one solution. In the example, we find

$$N_2(\mathbf{x}) = \{1, 0, 1, 0), (1, 1, 0, 0), (1, 1, 1, 1), (0, 0, 0, 0), (0, 0, 1, 1), (0, 1, 0, 1)\}.$$

In general, we can see that a kth-order 1-neighbourhood is the surface of a ball of radius k (using the Hamming metric) in the binary hypercube. For large n, the size of a kth-order neighbourhood is $\binom{n}{k}$ and searching such a large neighbourhood is likely to be computationally impracticable. Most NS methods use neighbourhoods of order 2 or 3 at the most.

In the case of a genetic algorithm, we usually have two parents which are recombined to give two offspring by a crossover operation. The basic NS framework can only fit this situation if we generalize the concept of a neighbourhood. To do this, we first define the concept of an *intermediate* vector.

Definition A vector $\mathbf{y}$ is *intermediate* between two vectors $\mathbf{x}$ and $\mathbf{z}$, written as $\mathbf{x} \diamond \mathbf{y} \diamond \mathbf{z}$, if and only if

$$\text{either } y_i = x_i \text{ or } y_i = z_i \text{ for } i = 1, \ldots, n.$$

We can now define a *2-neighbourhood* of $\mathbf{x}$ and $\mathbf{z}$ to be the set of all their intermediate vectors.

An equivalent definition of intermediacy is in terms of the Hamming distance between two vectors, given by

$$d(\mathbf{x}, \mathbf{y}) = \sum_{i=1}^{n} |x_i - y_i|.$$

Then $\mathbf{y}$ is intermediate between $\mathbf{x}$ and $\mathbf{z}$ if and only if

$$d(\mathbf{x}, \mathbf{z}) = d(\mathbf{x}, \mathbf{y}) + d(\mathbf{y}, \mathbf{z}).$$

(These and other geometrical properties of the n-dimensional hypercube are described in greater detail in Kanerva [23], who prefers the term 'between-ness' for what we have called intermediacy.)

This definition enables us to be more precise: we can define a kth-order *2-neighbourhood* as the set of all intermediate vectors at a Hamming distance of k from one end point. Formally,

$$N_k(\mathbf{x}, \mathbf{z}) = \{\mathbf{y} : (\mathbf{x} \diamond \mathbf{y} \diamond \mathbf{z}) \text{ and } (d(\mathbf{x}, \mathbf{y}) = k \text{ or } d(\mathbf{y}, \mathbf{z}) = k)\}.$$

Clearly, there will be $d \equiv d(\mathbf{x}, \mathbf{z})$ bits where $x_i \neq z_i$, and in the remaining cases (where $x_i = z_i$), the value of y_i cannot vary. The size of the 2-neighbourhoods is obviously

$$S_k = 2 \binom{d}{k} \quad \text{for } 1 \leq k \leq \lfloor d/2 \rfloor. \tag{1}$$

For example, if $\mathbf{x} = (0, 1, 1, 0)$ and $\mathbf{z} = (1, 0, 0, 0)$, then $d = 3$ and

$$N_1(\mathbf{x}, \mathbf{z}) = \{(1, 1, 1, 0), (0, 0, 1, 0), (0, 1, 0, 0), (0, 0, 0, 0), (1, 1, 0, 0), (1, 0, 1, 0)\}.$$

(There is no second-order neighbourhood here, since all points at a distance 2 from $\mathbf{x}$ are 1 unit distance from $\mathbf{z}$ and hence qualify as first-order points.)

A standard GA takes either one or two intermediate vectors which are generated by some form of crossover operator. Traditionally, the emphasis has been placed on the type of crossover used rather than on the vectors it generates, but it is evident even from the above simple example that we cannot generate a given neighbourhood by any type of n-point crossover operator, although, naturally, we can with uniform crossover.[2]

We can see this clearly if we use the concept of a crossover 'template'. Any crossover operator can be represented by a string defined on the alphabet $\{0, 1, \#\}$, where a 1 means the offspring uses $\mathbf{x}$, a 0 means the offspring uses $\mathbf{z}$, and a $\#$ means the bits are identical. As an example, the templates that generate the intermediate vectors in the example above (from $\mathbf{x} = (0, 1, 1, 0)$ and $\mathbf{z} = (1, 0, 0, 0)$) are as follows:

$$\{(0,1,1,\#),(1,0,1,\#),(1,1,0,\#),(1,0,0,\#),(0,1,0,\#),(0,0,1,\#)\}.$$

These templates may define one-point, 2-point and even 3-point crossovers, depending on whether the $\#$ symbol is interpreted as a 0 or a 1 in any particular case. (Note that these templates are not schemata—the $\#$ symbol should not be confused with the $*$ used to denote a schema in the traditional account of how GAs work.)

Thus the effect of traditional crossover operators is to *sample* the set of all intermediate vectors, but the distance of the offspring from its parents cannot be predicted, even if we restrict ourselves to one particular crossover, since the form of the offspring depends on how the operator interacts with the parents.

We note that it is quite possible for the traditional crossover to produce a 'degenerate' intermediate vector, i.e. a clone of one of the parents themselves. In the above example, a one-point crossover at position 3 would simply clone the parents. In traditional GA practice this is commonly accepted, on the grounds that it is quite reasonable to keep 'old' material in the population for future recombinations with different strings. In normal NS the re-visiting of points is also possible, In what follows, we ignore the possibility of degeneracy, since we can always constrain crossover to operate only on the 'interesting' section of the two strings.[3] We can easily identify the 'interesting' section as follows. Given two binary vectors $\mathbf{x}$ and $\mathbf{z}$, we can calculate the exclusive-OR (XOR) between them. For example, if $\mathbf{x} = (0, 1, 1, 0)$ and $\mathbf{z} = (1, 0, 0, 0)$, their XOR is $(1, 1, 1, 0)$.

---

[2] This point is also clearly demonstrated in the recent resurgence of interest in constructing mathematical models of a GA; in Vose [24] and Whitley [25] for example we find an explicit calculation of a 'mixing matrix' to characterize the probability of generating one vector from others by means of n-point crossover.

[3] This approach was first suggested by Booker [26] who termed it 'reduced surrogate' crossover. It has also been independently investigated by Fairley [27] who found it significantly improved the results obtained. Traditional 'generational' implementations of GAs commonly use a crossover probability of less than 1, which means clones may easily occur, and some have argued that this is beneficial. However, proponents of incremental or 'steady-state' implementations [28, 39] have argued persuasively *against* the generation of duplicates, and in this context the use of reduced surrogate crossover is an obvious strategy to reduce the occurrence of duplicates.

Having computed the XOR of the two parents, the *non-degenerate* crossover template will have the properties that

- it has a # wherever the XOR has a 0;
- it has a 1 or a 0 wherever the XOR has a 1;
- the number of 1s or 0s in these positions is between 1 and $d - 1$ (where d is the distance betwen the two parents, which is of course the same as the number of 1s in the XOR string.)

Clearly, there are $2^d - 2$ non-degenerate templates. For example, if $\mathbf{x} = (0, 1, 1, 0)$ and $\mathbf{z} = (1, 1, 0, 0)$, the XOR string is $(1, 0, 1, 0)$ and the only non-degenerate crossover templates are $(1, \#, 0, \#)$ or $(0, \#, 1, \#)$. Any other template will generate a clone of one of the parents.

We can characterize traditional crossovers by how they sample the space of intermediate vectors. We illustrate this below for the case of one-point and uniform crossover.

3.1 One-point crossover

One-point crossover[4] templates are of the form

$$(1, \ldots, 1, 0, \ldots, 0)$$

and there are clearly $(n - 1)$ of these for a string of length n. In general, suppose the XOR string has the leftmost and rightmost 1s at positions l and r respectively. Borrowing from the terminology of Bridges and Goldberg [29] (in a slightly different context), we call these the *sentry bits*. It is clear that only $r - l$ 'one-point' templates will produce a non-degenerate neighbour. If there are no #s between the sentry bits, the probability[5] of generating a kth-order neighbour is

$$P[k] = 2/(d - 1) \text{ for } 1 \leq k \leq \lfloor d/2 \rfloor,$$

if d is odd. If d is even, the distribution is

$$P[k] = 2/(d - 1) \text{ for } 1 \leq k \leq \lfloor d/2 \rfloor - 1,$$
$$= 1/(d - 1) \text{ for } k = \lfloor d/2 \rfloor.$$

If there are #s between the sentry bits, we have the possibility of generating some neighbours more than once. However, which neighbours these are depends on the nature of the parents, and in general it cannot be predicted. We will assume that there is no bias towards any one neighbour in particular, so that on the average the distribution is as characterized above.

---

[4] We consider only one-point crossover here for the sake of brevity. However, as pointed out by an anonymous referee, it is simple to derive two-point crossover as iterated one-point crossover.

[5] These are actually *conditional* probabilities—conditional on the offspring being at least 1 bit different from the parents. Thus we should really write $P[k|k > 0]$, but we will drop the condition for clarity.

Thus we see that one-point crossover samples these neighbourhoods at different rates. From equation 1, there are always more neighbours of order k than there are of order $k-1$, yet one-point crossover is indifferent to the size of the overall neighbourhood, giving equal chance to generating a neighbour at any order. However, as previously mentioned, one-point crossover will only be able to generate some of the neighbours at order k.

3.2 Uniform crossover

The situation with uniform crossover is quite different. In this case any (non-#) template position is equally likely to be a 0 or a 1, so the number of 1s (or 0s) is characterized by a Binomial distribution. If we again ignore the occurrence of #s between the sentry bits, the probability of a neighbour of order k is

$$P[k] = \frac{2}{(2^d - 2)} \binom{d}{k} \quad \text{for } 1 \le k \le \lfloor d/2 \rfloor \tag{2}$$

in the case d is odd. Again there is a small difference for the case when d is even, where

$$P[k] = \frac{2}{(2^d - 2)} \binom{d}{k} \quad \text{for } 1 \le k \le \lfloor d/2 \rfloor - 1,$$

$$= \frac{1}{(2^d - 2)} \binom{d}{k} \quad \text{for } k = \lfloor d/2 \rfloor.$$

Firstly, we see that in contrast to one-point crossover, it is possible to generate any of the neighbouring vectors at order k. Furthermore, because the probabilities are in exact proportion to the size of the neighbourhoods, we see that uniform crossover gives a much greater chance of sampling high-order neighbourhoods. Even for moderate values of d, this is a quite striking difference from one-point crossover. We now consider this aspect in more detail.

In order to characterize the effect of crossover more precisely, we need to know how large d is likely to be for a given value of n. We shall argue that, even for comparatively small values of n, most products of crossover are a long way from their parents.

We show in the Appendix that the distance between any two strings chosen at random which have respectively a and b 1-bits has a hypergeometric distribution. In principle we could use this result to compute the marginal distribution of k for given n but the computational demands would be excessive. Here we will simply use the case $n = 100$ (not particularly large for many applications of GAs) to illustrate the argument. On a simple binomial calculation, the values of a and b are overwhelmingly likely to lie between 30 and 70 (the probability of being outside these limits is 0.004%). For combinations of a and b within this range, we tabulate below approximate symmetric 99% confidence intervals for d as calculated from the hypergeometric distribution.

It is thus clear that it is highly unlikely that two random strings of length $n = 100$ will be less than 30 bits different, and even for $d = 30$, equation 2 shows

Table 1. 99% confidence intervals for distance between two random vectors

b	\multicolumn{9}{c}{99% confidence interval a}								
	30	35	40	45	50	55	60	65	70
30	30-52	31-55	34-58	35-59	38-62	39-63	42-66	43-67	46-68
35		32-56	33-57	36-60	37-61	38-62	41-65	42-66	43-67
40			34-60	35-61	36-62	37-63	38-64	41-65	42-66
45				36-62	35-61	36-62	37-63	38-62	39-63
50					36-62	35-61	36-62	37-61	38-62
55						36-62	35-61	36-60	35-59
60							34-60	33-57	34-58
65								32-56	31-55
70									30-52

that about 95% of all the offspring generated by uniform crossover lie at least 10 bits distant from one of the parents. (In contrast, for $d = 30$, only about 38% of offspring generated by one-point crossover would be that far from one of the parents.) Of course, d will usually be much greater than 30, so that it is clear that uniform (and even one-point) crossover generates offspring that are much further from a parent than would be the case with ordinary NS methods.

Inevitably, as the search develops, the parents resemble less and less a random sample of points of the hypercube, and the average distance between parents is likely to drop markedly. Louis and Rawlins [30] found that convergence to a single chromosome using a generational implementation was highly likely in 50 generations (on a problem with a 50-bit chromosome and a population of size 30). Implementations which use incremental reproduction typically converge faster, but this depends on the selection pressure and the mutation rate used. In practice we found that it may still take a relatively long time before the average distance between parents reaches a low level, even when cloning is permitted. In some experiments on a 100-bit function, we used an initial population of 30 strings, incremental reproduction with a variety of selection pressures, both one-point and uniform crossover, and mutation at a rate of 0 and 0.01. The rate at which the initial average inter-parent distance of 50 bits fell was measured, and it was found that at moderate selective pressures, with uniform crossover, and with a mutation probability (per bit) of 0.01, the average distance was still in excess of 5, even after 1000 function evaluations. (The distance between parents has to be less than 5 before order-1 neighbours become the most likely.)

It thus would seem to remain true that over a fairly extensive period of the search, uniform crossover in particular is likely to generate new points that are much further from their parents than is usual in traditional neighbourhood search methods.

4 Implications

The most prominent members of the NS family—simulated annealing and tabu search—share the characteristic that candidate moves are almost always to points in the *immediate* neighbourhood, i.e. to points whose Hamming distance from the current point is small. From the above analysis, we see that a GA is a very different form of NS, and its success in many diverse problems suggests that its differences are worth exploring in more detail.

These differences seem to be three-fold. Firstly, having a 2-neighbourhood essentially constrains the search to particular regions of the hypercube; the evidence of tabu search, which also restricts the search, although in a different manner, suggests that this is indeed a sensible approach.[6]

Secondly, the use of crossover operators entails taking large moves away from the points that are currently being considered. These are in direct contrast to normal NS, in which we only investigate the effect of small moves. (It is noteworthy that one GA implementation which is claimed to be extremely effective, the CHC algorithm introduced by Eshelman [31], uses a version of uniform crossover in which the offspring are deliberately constructed to be $d/2$ bits distant—as far as possible—from their parents.) This accounts for two commonly-observed features of GAs: they are less likely to get stuck in local optima (because they make large moves), but lack the 'killer instinct'—having found a region inhabited by good solutions they make heavy weather of finding the best in that region (because they do not make many small moves).

We may thus wonder how it is that GAs actually find optima at all! This brings us to the third point: as indicated above, the order of the NS actually varies dynamically. As the strings become more similar, the average distance between parents decreases and the average move length decreases with it. Thus, in the latter stages, a GA behaves rather more like a typical (although constrained) neighbourhood search, and will therefore exhibit a 'hill-climbing' type of behaviour.

Finally, we must point out that this analysis is predicated on the assumption that degenerate offspring (clones) are avoided—perhaps by using Booker's reduced surrogate crossover [26]. However, we should recognize that reported GA implementations do not always prevent the generation of clones. In these cases, our probabilities would need to be adjusted to reflect the possibility of 'zero-order' neighbourhoods. Clearly, as the search proceeds, and in the absence of something like reduced surrogates, the chance of generating clones rises considerably. This is certainly another major difference from normal NS—from a NS perspective, a detrimental one, as it is difficult to think of reasons why it

---

[6] It also suggests it may be worth a more systematic investigation of 'consensus' type operators, where we may have several 'parents'; this would also restrict the search more or less severely depending on the way in which the inheritance procedure is implemented. For example, we could fix a particular allele of a gene if a certain proportion ϕ or more of the parents shared its value. Increasing ϕ would make the restriction less severe.

should be allowed to occur. In a single descent search, for instance, we would never generate the same solution twice, while it is of the essence of tabu search to prevent this occurring as far as possible. In simulated annealing it is possible to re-visit points, but the probability is small.

4.1 The role of mutation

The above analysis shows that *in terms of the new vector(s) produced* traditional crossover actually behaves in a very similar way to mutation applied at a relatively high rate. The difference is that mutation can apply to *any* bit, while the new information from crossover is restricted to appearing at loci where the parents differ. This does raise an interesting question: would we get qualitatively the same results *if we applied mutation only*, but used global statistics on allele frequencies to determine the chance of mutation? The answer would appear to be in the affirmative: Syswerda's [32] idea of 'bit simulated crossover' (BSX) is very close to this idea, and was shown to produce comparable results to a traditional GA in a range of experiments. This approach was recently investigated further by Eshelman and Schaffer [33], who argued firstly that pairwise mating is the crucial ingredient missing from BSX, but secondly, that the situations where this matters are fairly rare.

Traditional GA mutation can obviously also be considered from an NS viewpoint in a similar way. We hope to give a more detailed consideration of this aspect in later work. It is clear, however, that mutation rates of about n^{-1} (which are fairly commonly used) imply that low-order neighbours are the most likely to be generated. In implementations which use crossover-OR-mutation at each recombination event, this provides a stochastic mechanism for exploring points which are close in Hamming space. In cases where the GA uses crossover-AND-mutation the effect will be to add some noise to the search, but not to change the expected size of the move very greatly.

5 Some research directions

The above analysis is to some extent simply stating more precisely some of the traditional 'folklore' about GAs. However, we believe that this analysis also suggests some further avenues to explore.

Firstly, we have argued that the GA's ability to 'jump' a long way from the regions currently occupied by the parents provides some explanation for its success in avoiding local optima. It is not of course the only explanation, but it seems plausible that to mimic such behaviour in more traditional NS might result in improved performance. This does of course depend on the nature of the fitness landscape, and it may be that what is beneficial in the case of a very 'rugged' landscape is less so in the case of smoother surfaces. We are currently investigating a NS approach based on this idea which we have called a 'simulated genetic algorithm'.

Secondly, it seems at least plausible that (as has been suggested by Rayward-Smith [20] for example), a more thorough exploration of 'close' intermediate points (at a distance of 1 or 2 bits, say) than is possible by mutation alone might be appropriate in implementing a GA, rather than relying on the very infrequent occasions on which this will happen by chance.

5.1 Neighbourhood search crossover

On this basis, we have investigated the effect of using a 'neighbourhood search crossover of order k' (NSX_k), which will examine all intermediate vectors of order k, and pick the best as the new offspring. Clearly, examining *all* the intermediate vectors would in many cases require far too much computation.

A simple procedure for applying NSX_1 is as follows: we first compute the exclusive-OR (XOR) vector of x and z. We then create a set of crossover templates for z by copying this string exactly except that we change one of the 1s between the sentry bits to a 0. There will be d different templates, where d is the distance between x and z . By reversing all the bits in each template we get the corresponding template for x. We then apply each template to the parents x and z to generate the neighbourhood $N_1(x, z)$; the extension to higher-order neighbourhoods is obvious. We note that this is different from the approaches referred to in section 1: the operator may still respect and/or assort schemata.

We have tried this out on some combinatorial problems. First, we investigated some instances of the multi-constrained 0/1 knapsack problem. The existence of constraints is often a problem to GAs: naive penalty-function methods seldom work well, as they tend either to search the 'interior' of the solution space, well away from the boundaries (using penalties which are too severe), or to generate many infeasible solutions (using penalties that are too light). In experimenting with GAs for this problem we observed that many solutions generated were often only 1 or 2 bits different from optimal, but ordinary crossovers failed to exploit this. We used NSX_1 in conjunction with one-point crossover and a penalty-based method that had previously given reasonable results. However, the new approach was only a little more consistent in finding high-quality solutions: perhaps the presence of constraints makes this an unsuitable test-bed.

This approach can also be extended to non-binary coded problems: a similar approach was taken to sequence-coded chromosomes in a set of flowshop sequencing problems. A report on this work is in preparation, but we can summarise some preliminary results here. These are more promising than those found in the binary case: embedding NSX_1 in a traditional GA produced slightly better results on average. Moreover, NSX_1 has a particular advantage in this application: it is possible to calculate the effect of a 1-bit change at small computational cost, whereas a many-bit change essentially needs a complete re-evaluation of the objective function. (The same is often true in general: a change in a single bit only affects one parameter, and leaves the others unchanged; in some cases the computational complexity of computing the effect of n 1-bit changes may actually be the same as computing the effect of one many-bit change.) Thus it proved possible to speed up the search by a factor of about 5 for problems with 20 jobs.

A report is also currently nearing completion [34] on another series of experiments in which NSX was applied to graph partitioning problems. Here it was found that NSX out-performed both a greedy hybrid approach and an implementation of Radcliffe's [35] RAR operator. These experiments cannot be claimed to be conclusive, and as indicated more comprehensive work is currently in progress. However, it at least suggests that NSX is worth considering for optimization problems where the objective can be structured in such a way that close neighbours can be evaluated cheaply.

As a final thought, we observe that this perspective on GAs also offers the chance to integrate some of the ideas used in recent developments of NS such as tabu search. We have argued elsewhere [36] that incorporating memory into a GA offers the potential for improving performance. This is one area in which TS has brought considerable gains, but building GA/TS hybrids has so far not been a common activity, perhaps because they seem so different. Using NSX might give more scope to such possibilities.

6 Conclusion

In this paper we have attempted an initial characterization of a GA as a form of neighbourhood search. We have shown that the concept of a neighbourhood has to be extended in order to interpret a GA in this way, and that this points up some of the ways in which GAs differ from standard NS. The main differences reside in the fact that in a GA the search is in one sense more restricted, in that it is constrained to a particular subset of genes, while at the same time it is much wider in the sense that moves are allowed to be much longer than is usual in NS methods. Both of these follow from the modified definition of a neighbourhood, which is necessary because GAs use pairwise mating of strings drawn from a population. (Naturally, it is also true that, on a higher level, the fact that GAs use populations does in itself constitute a major difference from traditional NS. We might argue that a GA is conducting many of its own brand of neighbourhood search and exchanging information between them. It is interesting that ideas of this type have recently been suggested under the name of 'vocabulary building' in the context of tabu search [18].)

Some research directions have been suggested, and some preliminary experimental work reported. However, more theoretical and computational research needs to be done in order to evaluate the effects of viewing GAs from this perspective. This paper does not pretend to be a definitive account of how a GA fits in to the NS framework, but we hope that this modest first step will stimulate more research into the potential benefits of integrating GAs and more traditional neighbourhood search methods.

Appendix

We derive here the hypergeometric distribution for the distance between two strings.

Theorem Suppose we have two binary strings **a** and **b** of length n, where the number of 1-bits in each string is a and b respectively. Suppose also, without loss of generality, that $a \geq b$. Then the random variable D—the distance between **a** and **b**—has the probability distribution

$$P[D = d = a - b + 2j] = \frac{\binom{a}{b-j}\binom{n-a}{j}}{\binom{n}{b}} \text{ for } 0 \leq j \leq \min(b, n-a).$$

Proof First, it is clear that the minimum value of d is $a - b$, which occurs when all the 1-bits in **b** match a 1-bit in **a**. For a distance of $d = a - b + 2j$, we require that $b - j$ of the 1-bits in **b** must match a 1-bit in **a**; this can be arranged in $\binom{a}{b-j}$ ways, and contributes $a - b + j$ to the distance. The remaining j 1-bits must match a 0-bit in **a**, which can be arranged in $\binom{n-a}{j}$ ways, and contributes j to the distance. To establish the probability distribution, we need the total number of arrangements of the b 1-bits in **b**, which is $\binom{n}{b}$.

Finally, the maximum value of d occurs when either all the 1-bits in **b** match a 0-bit in **a** ($b \leq n-a$), or when all the 0-bits in **a** have been matched with a 1-bit from **b** ($b \geq n - a$). This establishes the upper limit on j, and thus completes the proof. $\square$

References

1. E.Falkenauer and A.Delchambre (1992) A genetic algorithm for bin packing and line balancing. In *Proceedings of the IEEE International Conference on Robotics and Automation*.
2. C.R.Reeves (1993) Hybrid genetic algorithms for bin-packing and related problems. (submitted to *Annals of OR*).
3. C.R.Reeves (1993) A genetic algorithm for flowshop sequencing. To appear in *Computers & Ops.Res.*.
4. J.L.Blanton Jr. and R.L.Wainwright (1993) Multiple vehicle routing with time and capacity constraints using genetic algorithms. In [37].
5. P.Jog, J.Y.Suh and D. Van Gucht (1991) The effects of population size, heuristic crossover and local improvement on a genetic algorithm for the travelling salesman problem. In [38].
6. D.Whitley, T.Starkweather and D.Shaner (1991) The traveling salesman and sequence scheduling: quality solutions using genetic edge recombination. In [39].

7. A.Homaifar, S.Guan and G.E.Liepins (1993) A new approach on the travelling salesman problem by genetic algorithms. In [37].

8. J.H.Holland (1975) *Adaptation in Natural and Artificial Systems.* University of Michigan Press,Ann Arbor.

9. D.E.Goldberg (1989) *Genetic Algorithms in Search, Optimization, and Machine Learning.* Addison-Wesley, Reading, Mass.

10. C.R.Reeves (Ed.) (1993) *Modern Heuristic Techniques for Combinatorial Problems.* Blackwell Scientific Publications, Oxford.

11. C.R.Reeves (1994) Genetic algorithms and combinatorial optimization. In *Proceedings of the UNICOM Seminar on Adaptive Computing and Information Processing,* Brunel University, UK.

12. H.Mühlenbein (1991) Evolution in time and space—the parallel genetic algorithm. In [40].

13. N.L.J.Ulder, E.H.L.Aarts, H.-J. Bandelt, P.J.M.Laarhoven and E.Pesch (1991) Genetic local search algorithms for the travelling salesman problem. In H.-P.Schwefel and R.Männer (1991) (Eds.) *Parallel Problem-Solving from Nature.* Springer-Verlag, Berlin.

14. P.Prinetto, M.Rebaudengo and M.Sonza Reorda (1993) Hybrid genetic algorithms for the travelling salesman problem. In [41].

15. N.J.Radcliffe (1991) Forma analysis and random respectful recombination. In [42].

16. N.J.Radcliffe (1991) Equivalence class analysis of genetic algorithms. *Complex Systems,* 5, 183-205.

17. D.Orvosh and L.Davis (1993) Shall we repair? Genetic algorithms, combinatorial optimization and feasibility constraints. In [37].

18. F.Glover and M.Laguna (1993). Tabu Search. In [10].

19. R.J.M.Vaessens, E.H.L Aarts and J.K.Lenstra (1992) A local search template. In [43].

20. V.J.Rayward-Smith (1994) A unified approach to tabu search, simulated annealing and genetic algorithms. In *Proceedings of the UNICOM Seminar on Adaptive Computing and Information Processing,* Brunel University, UK.

21. J.Antonisse (1989) A new interpretation of schema notation that overturns the binary encoding constraint. *In* [38], 86-91.

22. N.J.Radcliffe (1992) Non-linear genetic representations. In [43].

23. P.Kanerva (1988) *Sparse Distributed Memory.* MIT Press, Cambridge, Mass.

24. M.D.Vose (1993) Modeling simple genetic algorithms. In [44].

25. L.D.Whitley (1993) An executable model of a simple genetic algorithm. In [44].

26. L.B.Booker (1987) Improving search in genetic algorithms. In [46].

27. A.Fairley (1991) *Comparison of methods of choosing the crossover point in the genetic crossover operation.* Dept. of Computer Science, University of Liverpool.

28. G.Syswerda (1991) A study of reproduction in generational and steady-state genetic algorithms. In [40].

29. C.L.Bridges and D.E.Goldberg (1987) An analysis of reproduction and crossover in a binary-coded genetic algorithm. In [45].

30. S.J.Louis and G.J.E.Rawlins (1993) Syntactic analysis of convergence in genetic algorithms. In [44].

31. L.J.Eshelman (1991) The CHC adaptive search algorithm: how to have safe search when engaging in non-traditional genetic recombination. In [40].

32. G.Syswerda (1993) Simulated crossover in genetic algorithms. In[44].

33. L.J.Eshelman and J.D.Schaffer (1993) Crossover's niche. In [37].

34. C.Höhn and C.R.Reeves (1994) Heuristic genetic search methods for graph parti-
    tioning. To be presented at the International Conference on Systems Engineering,
    Coventry University, September 1994.
35. N.J.Radcliffe and F.A.W.George (1993) A study in set recombination. In [37].
36. C.R.Reeves (1993) Diversity and diversification in genetic algorithms: some con-
    nections with tabu search. In [41].
37. S.Forrest (Ed.) (1993) *Proceedings of* 5th *International Conference on Genetic Al-
    gorithms.* Morgan Kaufmann, San Mateo, CA.
38. J.D.Schaffer (Ed.) (1989) *Proceedings of* 3rd *International Conference on Genetic
    Algorithms.* Morgan Kaufmann, Los Altos, CA.
39. L.Davis (Ed.) (1991) *Handbook of Genetic Algorithms.* Van Nostrand Reinhold,
    New York.
40. G.J.E.Rawlins (Ed.) (1991) *Foundations of Genetic Algorithms.* Morgan Kauf-
    mann, San Mateo, CA.
41. R.F.Albrecht, C.R.Reeves and N.C.Steele (Eds.) (1993) *Proceedings of the Interna-
    tional Conference on Artificial Neural Networks and Genetic Algorithms,* Springer-
    Verlag, Vienna.
42. R.K.Belew and L.B.Booker (Eds.) (1991) *Proceedings of* 4th *International Confer-
    ence on Genetic Algorithms.* Morgan Kaufmann, San Mateo, CA.
43. R.Männer and B.Manderick (Eds.) (1992) *Parallel Problem-Solving from Nature,
    2.* Elsevier Science Publishers, Amsterdam.
44. L.D.Whitley (Ed.) (1993) *Foundations of Genetic Algorithms 2,* Morgan Kauf-
    mann, San Mateo, CA.
45. J.J.Grefenstette(Ed.) (1987) *Proceedings of the 2nd International Conference on
    Genetic Algorithms.* Lawrence Erlbaum Associates, Hillsdale, NJ.
46. L.Davis (Ed.) (1987) *Genetic Algorithms and Simulated Annealing.* Morgan Kauff-
    mann, Los Altos, CA.

A Unified Paradigm for Parallel Genetic Algorithms

A. Kapsalis, G.D. Smith and V.J. Rayward-Smith

University of East Anglia
e-mail: ak@sys.uea.ac.uk

Abstract

Genetic Algorithms contain natural parallelism. There are two main approaches in parallelising GAs. The first one is parallelising individual functional components of a standard, sequential GA. The only difference with the sequential GA is in the computation speed. The second approach more closely resembles the real life simultaneous evolution of species, which is the central theme in GAs. Algorithms following this approach are still referred to as GAs but are different from Holland's standard GA. For these algorithms it is not the improvement in computation speed that is the driving factor, but the efficiency with which they search a given solution space. We describe a number of the most common parallel GA methods found in the literature and mention practical issues concerning their implementation in a Transputer based system. We go on to introduce the Unified Parallel GA system, based on our GA toolkit, GAmeter, which allows the user to select one or more of the GA methods described, by setting various parameters. Finally we present results for the Steiner tree Problem in Graphs (SPG).

1.   Introduction

The term Genetic algorithms (GAs) refers to a family of computational models inspired by Darwin's theory of evolution and based on principles derived from natural population genetics. The natural processes of reproduction, selection and survival of the fittest are simulated and used to give good approximate solutions to a wide range of optimisation problems. The foundations of the technique were described by Holland [Holl75], who also established much of the theory to explain the subsequent success of the application of GAs to a wide variety of problems.

Research on improving the performance of Holland's "canonical" GA has created a vast number of alternative algorithms based on the same principles. The algorithm for the standard/sequential GA (SGA), shown in figure 1, is a generalisation that covers most of the various instantiations of the sequential GAs. In fact, it was reported in [Kaps93a] and [Rayw94] that this algorithm can be used to describe other search methods such as hill climbing, simulated annealing and tabu search.

The GA starts by creating an initial (usually random) pool, P, of chromosomes or "genetically" described solutions to a problem. These chromosomes are *Evaluated* and a fitness score is associated with them. At each iteration a number of Parent chromosomes, Q, are *Selected* from the pool based on their fitness scores. The selected chromosomes *combine* using genetic operators like *crossover* and *mutation* to *Create* the set R of new chromosomes. The new chromosomes are evaluated and a call to function *merge* determines which strings will form the pool P for the next iteration. Iterations are usually referred to as generations.

topology for such an approach is the star topology. Because there are different demands in communication depending on the degree of parallelisation used we separate our investigation of the Parallel GA (PGA) to *Level1* and *Level2* .

Level1

In *Level1* only the *Evaluate* function is parallelised. The master is a GA that transmits all chromosomes that need to be evaluated to the available slaves. The slaves wait until they receive a chromosome at which point they evaluate it and return its fitness value to the master.

ParallelEvaluate()

```
Initialise Queue
Send (Chromosome, ID Number) to slave at the head of the queue
While not all fitness values collected do
  begin
        if (Send completed) & (queue not empty) &
                                    (more Evaluations required) then
            Send (Chromosome, ID Number) to slave at the head of the queue;
        if there exists message to receive then
                    Receive (fitness value, ID Number);
  end;
```

Slave

```
While message>0 do
  begin
        Wait to Receive (Chromosome, message);
        if message>0 then Evaluate (Chromosome,fitness value);
        Send (fitness value, message);
  end;
```

If $|R|$ is the number of chromosomes that need to be evaluated at each generation then the total size of messages exchanged between the master and the slaves is:

$$|R| * (StringSize+MessageSize) \text{ sent by the master and}$$
$$|R| * (fitnessValueSize + MessageSize) \text{ returned by each slave.}$$

Message is an integer used as a tag or identification number so that the fitness value returned by the slave is associated with the appropriate chromosome. This is needed because the slaves might not respond in the same sequence with which they receive chromosomes. Negative valued messages are used to control the operation of the slave. Depending on the *Message* value it can wait to receive new problem data or stop operating.

Level2

In *Level2* both *Create* and *Evaluate* are parallelised. The most common genetic operator used in create is *crossover*. Given two chromosomes this operator combines them and creates two new chromosomes. Every new chromosome created must be evaluated and therefore *Evaluate* can be seen as a component of *Create*. The master sends a pair of chromosomes from Q to each of the slaves. Once the slaves receive a pair of chromosomes they choose a genetic operator to apply and *Evaluate* each of the new strings created. If the operator selected by the slaves requires only one chromosome then the same operator is applied to the second chromosome. Finally the slave sends the two new chromosomes and their fitness values back to the master.

The duplication of operators that produce only one chromosome is done so that the number of new chromosomes created reflects the operators parameter settings. If, for example, *crossover* rate is 60% and *reproduction* (replication) rate 40% then we would expect 60% of the new chromosomes to have resulted from *crossover* and 40% of them from *reproduction*.

ParallelCreate()

```
Initialise Queue
Send (message,Chromosome1,Chromosome2) to slave at the head of the queue;
While not all children collected do
  begin
        if (Send completed) & (queue not empty) & (more parents to send) then
              Send (message,Chromosome1,Chromosome2) to
                                          slave at the head of the queue;
        if there exists message to receive then
              Receive (Chromosome1,fitness value1,Chromosome2,fitness value2);
  end;
```

Slave

```
Initialise
While message>0 do
 begin
    Wait to Receive (message,Chromosome1,Chromosome2);
        if message>0 then
            begin
                  Choose Genetic Operator (Crossover or Reproduction);
                  Selected Operator (Chromosome1, Chromosome2, Child1, Child2);
                  Evaluate (Child1,fitness value1);
                  Evaluate (Child2,fitness value2);
            end;
    Send (Child1, fitness value1, Child2, fitness value2);
 end;
```

If $|Q|$ is the number of chromosomes selected at each generation then $|Q|/2$ pairs of chromosomes must be sent to the slaves at each generation. The total size of messages exchanged between the master and the slaves is:

$$(|Q|/2)*(2*StringSize+MessageSize) \text{ sent by the master and}$$
$$(|Q|/2)*(2*fitnessValueSize + 2*StringSize) \text{ returned by each slave.}$$

These simplify to:

$$|Q|*(StringSize+MessageSize/2) \text{ messages sent by the master and}$$
$$|Q|*(fitnessValueSize+StringSize) \text{ returned by each slave.}$$

The decision as to which operator to use for a pair of chromosomes is taken in the slaves and for this decision a random sequence must be sampled. The slaves sample from the same sequence and all start sampling from the same position in the sequence. In the SGA operator decisions are taken in sequence. If we require the parallel GA to be a perfect emulation of the SGA then the slaves collectively would have to sample from exactly the same sequence as the SGA. Every time a decision is made the position from which we sample the sequence progresses by one. Hence the master can keep track where in the sequence the next sample should come. This position can be sent to the slave and the slave can then generate the random number and make the decision as to which operator to apply. Alternatively, the master can be responsible for all random number generations and can inform the slaves of the operators to be used.

In practice, a perfect emulation may not be required, in which case we can sample from local random number streams and this may give a satisfactory performance.

A finer grain parallelism can be achieved by distributing the 2 evaluations performed in the slaves. However, this is not practical for distributed memory systems because of the additional communication needed for the strings that are to be evaluated.

The number of chromosomes selected at each generation is equal to the number of chromosomes created, or $|Q|=|R|$. Comparing the total size of messages sent and received in one generation for *Level1* and *Level2* we notice that the increase in communication for the *Level2* parallelisation is due to the increase in the size of the messages received by the master. The amount by which this is increased is $|R|*(StringSize-2*MessageSize)$.

2.2   New Parallel search methods based on the GA concept (Distributed GAs)

Several approaches have been proposed to overcome the limitations of parallelising the SGA. Most of them can be thought to have resulted from distributing an SGA in which special functions have been incorporated to manage information exchange among the Distributed GAs (DGAs). A general algorithm for this approach is shown in figure 2. This algorithm is the basis of the Unified Parallel GA (UPGA) described in the next section.

```
type
      chromosome= sequence of elements from an alphabet;

P,Q,R,I : set of chromosome;
Neighbours: Array of Network Addresses;

Initialise (P);
Evaluate (P);

While not finish(P) do
   begin
        if Immigration then   Transmit (ImiSelect (P), Neighbours);
        Q := Select (P);
        R := Create (Q);
            Evaluate (R);
        I := ImiReceive (Neighbours);
        P := Merge (P,Q,R,I);
   end
        Synchronise;
end Search
```

Figure 2: The Distributed GA (DGA).

```
type
    chromosome= sequence of elements from an alphabet
                                        (usually {0,1});

P,Q,R : set of chromosome;

Initialise (P);
Evaluate (P);

While not finish(P) do
      begin
         Q := Select (P);
         R := Create (Q);
             Evaluate (R);
         P := Merge (P,Q,R);
      end
end Search
```

Figure 1: The Standard/Sequential GA (SGA).

2. Two Main Approaches to Parallelising GAs

In nature, populations of thousands or even millions of individuals evolve in parallel. The size of the population affects the evolutionary process but not the rate with which each individual participates in it. Individuals are created and die asynchronously. There is no clear cut division between *Creation* and *Merge* as in figure 1. Generations or iterations in our evolutionary history are not so easy to distinguish as is the case for the sequential computer simulations.

The advent of ever faster and ever cheaper parallel hardware has fostered the research on comparatively new, more natural Parallel Genetic Algorithms. However, there is a vast amount of experimental results and theory on sequential GAs, so there is still a need for parallel GAs which operate in the same way as the SGA but which take advantage of the performance increase resulting from parallel hardware. For that reason there are two basic types of parallel GAs. Parallel GAs modelled around the SGA (referred to as PGAs in this text) and new, distributed search methods based on the biological metaphor that motivates GAs (referred to as DGAs in this text).

2.1 Parallelisation of the standard sequential GA (PGA)

From the algorithm of figure 1 we see that the population $P(t+1)$ at iteration or time-step $t+1$ is dependant on $P(t)$, $Q(t)$ and $R(t)$. This means that only the functions *Select*, *Create* and *Evaluate* can be parallelised. We also notice that *Select* operates on the entire population and thus can be efficiently parallelised on a shared memory system, which is the system best suited for parallelising the SGA. For an MIMD system like the Transputer based system used, the communication costs are too high, since every processor would have to receive the entire population or at least each individual's fitness value at each generation. Parallelisation of *Create* and *Evaluate* is less costly and the performance of the resulting algorithm is independent of the size of the population provided that the cost of sending a message to 1 processor and to M processors is the same and assuming that *Evaluate* is more expensive than *Select* (as is usually the case).

To parallelise the SGA, a master-slave approach was used. The ideal

Neighbours: List of network addresses for all DGAs that are conceptually considered as *Neighbours* to this DGA. Physical neighbouring resulting from the network structure is not necessarily reflected by this list. This list must be specified for every DGA and can vary dynamically.

Transmit: This sends one or more chromosomes to all *Neighbours*.

ImiSelect :   A chromosome is selected from *P* with any one of the selection methods available to *Select* but not necessarily with the same method as used for *Select*.

ImiReceive :   Receive any chromosome sent to this DGA.

The algorithm shown in figure 2 is completely asynchronous. A DGA in the distributed network can *Transmit* or *Receive* chromosomes at any instance. To avoid problems occurring when a DGA has terminated its search but there are chromosomes sent to it before the sender DGA(s) have been informed, the function *Synchronise* is used. This implements a form of handshaking. The algorithm waits until all neighbours have reached an agreement in which case it is allowed to continue and terminate the search. Because no communications are pending after the *Synchronise* step, the algorithm can also restart the search. For certain models of distributed DGAs it is important that the search progresses in some orderly manner. For these cases the *Synchronise* function must be inserted in the iteration loop.

    A number of models have resulted from the Distributed GA approach to parallelising GAs. The most common ones are: the *Island* model, the *Cellular DGA* model and the *Fine Grained SGA* model.

The Island model

    In this model several DGAs are operating in parallel. Occasionally, information exchange/migration takes place between the neighbours. Each DGA evolves its own population and at specific times it sends one or more members of the population (immigrants) to its *Neighbours*. The *Neighbours* can either accept the immigrants into their populations or refuse them. In figure 3 an example for the *Island* model is shown. In this example 5 DGAs are used and each one has 4 neighbours thus creating a complete network.

    One of the first studies using this model was published by Pettey, Leuze and Grefenstette [Pett87]. Their motivation was in creating an algorithm that can perform as an SGA with a large population size, but without increasing the optimisation time. In their model, the best individual is exchanged between neighbours at each generation. They presented results for DeJong's set of functions [Jong75].

    At the same time Cohhon, Hedge, Martin and Richards [Coho87] inspired by the paleonthological theory of punctuated equilibria, presented a similar model. Each DGA proceeds independently for a number of generations judged adequate to reach equilibrium. After this number of generations has elapsed, and this is the same for all DGAs, an "epoch" has been completed and a change of the environment is forced on each DGA. Each DGA transmits randomly selected copies of its population to neighbouring processors. From the surplus of individuals that each DGA acquires, a number equal to the population size is probabilistically selected to form the initial population for the next epoch. They applied this model to the Optimal Linear Arrangement problem and reported an increased efficiency of the parallel search compared with that of the SGA. They also referred to a method for parallelising the SGA similar to that of *Level2* and suggested approximate calculation of the fitness value as an alternative scheme for changing the environment. In a later paper [Coho90], they proposed applying each GA in the network with different control parameters, such as *crossover* and *mutation* rates or *pool* size. In [Coho91] they applied their algorithm to the k-partition problem.

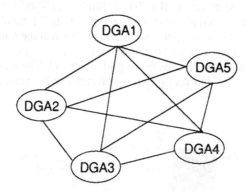

Figure 3: The Island model with each DGA having 4 neighbours.

A similar model was presented also by Tanese [Tane87] in which each DGA exchanges copies of its best individuals with one neighbour. The neighbour varies over time. Tanese reported results on a Walsh like function and showed that the rate at which exchanges (immigration) take place is important for the performance of the algorithm. Pettey and Leuze [Pett89] proposed a theoretical analysis on the allocation of trials to above average schemata. They concluded that the *Island* model of DGA tends to maintain the efficiency of the sequential algorithm.

An interesting alternative method to distributed problem-solving was proposed by Husbands and Mill [Husb91]. Their idea is to recast a complex problem in terms of simpler, co-operating sub-problems. Each DGA is solving one of the sub-problems. The genotype for each DGA is different, thus no migration is possible. However, the fitness of each individual in every DGA's population depends on the interactions with population members of other DGAs, thus letting sub-populations evolve in a shared world. The model they proposed includes an additional population of individuals, called the "arbitrators", that are responsible for resolving conflicts between members of the other populations. They successfully applied their model to a version of the manufacturing scheduling problem, previously regarded as too complex to tackle.

The Cellular DGA model

DGA models that fit into this category contain a notion of spatiality, in that the individuals of the DGAs population are assigned to a specific position in a given spatial environment. Usually a two dimensional grid, such as that shown in figure 4, is used. The grid is implemented as a toroidal array in order to avoid border effects. In terms of the DGA of figure 2, each square in the 2-dimensional grid can be thought of as a single DGA evolving the single string it contains. The second string needed for crossover is obtained from one of its four neighbours (North, South, East, West). Different shades for the squares in the grid of figure 4 represent different solutions. Initially the grid is populated with random solutions but after a few generations many small pockets of similar solutions are formulated. As the search progresses, the pockets merge creating larger groups of similar solutions. Eventually most grid positions are populated with the same solution. Although there are no explicit islands in the model, there is the potential for similar effects due to the "distance" or number of moves separating strings in different grid positions. This kind of separation is referred to as *isolation by distance* ([Coll91], [Mühl91a], [Gorg91]).

The model described above can be generalised to *N* dimensions in which each DGA has 2*N neighbours. It becomes similar to SGA when the number of neighbours to a cell in the grid is equal to the total number of cells in the grid. This can happen when, for example, there are 25 cells in the grid and *N*=12. The number of individuals in each DGA and therefore in every grid position can also be increased.

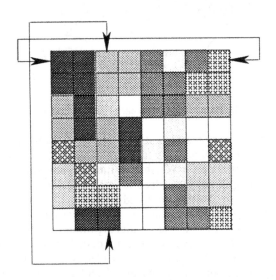

Figure 4: The *Cellular DGA*.

As in the case of the *Island* model, the concept of the *Cellular DGA* was independently developed by a number of researchers. Manderick and Spiessens [Mand89] compared the *Cellular DGA* with the SGA on De Jong's [Jong75] standard suite of functions. The *Cellular DGA* approach provided a more thorough exploration of the search space which resulted in a more effective algorithm for tougher problems. They concluded that there is no optimal population size when working on a parallel machine. The more powerful the machine, the larger the population can be and the better the results (see also [Gold89b]). On experimenting with different neighbourhood sizes they stated that the optimal size depends on the problem in hand (a neighbourhood size of 25 is the most robust setting), the larger the size the more similar is the performance of the *Cellular DGA* to the SGA.

Mühlenbein et al. [Mühl88, Mühl89] proposed a similar model in the ASPARAGOS system, inspired by Goldberg's [Gold89] polynation model. They used a hybrid SGA in which each individual does local *hill-climbing* and *crossover* is modified to a recombination operator based on voting. *Selection* is based on ranking of all individuals in the neighbourhood and of the global best. The neighbourhood size is 4. This system has been successfully applied to a range of problems. In [Mühl88] Mühlenbein, Gorges-Schleuter and Kramer tackled very large Travelling salesman problem instances. In [Mühl89] Mühlenbein reported a new optimum solution for the largest published problem instance of the quadratic assignment problem. Mühlenbein, Schomisch and Born in [Mühl91b] and [Mühl91c] concluded that ASPARAGOS performs at least as well as other SGAs in the optimisation of continuous functions and they were able to find the global minimum for functions of very high dimensions. The same system has also been applied by Gorges-Schleuter and von Laszewski [Gorg89, Gorg91, Lasz90] to Travelling Salesman and Graph Partitioning problems, using more standard genetic operators.

Davidor in [Davi91] described a similar model that he called the "ECO Genetic Algorithm". He presented an interesting reformulation of the schema theorem and experimental results showing how, during evolution, islands of similar individuals quickly emerge on the grid. These islands become larger according to how much their fitness outscores that of neighbouring islands. Similar results regarding the formation of "clusters" of similar solutions are reported by Spiessens and Manderick in [Spie91]. Comparison of the time complexities between the SGA and the *Cellular DGA* together with experimental results on various selection mechanisms and crossover operators are also reported. Tomassini [Toma93] introduced a diffusion-like process in the cellular model in which individuals take short random walks around their grid positions. If the individual arriving at a grid position is better than the individual already in that position, it replaces it. Experimental results from the optimisation of various functions show that diffusion might help avoid early loss of diversity and increase the likelihood with which good individuals combine.

A Fine Grained Sequential Like GA

The possibility of using the distributed GA approach to generate "fine grained" GAs having similar properties with the SGA have also been investigated. As was stated in section 2.1 the main obstacle in implementing such a model in a distributed memory system is the global information needed for the selection process. A solution has been proposed by Goldberg [Gold90] in the form of an alternative selection mechanism. *Tournament Selection* implements an approximate form of ranking. For each sub-population (one for every DGA) a tournament is held in the following way. A number of individuals are selected at random from the sub-population and the best one is the winner. Creation of the new individuals takes place locally with at most the communication of one individual from a neighbour. The subpopulations (or DGAs) are chained together. An individual is received and creation takes place for every other DGA. A DGA which does not perform any *Create* operations is responsible for sending the winner of the local tournament to its immediate neighbour.

This model can also be implemented for more than one tournament held in each DGA and the transmission of more than one individual from every other DGA. It can also be modified so that the immediate neighbour of a DGA varies from generation to generation.

Although a *Select* mechanism that preserves the panmictic i.e. global characteristic of the SGA and is relatively efficient for implementation in a distributed system has been described, an analogous *Merge* mechanism remains to be described. The *Merge* mechanism does not create any difficulty if the entire population is replaced at each generation by the newly created individuals. Problems occur in the case where we wish to have overlapping populations. An example of this is *steady state* GAs. In these, only two new individuals enter the population at each generation (this strategy was shown to provide better solutions but slower convergence than a non overlapping GA). If this model is to maintain the properties of the SGA a merge mechanism must be devised to handle the placement of the new individuals in the existing population. In the SGA the new individuals usually replace the worst individuals already in the population. This implies some central knowledge of the distribution of fitness values in the population which incurs a high communication cost when the population is broken up into many sub-populations which are far apart.

The problem with the *Merge* mechanism described above can be resolved by the use of an efficient mechanism to maintain sorted subpopulations. This will also enable the use of more standard selection mechanisms based on ranking.

3. The Unified Parallel GA

The discussion in the sections above of the various approaches to parallelising GAs based on the SGA of figure 1 has paved the way for the introduction of the *Unified Parallel GA*, (UPGA). As the name implies this combines all previously described algorithms (including the SGA) under a single unifying algorithm.

```
type
    chromosome= sequence of elements from an alphabet;

P,Q,R,I : set of chromosome;
Neighbours: Array of Network Addresses;
Plevel: in [0..2];  {1=Level1, 2=Level2, 0=Inactive}
Distr: Boolean;

Initialise (P);
Evaluate (P);

While not finish(P) do
 begin
   if Immigration & Distr then Transmit (ImiSelect (P), Neighbours);
   Q := Select (P);
   if Plevel=2 then R:= ParallelCreate(Q)
               else
                       begin
                               R := Create (Q);
                               if Plevel=1 Then ParallelEvaluate (R)
                                       else      Evaluate (R);
                       end;
   if Distr then      I := ImiReceive (Neighbours);
   P := Merge (P,Q,R,I);
 end
        Synchronise;

end Search
```

Figure 5: The Unified Parallel GA (UPGA).

A description of the various components of the UPGA can be found in the preceding sections. The only new parameters which need to be explained are *Plevel* and *Distr*. The *Plevel* and *Distr* parameters are used to decide on the parallel approach and model required. If *Plevel* has a value of 0 then no parallelisation of the *Evaluate* or *Create* functions will be performed. A value of 1 means that *Evaluate* will be parallelised and a value of 2 that both *Create* and *Evaluate* will be parallelised (PGA levels 1 and 2 respectively). If the value of the *Distr* parameter is true then the algorithm will behave as part of a distributed network of GAs (DGA). If no parallel approach is selected by any of the parameters then the algorithm will behave like an SGA. The UPGA can result in a new, finer grained parallel algorithm when both approaches are activated simultaneously. This will create a network of distributed PGAs or DPGAs.

With the UPGA, any of the various models for the DGAs described in the previous section can be implemented. To achieve this, the parameter values must be carefully decided upon and appropriate functions for *Transmit*, *Select*, *ImiReceive* and *Merge* must be used. The parallel version of the GAmeter toolkit developed at UEA provides all the necessary parameters and functions to allow use of any of the parallel models described. Because of limited space a description of the toolkit can not be included. More information on the SGA version of the toolkit can be obtained from [Kaps93a] and [Kaps92].

4. The Steiner Problem in Graphs (SPG)

Let $G=(V, E)$ be an undirected graph with a finite set of vertices, V, an edge set, E and a cost function assigning a positive, real cost, to each edge in E. Given a subset, S, of the vertices in G, the SPG is that of finding a subgraph, $G' =(V', E')$, of G such that V' contains all the vertices in S, G' is connected and the sum of the cost of all edges in E' is minimal.

The SPG has attracted considerable research interest especially over the last three decades. It has become one of the "classic" combinatorial optimisation problems alongside other famous problems such as the Travelling Salesman Problem and the Knapsack Problem. The decision problem associated with the SPG has been proven to be NP-complete [Karp93] and hence the problem is most unlikely to be solved by a polynomial time algorithm.

A comprehensive overview of the SPG is given by Winter in [Wint87] and Hwang and Richards in [Hwan92]. Applications of the problem to areas such as topological network design, printed circuit design, multiprocessor scheduling and phylogeny are described by Foulds and Rayward-Smith in [Foul83] and by Winter in [Wint87].

For some special cases, the SPG can be solved in polynomial time. In particular, if $|S| = 2$, SPG reduces to a simple shortest path problem and can be solved using Dijkstra's algorithm [Dijk59]. If $S = V$, the SPG reduces to finding the minimum spanning tree (MST) of G and we can use either the algorithm of Prim [Prim57] or Kruskal [Krus56]. The fundamental problem in the general case is that we do not know which vertices are Steiner vertices. Once we have determined the Steiner vertices, Z (say), the required Steiner tree is simply the MST of the subgraph of G induced by $V'=S \cup Z$.

We can represent a chromosomal solution to the SPG by a bit string of size equal to $|V-S|$ in which each bit position i corresponds to a node $v_i \in |V-S|$ and a 1 means that vertex $v_i \in |V\text{-}S|$, a 0 means that $v_i \notin |V\text{-}S|$.

Given a chromosome representing a set, U, of vertices, the evaluation proceeds as follows. Firstly, we construct the subgraph of G induced by $U \cup S$. Say this subgraph comprises k (≥ 1) components. We compute the MST of each component using Prim's algorithm [Prim57] and sum these values. If $k>1$, we add a large penalty linearly dependent on k. Each MST evaluation is $O(m^2)$, where m $(\leq n)$ is the number of 1s in the chromosome and n is the length of the chromosome. A more detailed description of the use of GAs to solve the SPG can be found in [Kaps93b].

5.   Experimental Results for the SPG

In this section we present results obtained for the SPG using some of the parallel GA models described above. The parallel version of GAmeter was used with the appropriate parameter settings for PGAs *Level1* and *Level2* and for the *Island* model of DGA. For the experiments a set of 18 problems was used. This set of problems has been used by many researchers in the field and therefore our results can be compared with those of other methods. This set of problems is known as Beasley's [Beas90] B-problem set and can be obtained through electronic mail or ftp to mscmga.ms.ic.ac.uk (155.198.66.4).

In all experiments, unless otherwise stated, the *Crossover* probability was set to 60%, *Mutation* probability to 2%, the *Selection* mechanism used was the common *Roulette wheel*, the *Merge* mechanism was GAmeter's *Replace all* mechanism. With this mechanism only the newly created individuals that have a fitness value better than 1.2 times the average fitness of all individuals in the pool are accepted. The size of the pool was 11 and the best solution always remained in the pool. For all experiments a Transputer system was used.

In tables 1, 2 and 4 the solutions obtained for each problem is shown in rows 1 to 18. The "Optimal" column shows the value of the known optimal solution for each of the problems. Each problem was solved 5 times and the average solution obtained in the 5 runs is shown in the "Average" column. The next column shows the percentage deviation of the best solution obtained in the 5 runs from the optimal. The "Times Optimal" column shows how many times out of the 5 runs the optimal solution was found. Finally the "Time" and "Evaluations" columns show the average time and evaluations needed to find the best solution (not necessarily the optimal). Table 4 shows the aggregate results obtained by each of the 5 DGAs used in the *Island* model and not the results from 5 runs of the model. This is done so that we can compare the efficiency of the two search strategies. For levels 1 and 2 the strategy used is: do many independent searches whereas, for the *Island* model it is: exchange information while doing the searches in parallel.

In section 2.1, where we described the *Level2* PGA model, we mentioned that the search can progress in a different way from that of the SGA or the *Level1* model if a message is not included in the information sent from the master to the slaves. The results shown in table 2 are for the *Level2* model without the inclusion of seed messages. Comparing the solutions from tables 1 and 2 we notice that the *Level2* method found the optimal more times than the *Level1* and also was able to find the optimal in all problems. *Level1* failed to find the optimal for problem 15 but the sub-optimal solution it discovered was within 0.62% of the optimal. The rate of success in finding the optimal solution is 74% for *Level1* and 82.2% for *Level2*. However, the

Level2 PGA required twice as much time and evaluations to improve the solutions by only a small factor.

Problems	Optimal	Average	Deviation of Min	Times Optimal	Time (sec)	Evaluations
1	82	82	0	5	11.6	1183
2	83	83	0	5	12.2	1337
3	138	138	0	5	6.4	481
4	59	59.8	0	4	115	17421
5	61	61	0	5	16.8	1861
6	122	122.4	0	4	12.4	965
7	111	111	0	5	115.2	9457
8	104	104.6	0	4	752.4	52185
9	220	220	0	5	20.6	767
10	86	86	0	5	47	3351
11	88	88.4	0	4	603	50653
12	174	174	0	5	18.8	781
13	165	172	0	2	888.6	46457
14	235	237.4	0	1	146	5195
15	318	320	0.62%	0	55.2	1249
16	127	134.8	0	1	409.4	19335
17	131	132.6	0	2	70.2	2219
18	218	218	0	5	145	3633
Total				67	3445.8	218530

Table 1: Steiner Tree solutions for *Level1* PGA using 10 slaves.

Problems	Optimal	Average	Deviation of Min	Times Optimal	Time (sec)	Evaluations
1	82	82	0	5	15	1929
2	83	83	0	5	14.4	1749
3	138	138	0	5	15.2	1327
4	59	59	0	5	26.6	4501
5	61	61	0	5	25	3477
6	122	122.4	0	4	16.2	1453
7	111	111	0	5	252.6	27143
8	104	105.8	0	2	69.6	4951
9	220	220	0	5	34	1449
10	86	87	0	4	150.4	14663
11	88	88.4	0	4	1382.8	136025
12	174	174	0	5	53	2417
13	165	167.8	0	3	623.4	35921
14	235	237	0	1	1060	47141
15	318	318.4	0	4	1164.6	33089
16	127	132.8	0	2	1005.4	59851
17	131	131	0	5	516.2	22495
18	218	218	0	5	180.8	4959
Total				74	6605.2	404540

Table 2: Steiner Tree solutions for *Level2* PGA using 5 slaves.

In order to compare the gain in speed produced by the various models, we use the number of function evaluations performed in every second. This number is shown in table 4, for the SGA, *Level1*, *Level2* and *Island* models. Measurements for problem 1, consisting of 50 vertices, and problem 18, consisting of 100, are shown in this table. The *Level1* model achieves a speed of between 3.4 and 4.4 times that of the SGA using 10 processors. On the other hand, the *Level2* model achieves a speed of between 4.2 and 4.9 times that of the SGA by using only half the number of processors. Clearly the *Level2* parallelisation performs better than the *Level1* and achieves an almost linear speed up. Of course neither of the two models can achieve a linear speed up because there are parts in the SGA that can not be parallelised efficiently on the transputers. Although the *Level1* parallelism performs poorly in this set of problems it improves rapidly as the size (or difficulty of the evaluation function) increases. For very large problems it is expected to outperform *Level2*.

Problems	SGA	Level1	Level2	Island
1	30	102	128	29
18	5.6	25	27.4	5.3

Table 3: Comparison of the number of evaluations per second performed by each of the models.

For the *Island* model more explanations on the set-up used for the experiment are required. Five DGAs were used in a complete network, shown in figure 3. DGA1 and DGA4 was set to send a chromosome to all other DGAs every 80 generations, DGA2 was set up to send a chromosome every 50 generations, DGA3 was set up to send a chromosome every 75 generations and DGA5 was set up to send a chromosome every 60 generations. Furthermore, each DGA was operating with slightly different mutation rates. For DGA1 the mutation rate was 1%, for DGA2 it was 2%, for DGA3 it was 0.5%, for DGA4 it was 1.5% and for DGA5 it was 2.5%. The chromosome transmitted was selected from each population using the *Roulette Wheel* mechanism.

The set-up and parameter settings were done arbitrarily and without any prior testing. The results obtained (shown in table 4) were not as good as expected. The *Island* model found the optimal solution for most of the problems but the exchange of good individuals made it converge to local optimal solutions more often than *Level1* or *Level2*. Because local optima are discovered quicker in the search, transmitting good individuals does not appear on the surface to be a good idea. A possible explanation is that once a DGA discovers a local optimal and transmits it to the other DGAs that have worse solutions, they tend to favour this sub-optimal solution more, even if most of the solutions are in other, more promising regions of the search space. The rate with which individuals are transmitted and the number of DGAs in a neighbourhood are important parameters for the island model. The way the comparison was conducted was unfair for the *Island* model. If it is viewed as a unit (i.e. each DGA in the model is viewed as an element of the "island" search algorithm) then the results were based on just 1 search compared to the 5 performed for *Level1* and *Level2*. In addition, there was a time limit for each of the experiments and therefore the model that was able to search more points in the space per unit of time had an unfair advantage with respect to the rest of them. Further research on this and other DGA models is in progress.

Because the *Island* model consists of many GAs running in parallel the number of evaluations per second performed in each DGA is almost equal to that of

Problems	Optimal	Average	Deviation of Min	Times Optimal	Time (sec)	Evaluations
1	82	82	0	5	37	1077
2	83	83	0	5	23.4	573
3	138	138	0	5	16	283
4	59	59	0	5	32.8	1050
5	61	61	0	5	18.6	496
6	122	122	0	5	22.8	440
7	111	111	0	5	90.6	1376
8	104	107	2.88%	0	62	694
9	220	220	0	5	78.8	687
10	86	86	0	5	2258.8	53029
11	88	89.6	0	1	243.4	5050
12	174	174	0	5	76.2	737
13	165	177	7.27%	0	186	1254
14	235	238	1.27%	0	225.2	1435
15	318	320	0.62%	0	783.6	4466
16	127	133	4.72%	0	412	3956
17	131	133	1.52%	0	893.4	8766
18	218	218	0	5	96.2	512
Total				56	5556.8	85881

Table 4: Steiner Tree solutions for the DGA with each processor having 4 Neighbours.

the SGA. However, since the Island model performs many searches in parallel a linear speedup of the total number of evaluations is obtained.

6.   Conclusions

Many parallel models for GAs have been proposed by researchers in recent years. The field has been shown to be a very promising one, with new and better models appearing all the time. The possibilities seem endless. The Distributed GA approach seems the most promising for the creation of more parallel GA variants, in the near future. Some new DGA models are currently under investigation here at UEA.

All models can be easily combined under a single Unified Parallel GA kernel. The benefits are enormous. A comparison of the various models can now be more easily realised. The merging of the models, by itself, provides new, hybrid parallel GA models. Combining this with the development of hybrid search techniques, including components of genetic algorithm, simulated annealing and tabu search, may lead to the development of parallel generic hybrid, parallel search techniques and toolkits to support these [Rayw94].

Experimental results have proven that GAs provide a remarkably successful method for finding solutions to the Steiner Problem in Graphs. From a practical point of view the *Level2* model performs better than the *Level1* for all problems tested. However, this should not be generalised since the results shown depend on the computational performance of the algorithm as well as the search efficiency. The performance of the *Level1* model increases more rapidly with the size of the problem and is expected to perform better than the *Level2* for large problems. Initial results for the *Island model* of DGAs are not so good and further experimentation is required before any definitive conclusions can be drawn.

References

Beas90. Beasley, J.E. OR-Library: distributing test problems by electronic mail. . *Opl. Res. Soc..41* (1990), 1069-1072.

Coho87. Cohoon, J.P., Hegde, S.U., Martin, W.N. and Richards, D. Punctuated Equilibria: A Parallel Genetic Algorithm. In *Genetic Algorithms and their Applications: Proceedings of the Second International Conference on Genetic Algorithms,* Grefenstette, J.J., Lawrence Earlbaum Associates, 1987, pp. 155-161.

Coho90. Cohoon, J.P., Martin, W.N. and Richards, D. Punctuated Equilibria: A Parallel Genetic Algorithm. In *Proceedings of the First International Workshop on Parallel Problem Solving from Nature,* 1990, pp. A-21.

Coho91. Cohoon, J.P., Martin, W.N. and Richards, D. A Multi-population Genetic Algorithm for solving the K-Partition Problem on Hyper-cubes. In *Proceedings of the fourth international conference on Genetic Algorithms,* Belew, R.K. and Booker, L.B., Morgan Kaufman Publishers, 1991, pp. 245-248.

Coll91. Collins, R. and Jefferson, D. Selection in Massively Parallel Genetic Algorithms. In *Proceedings of the fourth international conference on Genetic Algorithms,* Belew, R.K. and Booker, L.B., Morgan Kaufman Publishers, 1991, pp. 249-256.

Davi91. Davidor, Y. A Natural Occuring Niche & Species Phenomenon: The Model and First Results. In *Proceedings of the fourth International Conference on Genetic Algorithms,* Belew, R.K. and Booker, L.B., Morgan Kaufmann, San Mateo, CA, 1991, pp. 257-263.

Dijk59. Dijkstra, E.W. A note on two problems in connection with graphs. *Numer.Math. 1*(1959), 269-271.

Foul83. Foulds, L.R. and Rayward-Smith, V.J. Steiner problems in graphs: Algorithms and applications. *Engineering Optimization* 7(1983), 7-16.

Gold89. Goldberg, D.E. *Genetic Algorithms in Search, Optimization, and Machine Learning,* Addison-Wesley, Reading, Mass. (1989).

Gold89b. Goldberg, D.E. Sizing Populations for Serial and Parallel Genetic Algorithms. In *Proceedings of the Third International Conference on Genetic Algorithms,* Schaffer, J.D., Morgan Kaufmann, Los Altos, CA, 1989, pp. 70-79.

Gold90. Goldberg, D.E., "A Note on Boltzmann Tournament Selection for Genetic Algorithms and Population-oriented Simulated Annealing," , Deparment of Engineering Mechanics, TCGA Report 90003, 1990.

Gorg89.   Gorges-Schleuter, M. ASPARAGOS An Asynchronous Parallel Genetic Optimization Strategy. In *Proceedings of the Third International Conference on Genetic Algorithms,* Schaffer, J.D., Morgan Kaufmann, Los Altos, CA, 1989, pp. 422-427.

Gorg91.   Gorges-Schleuter, M. Explicit Parallelism of Genetic Algorithms through Population Structures. In *Parallel Problem Solving from Nature,* Springer Verlag, 1991, pp. 150-159.

Holl75.   Holland, J.H. *Adaptation in Natural and Artificial Systems,* University of Michigan Press, Ann Arbor, MI (1975).

Husb91.   Husbands, P. and Mill, F. Simulated Co-Evolution as The Mechanism for Emergent Planning and Scheduling. In *Proceedings of the fourth international conference on Genetic Algorithms,* Belew, R.K. and Booker, L.B., Morgan Kaufman Publishers, 1991, pp. 264-270.

Hwan92.   Hwang, F.K. and Richards, D.S. Steiner tree problems. *Networks* 22(1992), 55-89.

Jong75.   Jong, K.A.D. *An analysis of the behavior of a class of genetic adaptive systems,* Ph.D. dissertation, University of Michigan, 1975.

Kaps92.   Kapsalis, A. and Smith, G.D., "The GAmeter Toolkit Manual," , School of Information Systems, Computing Science Technical Report, 1992.

Kaps93a.   Kapsalis, A., Rayward-Smith, V.J. and Smith, G.D. Fast Sequencial and Parallel Implementation of Genetic Algorithms using the GAmeter Toolkit. In *Proceedings of the Int. Conf. on Artificial Neural Nets and Genetic Algorithms,* Springer Verlang, 1993.

Kaps93b.   Kapsalis, A., Rayward-Smith, V.J. and Smith, G.D. Solving the Graphical Steiner Tree Problem using Genetic Algorithms. *J. Oper. Res. Soc.* 44(1993), 397-406.

Karp93.   Karp, R.M. Reducibility among combinatorial problems. In *Complexity of computer computations ,* Miller, R.E. and Thatcher, J.W., Springer Verlang, 1993, pp. 85-103.

Krus56.   Kruskal, J.B. On the shortest spanning subtree of a graph and the travelling salesman problem. *Proc. Amer. Math. Soc.* 7(1956), 48-50.

Lasz90.   von Laszewski, G. and Mühlenbein, H. A Parallel Genetic Algorithm for the Graph Partitioning Problem. In *Proceedings of the First International Workshop on Parallel Problem Solving from Nature,* 1990, pp. B-12.

Mand89.   Manderick, B. and Spiessens, P. Fine-Grained Parallel Genetic Algoritms. In *Proceedings of the Third International Conference on Genetic Algorithms,* Schaffer, J.D., Morgan Kaufmann, Los Altos, CA, 1989, pp. 428-433.

Mühl88.   Mühlenbein, H., Gorges-Schleuter, M. and Kramer, O. Evolution Algorithms in Combinatorial Optimization. *Parallel Computing* 7(1988), 65-88.

Mühl89.   Mühlenbein, H. Parallel Genetic Algoritms, Population Genetics and Combinatorial Optimization. In *Proceedings of the Third International Conference on Genetic Algorithms*, Schaffer, J.D., Morgan Kaufmann, Los Altos, CA, 1989, pp. 416-421.

Mühl91a.   Mühlenbein, H. Evolution in Time and Space - The Parallel Genetic Algorithm. In *Foundations of Genetic Algorithms*. Morgan-Kaufmann, Rawlins, G., pp. 316-337, 1991.

Mühl91b.   Mühlenbein, H., Schomisch, M. and Born, J. The Parallel Genetic Algorithm as Function Optimizer. *Parallel Computing* 17(1991), 619-632.

Mühl91c.   Mühlenbein, H., Schomisch, M. and Born, J. The Parallel Genetic Algorithm as Function Optimizer. In *Proceedings of the 4th International Conference on Genetic Algorithms*, Belew, R.K. and Booker, L.B., Morgan Kaufmann, San Mateo, CA, 1991, pp. 271-278.

Pett87.   Pettey, C.B., Leuze, M.R. and Grefenstette, J.J. A Parallel Genetic Algorithm. In *Genetic Algorithms and their Applications: Proceedings of the Second International Conference on Genetic Algorithms*, Grefenstette, J.J., Lawrence Earlbaum Associates, 1987, pp. 155-161.

Pett89.   Pettey, C.B. and Leuze, M.R. A Theoretical Investigation of a Parallel Genetic Algoritm. In *Proceedings of the Third International Conference on Genetic Algorithms*, Schaffer, J.D., Morgan Kaufmann, Los Altos, CA, 1989, pp. 398-405.

Prim57.   Prim, R.C. Shortest connection networks and some generalizations. *Bell System Tech. J.* 36(1957), 1389-1401.

Rayw94.   Rayward-Smith, V.J. A Unified Approach to Tabu Search, Simulated Annealing and Genetic Algorithms. In *Unicom conference proceedings: Adaptive Computing and Information Processing*, Unicom Seminars Ltd, Brunel Science Park, Cleveland Rd., Uxbridge, Middlesex, UB8 3PH, 1994.

Spie91.   Spiessens, P. and Manderick, B. A Massively Parallel Genetic Algorithm: Implementation and First Results. In *Proceedings of the fourth International Conference on Genetic Algorithms*, Belew, R.K. and Booker, L.B., Morgan Kaufmann, San Mateo, CA, 1991, pp. 279-287.

Tane87.   Tanese, R. Parallel Genetic Algorithm for a Hypercube. In *Genetic Algorithms and their Applications: Proceedings of the Second International Conference on Genetic Algorithms*, Grefenstette, J.J., Lawrence Earlbaum Associates, 1987, pp. 177-183.

Toma93.   Tomassini, M. The Parallel Genetic Cellular Automata: Application to Global Function Optimization. In *Proceedings of the Int. Conf. on Artificial Neural Nets and Genetic Algorithms,* Springer Verlang, 1993, pp. 383-391.

Wint87.   Winter, P. The Steiner problem in networks: A survey. *Networks 17*(1987), 129-167.

Distributed Coevolutionary Genetic Algorithms for Multi-Criteria and Multi-Constraint Optimisation

Phil Husbands
School of Cognitive and Computing Sciences
University of Sussex
Brighton, UK, BN1 9QH
email: philh@cogs.susx.ac.uk

Abstract

This paper explores the use of coevolutionary genetic algorithms to attack hard optimisation problems. It outlines classes of practical problems which are difficult to tackle with conventional techniques, and indeed with standard 'single species' genetic algorithms, but which may be amenable to 'multi-species' coevolutionary genetic algorithms. It is argued that such algorithms are most coherent and effective when implemented as distributed genetic algorithms with local selection operating. Examples of the successful use of such techniques are described, with particular emphasis given to new work on a highly generalised version of the job shop scheduling problem.

1 Introduction

The vast majority of genetic algorithm (GA) work involves a single 'species'. That is, a single genetic encoding aimed at finding solutions to a single problem. The GA machinery may be configured to work with a single population or, in the case of 'island' models [9], a number of interacting populations. But in either case there is just one evaluation function, and just one solution encoding.

This paper shows how coevolutionary genetic algorithms, employing more than one interacting 'species' evolving under different evaluation functions, can be used to tackle certain types of hard optimisation problems in a more efficient way than single species GAs.

Particular stress is put on multi-criteria and multi-constraint problems. These are often among the most demanding of optimisation problems and severely test all search techniques. From this class two types of problems are deemed potentially most amenable to coevolutionary techniques.

- Well defined problems with well defined evaluation functions, but which are hugely complex. Sometimes such problems can be redefined in terms of interacting sub-problems. In this paper a scheduling problem is given as an example of a such a task. It is shown how coevolutionary techniques can be usefully applied to it.

- Problems with evaluation functions which are not well defined. Examples are where solutions are desired to perform well over a potentially infinite set of test cases; or when there is no way of weighting different, irresolvable, criteria relative to each other. Hillis's work is given as an example of a coevolutionary approach to such problems [4].

Since they are seen as the basic techniques underpinning sensible coevolutionary GAs, the paper begins with descriptions of distributed and coevolutionary distributed GAs. Hillis's host-parasite coevolutionary system is explained, and its application to practical problems other than those he has explored is outlined. The remainder of the paper looks in detail at the coevolutionary aspects of my work on a highly generalised version of the job shop scheduling problem. Finally general conclusions are drawn.

2 Parallel Distributed GAs

From the very earliest days of its development the GA's potential for parallelisation, with all its attendant benefits of efficiency, has been noted. The availability of hardware has recently allowed significant progress in this direction. The standard sequential GA uses global population statistics to control selection, so the processing bottleneck is evaluation. The earliest parallel models simply parallelised this phase of the sequential algorithm, see, for instance, the paper by Grefenstette [2]. Recently more sophisticated parallel GAs have started to appear in which population can be thought of as being spread out geographically, usually over a 2D toroidal grid. All interactions, e.g. selection and mating, are local, being confined to small (possibly overlapping) neighbourhoods on the grid. Such GAs will be referred to as *distributed* in the remainder of this paper. By doing away with global calculations, it is possible to develop fine-grained highly parallel asynchronous algorithms. There is mounting evidence to suggest that such systems are more robust and faster (in terms of solutions evaluated) than other implementations, e.g. see the articles by Collins & Jefferson [1] and

Husbands [5]. Highly parallel models can also result in powerful new ways of approaching optimisation problems at the conceptual level, as will be seen later in this paper.

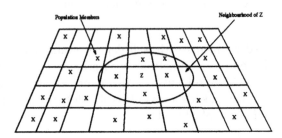

Figure 1: Fraction of typical 2D grid used for distributed GAs.

Typical fine grained distributed GAs use a grid like that shown in Figure 1. Population members are scattered across the grid with no more than one member per cell. Potential mates for a given individual will be found in a neighbourhood centred on the individual. Local selection rules ensure that the most fit members of the population in the neighbourhood are the most likely to be chosen as mates. After breeding, the offspring produced are placed in the neighbourhood of their parents, so genetic material remains spatially local. A probabilistic replacement of the weaker members of the neighbourhood may be employed. By using overlapping neighbourhoods the resulting algorithms generate extremely fierce local selection but allow any improved solutions found to flow around the grid. The effect is to continually stimulate the search process and prevent convergence of the population. A particular distributed algorithm will be described in more detail in Section 5. Further discussion of the parallel implementation of distributed GAs, and their comparison with other forms of parallel GAs, can be found in [5].

3 Coevolutionary Distributed GAs

The algorithms described thus far have, at least implicitly, referred to a single population (or 'species') searching for solutions to a single problem. Coevolutionary algorithms involve more than one 'species' breeding separately, from their own gene pools, but interacting at the phenotypic level.

In Nature the environment of most organisms is mainly made up of other organisms. Hence a fundamental understanding of natural evolution must take into account evolution at the ecosystems level. There are competing theories from evolutionary biology [11], but many are based on Van Valen's Red Queen Hypothesis [12] which states that any evolutionary change in any species is experienced by coexisting species as a change in their environment. Hence there is a continual evolutionary drive for adaptation and counter adaptation. As the Red Queen explained to Alice in Wonderland:

> "... it takes all the running you can do, to keep in the same place. If you want to get somewhere else, you must run at least twice as fast as that!" [From L. Carroll, *Through The Looking Glass*]

It is possible to exploit this phenomenon in GA-based artificial evolution to develop more and more complex competitive behaviours in animats (artificial animals [13]), without having to specify complicated evaluation functions [8]. As we will see, coevolutionary GAs can also be used to tackle difficult optimisation problems of the kinds outlined in the introduction to this paper.

How coevolutionary GAs should be implemented must, of course, at least in part depend on the application. But there are three obvious contenders.

- Separate sequential (or parallel) GAs for each species where the evaluation functions used in each somehow takes into account interactions with the other populations.

- A parallel 'island' implementation in which each population evolves in isolation with occasional interactions with members from other species.

- Distributed GAs for each species where the different populations are spread out over the *same* grid. Interactions between populations are (spatially) local.

In the first model it is difficult to maintain the kind of coherent coevolution required. Which members of any given population are coevolving with given members of another population? All with all? All members of any given population with randomly chosen members of all other populations? How can the population dynamics from generation to generation be controlled to maintain a consistent coevolutionary system? Negative experiences with such an implementation will be described later (see Section 5.1).

The second model will be adequate only if the inter-population interactions desired really are weak, otherwise it will give a misleadingly benign picture of the effects of the different species on each other's environments.

However, the third model gives a very clear and coherent implementation of coevolution. Each species interact locally with its own population but also with the members of the other species in its neighbourhood. Since all offspring appear in their parents' neighbourhoods, a consistent coevolutionary pressure emerges. Examples of such algorithms will be discussed in the following sections.

4 Parasites and Sorting Networks

Danny Hillis, who lead the team that developed the Connection Machine [3] , was the first to significantly extend the parallel GA paradigm by showing how to develop a more powerful optimisation system by making use of coevolution [4]. Using a distributed coevolutionary GA he had considerable success in developing sorting networks. In this extended model there are two independent gene pools, each evolving according to local selection and mating. One population, the hosts, represents sorting networks, the other, the parasites, represents test cases. Interaction of the populations is via their fitness functions. The sorting networks are scored according to the test cases provided by the parasites in their immediate vicinity. The parasites are scored according to the number of tests the network fails on.

Hillis's two species technique can be applied directly to many engineering optimisation problems where the set of test cases is potentially infinite. The most appropriate range and difficulty of evaluation tests can be coevolved with problem solutions as in the work described here. Another use of the technique, being explored at Sussex, is in the coevolution of problem solutions and sets of constraints to apply in the evaluation of these solutions. This is applicable where it is difficult, or impossible, to weight constraints relative to each other, and where there are too many constraints for it to be feasible to have them all active in all parts of the search space.

5 Coevolution, Arbitrators and Emergent Scheduling

This section outlines the use of a multi-species coevolutionary GA to handle a highly generalised version of Job Shop Scheduling (JSS). It is based on the notion of the manufacturing facility as a complex dynamical system akin to an ecosystem. Space does not allow a full description of the cost functions and encodings, such engineering and mathematical details can be found in an earlier paper [6]. This paper concentrates on previously unpublished details of the distributed genetic algorithms used.

The traditional academic view of JSS is shown in Figure 2. A number of *fixed* manufacturing plans, one for each component to be manufactured, are interleaved by a scheduler so as to minimise some criteria such as the total length of the schedule. However, a problem that would often be more useful to solve is that illustrated in Figure 3. Here the intention is to optimise the individual manufacturing plans *in parallel*, taking into account the numerous interactions between them resulting from the shared use of resources. This is a much harder

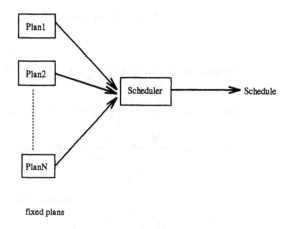

Figure 2: Traditional academic approach to job shop scheduling.

and far more general problem than the traditional JSS problem. In many manufacturing environments there is a vast number of legal plans for each component. These vary in the number of manufacturing operations, the ordering of the operations, the machines used for each operation, the tool used on any given machine for a particular operation, and the orientation of the work-piece (setup) given the machine and tool choices. All these choices will be subject to constraints on the ordering of operations, and technological dependencies between operations. Optimising a single process plan is an NP-hard problem. Optimising several in parallel requires a powerful search technique. This section presents a promising GA-based method for tackling the problem. This paper will not delve very deeply into the problem-specific technbical details, instead it will concentrate on GA issues. Much very useful work has been done in the application of GAs to scheduling problems. This will not be discussed here, since this paper is not intended to be specifically about scheduling. However, see [6, 10] for discussions of related work.

The idea behind the ecosystems model is as follows. The genotype of each specie represents a feasible manufacturing (process) plan for a particular component to be manufactured in the machine shop. Separate populations evolve under the pressure of selection to find near-optimal process plans for each of the components. However, their fitness functions take into account the use of shared resources in their common world (a model of the machine shop). This means that without the need for an explicit scheduling stage, a low cost schedule will emerge at the same time as the plans are being optimised.

Data provided by a plan space generator (complex and beyond the scope of this paper, see [6]) is used to randomly construct initial populations of structures representing possible plans, one population for each component to be manufactured. An important part of this model is a population of Arbitrators, again initially randomly generated. The Arbitrators' job is to resolve conflicts between

Parallel Plan Optimisation

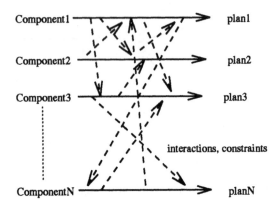

Figure 3: Parallel plan optimisation leading to emergent scheduling.

members of the other populations; their fitness depends on how well they achieve this. Each population, including the Arbitrators, evolve under the influence of selection, crossover and mutation.

Each process plan 'species' uses the plan encoding described in detail in [6], that paper also describes the special genetic operators used with the encoding, and the machining cost functions used in the algorithms given later.

The Arbitrators are required to resolve conflicts arising when members of the other populations demand the same resources during overlapping time intervals. The Arbitrators' genotype is a bit string which encodes a table indicating which population should have precedence at any particular stage of the execution of a plan, should a conflict over a shared resource occur. A conflict at stage L between populations K and J is resolved by looking up the appropriate entry in the Lth table. Since population members cannot conflict with themselves, and we only need a single entry for each possible population *pairing*, the table at each stage only needs to be of size $N(N-1)/2$, where N is the number of separate component populations. As the Arbitrators represent such a set of tables flattened out into a string, their genome is a bit string of length $SN(N-1)/2$, where S is the maximum possible number of stages in a plan. Each bit is uniquely identified with a particular population pairing and is interpreted according to the function given in Equation 1.

$$f(n_1, n_2, k) = g\left[\frac{kN(N-1)}{2} + n_1(N-1) - \frac{n_1(n_1+1)}{2} + n_2 - 1\right] \quad (1)$$

Where n_1 and n_2 are unique labels for particular populations, $n_1 < n_2$, k refers to the stage of the plan and $g[i]$ refers to the value of the ith gene on the Arbitrator genome. If $f(n_1, n_2, k) = 1$ then n_1 dominates, else n_2 dominates.

By using pair wise filtering the Arbitrator can be used to resolve conflicts between any number of different species. It is the Arbitrators that allow the scheduling aspect of the problem to be handled. In general, a population of coevolving Arbitrators could be used to resolve conflicts due to a number of different types of operational constraint, although their representation may need to increase in complexity.

It should be noted that in early versions of the work to be described, the Arbitrators were not used. Instead fixed population precedence rules were applied. Not surprisingly, this and similar schemes were found to be too inflexible and did not give good results. Hence the Arbitrator idea was developed and has proved successful.

5.1   Early MIMD Implementation

The first implementation of the basic ecosystems model was described in [7]. It used a set of interacting sequential genetic algorithms and was implemented on a MIMD parallel machine as well as on a conventional sequential machine. On each cycle of the algorithm each population was ranked according to cost functions taking into account plan efficiency. The concurrent execution of *equally ranked* plans from each population was then simulated. The simulation provided a final cost taking into account interactions between plans for different components. Conflicts were resolved by an equally ranked Arbitrator. This final cost was used by the selection mechanism in the GAs.

Although promising results were achieved with this model, it suffered from population convergence and little progress was made after a few hundred generations, despite many attempts to cure this problem. It was also felt that the implementation was over complicated and lacked coherence at some levels, resulting in some of the problems discussed earlier in Section 3. The ranking process appears to facilitate coevolution to some extent. But, since the populations are continually reordered there is little continuity, from generation to generation, in the members being costed together until population members are very similar. This may have indirectly been one of the causes of strong convergence. For these reasons this implementation was abandoned and another much more coherent version developed. This is based on the kind of geographically distributed GA mentioned earlier.

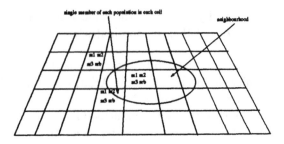

Figure 4: Distributed interacting populations.

5.2 Distributed Implementation

In the second implementation, the cost, hence selection, functions for plan organisms again involve two stages, but for Arbitrators now just one. The first stage involves population specific criteria (basic machining costs), as before, and the second stage again takes into account interactions between populations. Arbitrators are only costed at the second stage. Again the second phase of the cost function involves simulating the simultaneous execution of plans derived from stage one. The process plans and Arbitrators are costed as before, it is the way in which the cost functions are used within the GA machinery which is now quite different.

This second, more satisfactory, implementation spreads each population 'geographically' over *the same* 2D toroidal grid, this is illustrated in Figure 4. Each cell on the grid contains exactly one member of each population. Selection is local, individuals can mate only with those members of their own species in their local neighbourhood. Following Hillis [4], the neighbourhood is defined in terms of a Gaussian distribution over distance from the individual; the standard deviation is chosen so as to result in a small number of individuals per neighbourhood. Neighbourhoods overlap allowing information flow through the whole population without the need for global control. Selection works by using a simple ranking scheme within a neighbourhood: the most fit individual is twice as likely to be selected as the median individual. Offspring produced replace individuals from their parents' neighbourhood. Replacement is probabilistic using the inverse scheme to selection. In this way genetic material remains spatially local and a robust and coherent coevolution (particularly between Arbitrators and process plan organisms) is allowed to unfold. Interactions are also local: the second phase of the costing involves individuals from each population *at the same location on the grid*. This implementation consistently gave better faster results than the first. The notion of coevolution is now much more coherent; by doing away with the complicated ranking mechanism, and only using local selection, based on a concrete model of geographical neighbourhood, the problems and inconsistencies of the first implementation are swept away.

The overall algorithm is quite straightforward. It can be implemented sequentially or in a parallel asynchronous way, depending on hardware available.

Overall()

1. **Randomly generate each population, put one member of each population in each cell of a toroidal grid.**

2. **Cost each member of each population (phase1 + phase2 costs). Phase2 cost are calculated by simulating the concurrent execution of all plans represented in a given cell on grid, any resource conflicts are resolved by Arbitrator in that cell.**

3. **i = 0.**

4. **Pick random starting cell on the toroidal grid.**

5. **Breed each of the representatives of the different populations found in that cell.**

6. **If all cells on the grid have been visited Go to 7. Else move to next cell, Go to 5.**

7. **If $i <$ MaxIterations, i = i + 1, Go to 4. Else Go to 8.**

8. **Exit.**

The breeding algorithm, which is applied in turn to the members of the different populations, is a little more complicated.

Breed(current_cell, current_population)

1. **i = 0.**

2. **Clear NeighbourArray**

3. **Pick a cell in neighbourhood of current_cell by generating x and y distances (from current_cell) according to a binomial approximation to a Gaussian distribution. The sign of the distance (up or down, left or right) is chosen randomly (50/50).**

4. **If the cell chosen is not in NeighbourArray, put it in NeighbourArray, i = i+1, Go to 5. Else Go to 3.**

5. **If $i <$ LocalSelectionSize, Go to 3. Else Go to 6.**

6. **Rank (sort) the members of current_population located in the cells recorded in NeighbourArray according to their cost. Choose one of these using a linear selection function.**

7. Produce offspring using the individual chosen in 6 and current_population member in current_cell as the parents.

8. Choose a cell from ranked NeighbourArray according to an inverse linear selection function. Replace member of current_population in this cell with offspring produced in 7.

9. Find phase one (local) costs for this new individual (not necessary for Arbitrators).

10. Calculate new phase two costs for all individuals in the cell the new individual has been placed in, by simulating their concurrent execution. Update costs accordingly.

11. Exit.

The binomial approximation to a Gaussian distribution used in step 3, falls off sharply for distances greater than 2 cells, and is truncated to zero for distances greater than four cells.

In the results reported here a 15×15 grid was used, giving populations of size 225.

5.3   Results of Distributed Implementation

Results of a typical run for a complex 5 job problem shown in Figure 5. Each of the components needed between 20 and 60 manufacturing operations. Each operation could be performed, on average, on six different machines using 8 possible setups. There were many constraints on and between plans, but there were still a very large number of possible orderings of the operations within each plan. The graph at the top shows the best 'total factory cost' found plotted against the number of plan evaluations on the X axis. The graph represents the average of ten runs. The 'total factory cost' is calculated per grid cell by summing the costs of all the process plans (exactly one per component), including the waiting-time costs, represented at that cell. The lower left Gantt chart shows the loading of the machines (M0-M12) in the job shop early on in a typical run, and that at the right shows the loading towards the end of the same run. Time, in arbitrary units, is shown on the horizontal axis. A very tight schedule *and* low cost plans for each of the components are obtained. Figure 6 shows the state of the geographical grid in terms of the 'total factory cost' at each cell. The aim of the overall simultaneous plan optimisation problem is to find the best cell, i.e. the cell with the lowest 'total factory cost'. This will not necessarily contain copies of the *individually* lowest cost members of each population, but it will contain that set of plans, one from each population, that interact in the most favourable way. It can be seen from the figure that the initial random

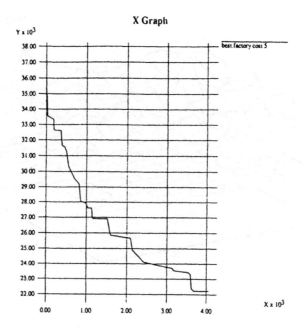

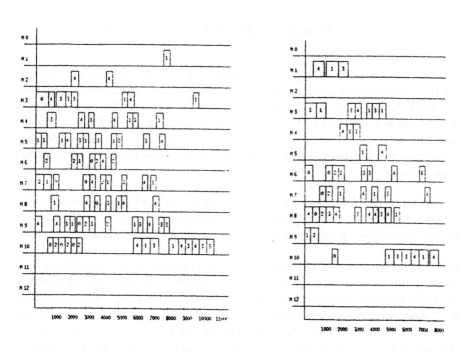

Figure 5: Results of distributed coevolution model with large problem. See text for explanantion.

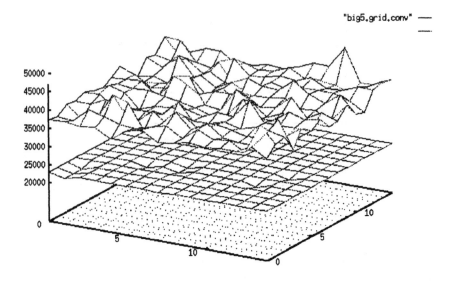

"big5.grid.conv"

Figure 6: Geographical grid states for large problem.

populations give a spread of poor 'total factory costs', this is rapidly reduced (a few thousand function evaluations) to a set of very good costs spread throughout the whole grid. These are shown more clearly in Figure 7, but note that grid states have not converged. In much longer runs this was still found to be the case. Results very close to those shown here were found on each of 50 runs of the system. Far more consistent behaviour than was found with the earlier implementation.

By replacing the Arbitrators in low cost cells with randomly generated ones, higher cost schedules were produced demonstrating that the Arbitrators are coevolving to make good decisions over all the stages of the plans.

Very promising preliminary results have been obtained for this complex optimisation problem. Although the search spaces involved are very large, this method has exploited parallelism sufficiently to produce good solutions. Most scheduling work deals with a problem rather different from that tackled here, so a direct comparison is very difficult. However, Palmer [10] has recently tackled the generalised problem using a simulated annealing based technique. He demonstrated that his technique had significant advantages over traditional scheduling for a class of manufacturing problems a little simpler than those used here. A direct comparison of the GA and simulated annealing methods is currently underway.

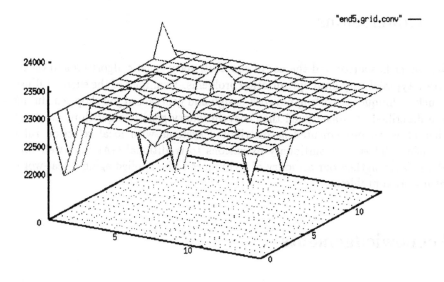

Figure 7: Expanded view of final geographical grid states for large problem.

5.4 Dynamics and Noise

Very important properties of manufacturing environments, namely dynamics
and noise, are not handled at all well by most classical JSS techniques. The
distributed coevolutionary method can handle both in a natural way. The dynamics of a job shop include such things as machines breaking down, job prior-
ities changing, jobs starting at different times, job due dates changing. These
changes are modelled directly in the cost functions and the whole 'ecosystem' re-
configures. The distributed implementation allows the rapid flow of re-adapted
solutions throughout the grid, the unconverged state of the populations facilitates the re-adaptation. Noise arises from the fact that the manufacturing
processes are not perfectly reliable and deterministic. This means that in reality
all cost functions must have stochastic elements. Again this can be handled
in a straightforward manner by using evaluation functions which require good
performance over distributions of random variables representing such things as
machining times. Hence a solution which is very sensitive to *exact* timings of
manufacturing processes will be selected against.

6 Conclusions

This paper has advocated the use of coevolutionary genetic algorithms to tackle certain types of very hard optimisation problems. Examples of the successful use of such techniques to sorting network design and generalised job shop scheduling were described. It was argued that coevolutionary genetic algorithms are most coherent when implemented as distributed systems with local selection rules operating. These coevolutionary extensions to traditional GAs may well considerably strengthen our armoury of techniques to be applied against real-world optimisation problems.

Acknowledgements

This work was supported by EPSRC grant GR/J40812. Thanks to two anonymous referees for some useful comments.

References

[1] R. Collins and D. Jefferson. Selection in massively parallel genetic algorithms. In R. K. Belew and L. B. Booker, editors, *Proceedings of the Fourth Intl. Conf. on Genetic Algorithms, ICGA-91*, pages 249–256. Morgan Kaufmann, 1991.

[2] J. Grefenstette. Parallel adaptive algorithms for function optimisation. Technical Report CS-81-19, Compt. Sci., Vanderbilt University, 1981.

[3] W.D. Hillis. *The Connection Machine*. MIT Press, 1985.

[4] W.D. Hillis. Co-evolving parasites improve simulated evolution as an optimization procedure. *Physica D*, 42:228–234, 1990.

[5] P. Husbands. Genetic algorithms in optimisation and adaptation. In L. Kronsjo and D. Shumsheruddin, editors, *Advances in Parallel Algorithms*, pages 227–277. Blackwell Scientific Publishing, Oxford, 1992.

[6] P. Husbands. An ecosystems model for integrated production planning. *Intl. Journal of Computer Integrated Manufacturing*, 6(1&2):74–86, 1993.

[7] Philip Husbands and Frank Mill. Simulated co-evolution as the mechanism for emergent planning and scheduling. In R. K. Belew and L. B. Booker, editors, *Proceedings of the Fourth Intl. Conf. on Genetic Algorithms, ICGA-91*, pages 264–270. Morgan Kaufmann, 1991.

[8] Geoffrey F. Miller and D. T. Cliff. Protean behavior in dynamic games: Arguments for the co-evolution of pursuit-evasion tactics. In D. Cliff, P. Husbands, J.A. Meyer, and S. Wilson, editors, *Animals to animats 3: Proceedings of the third international conference on simulation of adaptive behavior.* MIT Press, 1994.

[9] H. Muhlenbein, M. Schomisch, and J. Born. The parallel genetic algorithm as function optimizer. *Parallel Computing*, 17, 1991.

[10] G. Palmer. *An Integrated Approach to Manufacturing Planning.* PhD thesis, University of Huddersfield, 1994.

[11] N.C. Stenseth. Darwinian evolution in ecosystems: the red queen view. In P. Greenwood, P. Harvey, and M. Slatkin, editors, *Evolution: Essays in honour of John Maynard Smith*, pages 55–72. Cambridge University Press, 1985.

[12] L. Van Valen. A new evolutionary law. *Evolutionary Theory*, 1:1–30, 1973.

[13] S.W. Wilson. Knowledge growth in an artificial animal. In J. Grefenstette, editor, *Proceedings of the First International Conference on Genetic Algorithms and their applications.* Lawrence Erlbaum Assoc., 1985.

Inductive Operators and Rule Repair in a Hybrid Genetic Learning System: Some Initial Results

A. Fairley[1] and D.F. Yates

The Bio-Computation Group
Dept of Computer Science
University of Liverpool
Liverpool, L69 3BX.

Abstract

Symbolic knowledge representation schemes have been suggested as one way to improve the performance of classifier systems in the context of complex, real-world problems. The main reason for this is that unlike the traditional binary string representation, high-level languages facilitate the exploitation of problem specific knowledge. However, the two principal genetic operators, crossover and mutation, are, in their basic form, ineffective with regard to discovering useful rules in such representations. Moreover, the operators do not take into account any environmental cues which may benefit the rule discovery process. A further source of inefficiency in classifier systems concerns their capacity for forgetting valuable experience by deleting previously useful rules.

In this paper, solutions to both of these problems are addressed. First, in respect of the unsuitability of crossover and mutation, a new set of operators, specifically tailored for a high level language, are proposed. Moreover, to alleviate the problem of forgetfulness, an approach based on the way some enzyme systems facilitate the repair of genes in biological systems, is investigated.

1.0 Introduction

Research into the operation of classifier systems is still in its infancy, and as a result, there are many possible sources of inefficiency within such systems. One such source is the set of genetic operators used to generate the new rules for a system. Whilst they have been shown to be effective for simple representations such as the binary strings used in most GA applications, the two central genetic operators,

---

[1]Current Address: Naval Studies Department, Defence Research Agency, Southwell, Portland, Dorset, DT5 2JS

crossover and mutation, appear to be ineffective when high-level symbolic representation schemes are used. A second problem central to the operation of a classifier system concerns their capacity for rapidly forgetting valuable experience by deleting previously useful rules. Two possible solutions to this problem immediately suggest themselves, namely increasing the size of the rule base and incorporating a long-term memory of inactive, well-trained rules into the system. However, these solutions raise problems concerning the running time of the system and deciding when a rule has been useful, respectively. Moreover, as both the rule base and the memory have a finite size, then there is still no guarantee that useful information will be retained.

In this paper, solutions to both of these problems are addressed. First, in respect of the unsuitability of crossover and mutation, a new set of operators based on ideas from the field of inductive learning [1,2] are proposed. Moreover, to alleviate the problem of forgetfulness, an approach based on the way some enzyme systems facilitate the repair of genes in biological systems, is investigated.

The vehicle used for this work is a hybrid genetic learning system, SAGA, which combines a number of novel features in a co-evolutionary architecture to model a 2-player game. Although this paper necessarily gives a brief overview of SAGA, only one part of the hybrid architecture, namely that of a Component Classifier System, is used for the trials.

2.0 The Test Model

The problem used to investigate the effectiveness of these features was a simplified version of the *Two Tanks Problem* (TTP). This problem requires a tank, *T1*, armed with but a single gun, to destroy another tank, *T2*, possessing identical capabilities. In the simplified version of the problem adopted for this work, the setting was restricted to a 'one-dimensional board' composed of an infinite strip of identical contiguous squares.

An *episode* of the TTP commences with the two tanks in their initial positions, placed randomly at a distance of $19 \leq x \leq 21$ units apart, and with the range of each gun set to 15. The tanks take alternate moves (*T1* takes the first), and on each of its moves, a tank, subject to constraints, may:

- move forward 1 square;
- fire;
- increase gun elevation by 1 notch (equivalent to increasing gun range by one square).

These moves are subject to the constraints that:

a. each tank can fire at most $(x-18)$ times;
b. each gun can only be elevated at most 3 times or notches.

An episode ends when one of the tanks hits its opponent (achieved by firing when the inter-tank gap equals the range of the guns), or when t moves (the maximum length selected for an episode) have been made without a hit being scored.

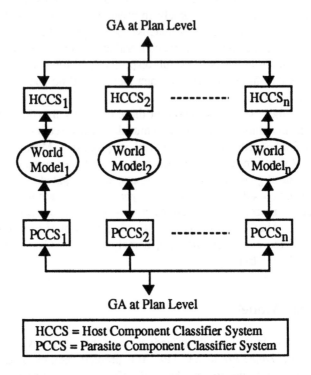

Figure 1: Outline of the SAGA model

3.0 The SAGA Model

3.1 Overview

SAGA (Strategy Analysis by Genetic Algorithm) (Figure 1) is a genetic learning system which combines aspects of both the Michigan and Pittsburgh approaches in a co-evolutionary architecture in an attempt to model a learning agent and its adversary in the TTP. It is composed of a number of *Component Classifier Systems* (CCSs), each of which has essentially the same architecture as a traditional classifier system[2]. The CCSs are partitioned into two populations of equal size, each CCS being paired with a partner in the other population. As distinct from the traditional view wherein

---

[2]Although no "standard" classifier system has ever been defined, central features, such as message passing, the existence of an internal message list and the BBA for credit assignment remain common to most applications.

each CCS has an environment with which it alone interacts, SAGA provides each CCS and its partner with a shared environment in which one CCS combatant acts for the first player and the other CCS operates as its adversary. As both are capable of evolving, they combine in a co-evolutionary manner in which each player is constantly trying to beat their opponent. This co-evolutionary effect is often referred to as the Red Queen effect [3] and is synonymous with the relationship between a 'parasite' and its 'host'. As a result, the two populations of CCSs in SAGA are called the *host* and *parasite* populations, with the host and parasite CCSs playing T1 and T2 respectively.

For rule discovery, SAGA uses two GAs, one which operates at the rule (classifier) level and the other at the plan level. The GA that is active at the rule level operates within each CCS and uses the strength of the extant rules to guide its search for improved rules. The second GA, that operating at the plan level, is applied less often and its effect is to combine the most profitable plans.

As this paper only investigates learning within a single CCS, both the parasite population and the GA which operates at the plan level were disabled. Thus, the system used for the work reported here employed but a single CCS, and the role of the adversary was taken by a plan employing a fixed strategy which entailed T2 firing on every move.

3.2 Knowledge Representation

In common with most classifier systems, each CCS in SAGA represents its knowledge as a set of production rules. However, as a departure from the norm, the rules in SAGA are expressed in a high-level (symbolic) language which facilitates the transfer of knowledge both to and from the system. Specifically, all conditions and actions are represented by tuples, each possessing an identifier and a set of arguments. Each identifier belongs to one of the six possible data types (Table 1) which together cover a wide variety of variables that need to be modelled in any problem domain.

The number of arguments in each condition tuple is dependent upon the identifier type, with two arguments for linear/cyclic identifiers (representing the lower and upper bounds respectively) and a single argument for tree structured identifiers (representing the root of the subtree of matching values). Identifiers of list type can have k, $k \geq 1$, arguments up to the maximum number of values for that identifier, whilst identifiers with pattern type have a single string argument which has the same length as an internal message.

SAGA's knowledge representation scheme is further enhanced by permitting conditions to be negated, and variable names, as well as constant arguments, to be used in conditions. For example, a condition such as (*distance_to_target, range, range*), where the two arguments represent the lower and upper bounds respectively, can be used to check if the distance from a tank to a particular target is the same as the range of its gun.

Each rule consists of $n > 0$ conditions and a single action to be taken when all n conditions are satisfied. Moreover, there is an internal message associated with every rule which is posted to the message list if the rule is active after conflict resolution. The only condition which must always be present in a rule is the internal condition, which, as its name suggests, refers to the state of the internal message list. All other conditions refer to various properties of the world model.

Type	Use	Example Condition
Linear	Used to represent integer or real scalar values	(speed, 100, 200)
Cyclic	A special form of linear variable which may assume only values which are cyclical in nature	(bearing, 330, 90)
Tree Structured	Used to represent hierarchical relationships	(weather, cloudy)
List	Used to represent non-hierarchical relationships	(target_type, MkI, MkII)
Pattern	Used to represent internal messages. They are strings over the alphabet {0,1,#} where # is a 'don't care' symbol	(detector1, 0##1)
Boolean	Used for variables which can assume the values TRUE or FALSE	(target_in_range, TRUE)

Table 1: Data Types supported by SAGA

3.3 CCS operation

Each CCS in SAGA operates cyclically - one cycle per 'episode move'. On each cycle, it inspects its world model (used to represent the extant state of the episode) and then determines the next move based on the knowledge in its tactical plan. This is achieved by selecting those rules in which all the conditions (both environmental and internal) are satisfied, and passing these 'matching rules' to a conflict resolution mechanism to ensure consistency in the action ultimately suggested. The action from the selected rule(s) is then applied to the world model which is subsequently updated. Subsequently, the opponent makes its move and again the world model is correspondingly updated. This marks the end of a cycle.

SAGA supports a number of alternative credit assignment schemes including the Bucket Brigade Algorithm (BBA) [4], epochal schemes such as the Profit Sharing Plan (PSP) [5], hybrids of the these, as well as a self-adapting scheme in which the distribution of credit between active rules varies depending upon the success of the tactical plan. It is not the purpose of this paper to focus on the capabilities of these alternative schemes and thus, for simplicity, only a simple BBA was used for the work reported. In respect of the BBA, given that classifiers i and j are active at times t and $t+1$ respectively, then the strength of classifier i is updated using

$$S_i(t+1) = S_i(t) + b.spec_j.S_j(t) + P(t) - b.spec_i.S_i(t) - tax.S_i(t)$$

where $S_i(t)$ is the strength of classifier i at time t, $spec_i$ is the specificity of i, and b and tax are the bid the taxation co-efficients respectively.

4.0 Genetic Operators

Genetic Operators are employed by classifier systems as agents for the production of new rules in the search for a high performance rule set. Whilst the genetic operators of crossover and mutation can be applied in their 'general form' to the rule base of any system, improved system performance may be achieved if other genetic operators are specifically tailored for a given knowledge representation. With this end in mind, the genetic operators made available in SAGA have been thus tailored.

In SAGA, genetic operators can be applied: at the end of an episode when its result is known, and during an episode when the occurrence of a special event acts as a trigger.

4.1 Triggered Operators

4.1.1 NO_MATCH
During the course of an episode, several different situations can occur wherein it may be profitable to invoke a genetic operator. One such situation is failing to find a matching rule at the start of a cycle. In this case, SAGA employs an operator, NO_MATCH, to create a new rule from the best of the candidate rules which are partially matched. The new rule possesses the conditions of the best candidate with each extended by a minimum amount so as to achieve a match. Usually (95% of the time), this rule is assigned the actions of the best candidate rule but on the other 5% of occasions its action is set at random to that of one of the other candidate rules. The reason for the adoption of this scheme in respect of a rule's actions is as follows. When the action adopted by the new rule is always that of the best candidate, α say, there will be occasions when the action adopted is a poor one. When this is the case, and the new rule is involved in an unsuccessful episode, the rule is likely to be specialised (see later) and yet another new rule will be needed if the situation recurs. Again however, rules suggesting α are likely to be the closest matches and hence a possibly infinite loop develops where α is the only action suggested for the unmatched situation. However, by adding the small random element, this problem will be avoided.

4.1.2 CLOSING_INTERVAL
Genetic operators can also be used to advantage when two rules which advocate the same action are active simultaneously. In such circumstances, it is likely that the rules apply to a similar set of situations, and consequently they can be combined. SAGA performs the combination according to the relative strengths of the rules. If both active rules are of high strength, then SAGA generalises a single condition

common to both rules using the operator CLOSING_INTERVAL, and invests a new rule with the generalised condition. This operator uses a condition in one rule to extend the corresponding condition in the other. For example, given the active rules R1 and R2 containing only the common linear conditions (*range*, 100, 200) and (*range*, 150, 250) respectively, then the effect of applying CLOSING_INTERVAL results in the '*range condition*' of the new rule being generalised to (*range*, 100, 250).

4.1.3 OPENING_INTERVAL

If two simultaneously active rules advocate the same action and both have low strength, then it is possible that there is an element common to both rules which is causing the poor performance. The operator OPENING_INTERVAL uses this idea to specialise one of the conditions in a rule using the corresponding condition in the other rule. For example, if rules R3 and R4 contain only the following common conditions involving *speed*: (*speed*, 100, 220) and (*speed*, 175, 290) respectively, the corresponding specialised condition would be either (*speed*, 100, 175) or (*speed*, 220, 290), and this will be invested in either R3 or R4 respectively.

4.1.4 CROSSOVER

There are two types of situation wherein SAGA uses the crossover operator to generate a new rule from two parent rules. The first occurs when two high strength rules are active simultaneously. As OPENING_INTERVAL is also used in this situation, significant disruption may occur in the make-up of the rules. Consequently, SAGA employs both CROSSOVER and OPENING_INTERVAL each with a 50% probability. The second situation in which CROSSOVER is used is that in which one of the two active rules has high strength and the other low strength. Here, the resulting rule replaces rule with lower strength.

4.2 End of Episode Operators

4.2.1 Generalisation Operators

SAGA employs the result of an episode to select appropriate operators to aid in the search for improved rules. If an episode is successful (that is T1 has won), it may be profitable to try to generalise one of the rules used during that episode. With this end in mind, SAGA records one of the rules used during an episode and if the episode is successful then, with a given probability, one of the following operators is applied to it:

(a)   CONJUNCTION_TO_DISJUNCTION
       This operator generalises rules which are both very specific and contain many conditions. This it achieves by dividing the selected rule's condition set into two subsets at random and creating two new rules, each possessing one of the subsets as its condition set. The action part of the new rules is that of the original. The original rule is deleted. For example, the rule C1 C2 C3 C4 C5 : A1 might be generalised to C1 C2 : A1 and C3 C4 C5 : A1.

(b)   CONDITION_DROPPING

This operator merely discards a condition from the selected rule in order to form a rule that is slightly more general.

(c)   EXTEND_BOUNDS

The action of this operator depends on the data type of a randomly selected condition in the rule. For linear and cyclic data types, the operator selects one of the bounds in the condition at random and mutates it. In such a situation, the amount by which the bound is altered must be determined. For example, considering the condition (distance_to_target, 100, 120) with an associated valid range of say 0 to 10000, by what amount should 120 (this has been selected) be altered in respect of the mutation? Certainly one of the factors that must be taken into consideration in such cases is the range of the corresponding variable in the condition. Another factor is the current strength of the rule - if a rule has a high strength, it is likely that it has been successful in the past, and altering the bound too much could severely degrade its performance. However, if a rule has low strength, the opposite applies. Consequently, SAGA combines all these factors in a Lamarckian mutation operator taking the form of equation 1.

$$New\ Bound = Old\ Bound \pm \frac{current\ range \,.\, C \,.\, Rand}{relative\ strength\ of\ rule} \qquad (1)$$

where C is a constant and $Rand$ is a random number such that $0 \leq Rand < 1$.

The analogue of this operator for tree structured identifiers involves replacing the bound of the most specific condition by either its parent or grandparent (selected at random). For list type identifiers, the tuple is generalised by adding another argument to the list (selected at random from those not already included), whilst for identifiers with a pattern type, one of the non-# bits in the argument is mutated to a #. No form of bound extension is applied to boolean conditions as this would lead to the condition being dropped.

(d)   COUPLING

This operator forges tighter links between a rule that was active during an episode and the rule which activated it. For example, if R5 is active at time t (posting the message M) and this activates R6 at time $(t+1)$, then the argument of the internal condition of R6 is changed to M, thereby linking it more closely with R5.

(e)   CONSTANT_TO_VARIABLE

For each pair of variables of the same type in different conditions of a rule which has been active during an episode, this operator first inspects the values that the variables possessed in the world model at the time when the rule was active. If the values are found to be equal then a new rule is created which expresses this fact. For example, if the rule:

IF ... (*range_of_T1*, 15, 18) (*dist_to_target*, 16, 20) ... THEN (*Fire*, TRUE)

was active during a successful episode, and at that time, the following variable-values were observed in the world model: range_of_T1 = 17, dist_to_target = 17, then the operator CONSTANT_TO_VARIABLE would create a new rule with the conditions:

(*range_of_T1*, *dist_to_target*, *dist_to_target*) AND (*dist_to_target*, 16, 20).

4.2.2 Specialisation Operators

When a particular sequence of actions leads to an unsuccessful episode (that is T1 failed to win), this is likely to be the result of 'overgeneralisation', and therefore, specialising one or more of the rules used during the episode may be profitable. Consequently, SAGA employs the following suite of specialisation operators:

(a)   CONDITION_ADDING

This operator is used when the selected rule, chosen randomly from those used during the episode, has only a small number of conditions. The effect of CONDITION_ADDING is to introduce into the rule a new condition which satisfies the recorded environmental state, provided it is not already present in the rule.

(b)   REDUCE_BOUNDS

This operator is employed when the selected rule has a fairly large number of conditions. The operator functions by using the state of the world model at the time the rule was active to exclude an unsuccessful value from a condition. For example, for conditions involving linear and cyclic identifiers, if the rule containing the condition (*bearing*, 100, 200) is active during an unsuccessful episode when the bearing in the world model is 180, then SAGA would specialise the condition in that rule to (*bearing*, 100, 179). In the case of conditions involving tree structured identifiers, REDUCE_BOUNDS operates by changing the relevant bound to the other child of the current bound's parent. For conditions involving pattern identifiers, the operator alters the message in such a way that a rule that is active at time ($t+1$) cannot be activated if the pattern is observed in the world model at time t. Meanwhile, if the identifier is of a list type, then the value in the list which matched the world model when the rule was used, is removed. Once again, there is no analogous operation for conditions involving identifiers of type boolean.

4.2.3 Other End of Episode Operators

SAGA employs two further operators at the end of an episode, irrespective of its outcome. These are:

(a)   DELETE
        This operator is used in an attempt to rid the system of old and ineffective
        rules by removing rules if their strength has fallen below a certain value (the
        deletion threshold).

(b)   GARBAGE_COLLECTOR
        This operator inspects every condition (except the internal condition) in
        every rule and if any condition is found to be maximally general (that is, it
        cannot be generalised), then it is deleted.

5.0 The Repair Mechanism

Due to the limited size of its population (rule base), a CCS may rapidly forget
valuable experience as a result of the deletion of previously useful rules. Two
possible solutions to this problem are readily suggested. The first is to consider
using a much larger rule base. However, this severely increases the running time of
the system without guaranteeing the retention of useful information. The second is
to maintain some form of memory. Zhou [6] suggests the use of a long term
memory of inactive, well-trained rules. However, this will also attract an overhead in
terms of time, as well as creating the difficult problem of deciding whether or not a
rule has been useful.

In SAGA, a rather different approach is adopted. Based upon the observation
that, although genes undergo change in biological systems, there are also enzyme
systems which facilitate their repair, SAGA employs a *repair mechanism*. Rather
than 'repairing' only selected genes (conditions) as in biological systems, SAGA's
repair mechanism restores the whole chromosome (rule) to a previous state. It
functions by comparing the current strength of each rule with a previous best strength
which is stored along with the version of the rule which achieved that strength. If the
current strength becomes significantly higher than its previous best, the strength and
the corresponding rule are recorded. If however, a rules current strength falls
significantly below its previous best, then the rule can be repaired by supplanting the
current rule by the earlier superior version. Such an operator is of great advantage in
situations for which a generalisation significantly degrades the performance of a rule.
The operator does not however reduce the efficiency of the system because the stored
copy is not used during any of the processor intensive phases of system operation
such as matching and conflict resolution. The only cost to the system is a doubling
of storage in respect of the rules and as storage is unlikely to be a real restriction on
most modern machines, then this is not a problem. However, it should be noted that
the operations underlying the repair mechanism, unlike those corresponding to such
ideas as Directed Mutation [7], is fixed and is not influenced by the environment.

6.0 Evaluation of Performance

To investigate the effectiveness of the operators described above, a number of
experiments were undertaken, with each experiment examining a different

combination of operators. In the experiments, two slightly different 'versions of the system' were used. In the first version, the value of the distance from T1 to T2 (*distance_to_target*) was omitted from the environmental message. Therefore, T1 was effectively blind and had to rely on the state of the internal message list to decide the appropriate action to take (that is it had to learn a sequence of actions). However, in the second version, this knowledge was made available, and so T1 was able to use environmental information, as well as internal messages, to decide upon its actions.

Each experiment consisted of 40 trials of up to 200 games apiece. At the start of each trial, T1 was invested with a plan consisting of six rules. Three of these could only become active on the first move of a game (with each proposing a different one of the three possible actions); the other three being active on all subsequent moves and again advocating a different one of the three possible actions. In the course of a trial, the operators described above were used to modify the plan during and after each episode as appropriate. The success or failure of T1 was recorded at the end of each episode. If it had been successful on 10 consecutive episodes, it was assumed to have found a winning strategy, and the trial was regarded as successful. If however T1 had failed to win on any 10 consecutive episodes during a trial, then the trial was deemed a failure. To gauge the success of a specific set of operators, two values were used - the number of successful trials out of the 40 and also the average number of episodes it took to find a winning strategy (the *learning rate*).

As the system commences each trial with a very general plan, the minimum set of operators (*base operators*) that the system needs to enable it to adapt favourably are the specialisation operators (used after an unsuccessful episode) and the triggered operator NO_MATCH. Whilst not essential for learning, only for reasons of efficiency , the DELETE and GARBAGE_COLLECTOR operators were also included in the set of base operators. Using first the 'blind' version of the system and then the 'sighted', bench-mark results were recorded using only these base operators. Subsequently, various combinations of the remaining operators were tested to determine whether they could improve the performance of the system.

7.0 Results

With over 20 run-time parameters, the number of possible system configurations is very large and so the results reported here are by no means the optimal results for a CCS in SAGA. However, it is hoped that they are a fair reflection of the relative performance of each operator.

The first set of results are shown in Table 2 and relate to the 'blind' system. When only the base operators are employed, a winning strategy is found on 12 out of 40 trials. The addition of either the repair mechanism, the triggered operators or the generalisation operators lead to a very slight improvement but the most considerable increase in performance was observed with the addition of the COUPLING operator. This is what would be expected, for in such a 'blind' system, the opportunities afforded the other operators to improve the plan would be few and far between because of the lack of environmental information. However, the system can build a winning strategy by finding appropriate rule sequences, and it is here that the COUPLING operator is particularly useful. Moreover, as expected, the addition of the CONSTANT_TO_VARIABLE operator had no impact upon system performance because there is insufficient information for it to form useful relationships. When several operators were added to the system simultaneously, the results became less predictable. Here, the generalisation operators tended to perform poorly, but the best performance was observed when the repair mechanism, COUPLING, and the triggered operators were all used simultaneously.

Operators	Times Successful (out of 40)	Learning Rate
Base Operators	12	118.7
COUPLING	16	88.2
Repair Mechanism	13	92
Triggered Operators	13	116.1
Generalisation Operators	13	90.6
CONSTANT_TO_VARIABLE	12	118.7
COUPLING and Repair Mechanism	17	65.6
COUPLING, Triggered Ops and Repair Mechanism	22	98.1
COUPLING, Generalisation Ops and Repair Mechanism	18	97.2

Table 2: Results using 'blind' system

Table 3 contains the results derived from the 'sighted' system. Clearly, knowledge of the distance to the target should improve the ability of the system to learn. This was the case and in fact, the system was able to find a successful strategy on 22 out of the 40 trials using only the base operators. Again the repair mechanism, the COUPLING and triggered operators all improved system performance, however the generalisation operators degraded the performance quite significantly. As was expected, the CONSTANT_TO_VARIABLE operator performed best of all due to its ability to link together *distance_to_target* and *range* so that firing occurred whenever the target was in range. Significantly though, having more than one operator did not improve performance further and in most cases, system performance was degraded.

Operators	Times Successful (out of 40)	Learning Rate
Base Operators	22	87.6
COUPLING	28	107.6
Repair Mechanism	30	74.3
Triggered Operators	31	92.5
Generalisation Operators	15	79.3
CONSTANT_TO_VARIABLE	38	48.1
COUPLING and Repair Mechanism	22	87.6
COUPLING, Triggered Ops and Repair Mechanism	29	89.1
COUPLING, Generalisation Ops and Repair Mechanism	15	81

Table 3: Results using 'sighted' system

8.0 Conclusions

In this paper, problems concerning (a) the unsuitability of 'simple' crossover and mutation for symbolic representations, and (b) forgetting previously useful rules, have been addressed. In respect of the first problem, a number of operators, especially tailored for use with high-level languages, have been investigated. These include Lamarckian-type operators which utilize environmental information to facilitate the rule discovery process, and triggered operators which use the occurrence of special events to select appropriate times to invoke specific operators.

With regard to the operators employed in SAGA, the following initial conclusions were reached:

- for situations in which the environment provides insufficient data, the COUPLING operator proved particularly useful;

- the CONSTANT_TO_VARIABLE operator is necessary to exploit common environmental features;

- the addition of the triggered operators (CLOSING_INTERVAL, OPENING_INTERVAL and CROSSOVER) improved system performance, especially when more environmental data was available;

- the generalisation operators (CONJUNCTION_TO_DISJUNCTION, CONDITION_DROPPING, and EXTEND_BOUNDS) may degrade performance by over-generalising successful rules.

Moreover, in order to alleviate the problem of forgetfulness, it was found that the inclusion of a repair mechanism improved the learning ability of the system for a range of different operator combinations. Detailed research of this particular mechanism is on-going and it is hoped to report in more depth on its capabilities in due course.

Acknowledgment

This research was performed under the CASE studentship scheme (SERC Award Ref. No. 9030996X) in co-operation with the Defence Research Agency, Portland.

References

[1]   Michalski, R.S. *A Theory and Methodology of Inductive Learning*, Artificial Intelligence, 20, 1983.

[2]   Holland, J.H, Holyoak, K.J, Nisbett, R.E, and Thagard, P.R. '*Induction: Processes of Inference, Learning and Discovery*' Cambridge: MIT Press, 1986.

[3]   Maynard Smith, J. *Evolutionary Genetics*, Oxford University Press, 1989.

[4]   Holland, J.H. *Properties of the Bucket Brigade Algorithm*, Proc. 1st Int. Conf. on Genetic Algorithms, 1985.

[5]   Grefenstette, J.J. *Credit Assignment in Rule Discovery Systems Based on Genetic Algorithms*, Machine Learning, 3, Kluwer Academic Publishers, 1988.

[6]   Zhou, H.H. *CSM: A Genetic Classifier System with Memory for Learning by Analogy*, PhD Thesis, Dept. of Computer Science, Vanderbilt University, Nashville, TN, 1987.

[7]   Paton, R.C. *Some Perspectives on Adaptation and Environment*, Internal Working Paper, Dept. of Computer Science, University of Liverpool, 1992.

Adaptive Learning of a Robot Arm

Mukesh J. Patel[*] and Marco Dorigo[*,#]

[*] *Progetto di Intelligenza Artificiale e Robotica, Dipartimento di Elettronica e Informazione, Politecnico di Milano, Piazza Leonardo da Vinci 32, 20133 Milano, Italy. patel@elet.polimi.it, dorigo@elet.polimi.it*

[#] *IRIDIA, Université Libre de Bruxelles, Avenue Franklin Roosevelt 50, CP 194/6, 1050 Bruxelles, Belgium, mdorigo@ulb.ac.be.*

ABSTRACT

ALECSYS, an implementation of a learning classifier system (LCS) on a net of transputers was utilised to train a robot arm to solve a light approaching task. This task, as well as more complicated ones, has already been learnt by ALECSYS implemented on AutonoMouse, a small autonomous robot. The main difference between the present and previous applications are, one, the robot arm has asymmetric constraints on its effectors, and two, given its higher number of internal degrees of freedom and its non anthropomorphic shape, it was not obvious, as it was with the AutonoMouse, where to place the visual sensors and what sort of proprioceptive (the angular position of the arm joints) information to provide to support learning. We report results of a number of exploratory simulations of the robot arm's relative success in learning to perform the light approaching task with a number of combinations of visual and proprioceptive sensors. On the bases of results of such trials it was possible to derive a near optimum combination of sensors which is now being implemented on a real robot arm (an IBM 7547 with a SCARA geometry). Finally, the implications these findings, particularly with reference to LCS based evolutionary approach to learning, are discussed.

Introduction

Learning classifier systems (LCS's) have been implemented in autonomous agents and machine learning simulations to solve a variety of complicated problems or learn new behaviours. A classifier system is particularly useful in rule based problem solving tasks, and, combined with an evolutionary approach to generate new rules it is a powerful search technique (Booker, Goldberg, Holland, 1989). Here we report a fairly novel implementation of an LCS on an industrial robot (IBM 7547 with a SCARA geometry). Briefly, the robot arm had to learn to detect and approach a light source. The main motivation for this research was to generalise the application of a specific version of LCS - called, ALECSYS - to that of control of a broader class of autonomous agents which had been trained to solve these sorts of tasks (Dorigo, 1994).

We report results of a number of preliminary exploratory simulations of the robot arm's relative success in learning to perform this task. The main interest lies in illustrating the nature of an evolutionary approach to adaptive learning in a fairly practical engineering domain. In particular, this paper focuses attention on issues related to the actual use of ALECSYS to learn the control system of the robot arm. Given the enormous difference between a small autonomous agent and an industrial robot arm this was not a straightforward task. For instance, even something as simple as the number and location of visual sensors, which on the AutonoMouse was a matter of attaching a pair on the "head," was not obvious. Nor was it clear how to provide the LCS information about the effectors to enable it to learn to solve the problem by taking a series of appropriate actions. In order to determine the best combination, that would support the solution of this task with a fair degree of success and within a reasonable period of time, a number of simulations were carried out. On the basis of results of such trials it was possible to derive a near optimum combination which is now being implemented on a real robot arm with the appropriate combination of sensory-motor capabilities. The outcome, as will become clear in the rest of this paper, is an implementation that is certainly not intuitive at first glance. Further, a selection of results of simulations are presented and evaluated, and some of their implications relating to learning and problem solution as a product (or outcome) of the agent-environment interaction are also discussed.

Evolutionary approach to adaptive learning

In this paper we focus on two related issues. First, the most important and relevant in the domain of evolutionary computation, is to see if the learning algorithm implemented on an autonomous agent - AutonoMouse - is generalisable to different sorts of learning agents. There is a growing research activity (see Meyer & Wilson, 1991; Varela & Bourgine, 1992; Meyer, Roitblat & Wilson, 1993; Fogel, 1994) that illustrates the usefulness of evolutionary computation to support adaptive learning and solve complex problems. In the case of adaptive learning, evolutionary computation techniques enhance the possibility of deriving solutions (or rules in the case of LCS) which collectively can enable the autonomous agent/system to function adequately in a fluctuating, uncertain and overall unpredictable (noisy) environment (Patel 1994a; 1994b) In other words, evolutionary computation supports learning (or task solution) that needs to have a high level of flexibility or adaptivity. However, much of this work remains confined to implementations (simulations or otherwise) of one type or sort of learning agent. Indeed, usually it is not even implemented on more than one specific (real) robot, assuming that the research work goes beyond the simulation stage. It is our contention that evolutionary or adaptive learning algorithms (i.e., GA's or LCS's or NN's etc.) should be generalisable to different task domains *and* autonomous (artificial) agents/systems. Otherwise they are little more than interesting models of specific types of learning confined to individual implementations and therefore of limited value for a more general insight into the process of adaptive learning as mediated by evolutionary computational techniques.

The other point of interest is to investigate the nature of the sensory information that would be necessary to support adaptive learning of an industrial robot. The use of an evolutionary computational approach to train a robot arm to learn certain sorts of goal seeking behaviour is not common, as opposed to a plethora of self-learning or adapting autonomous agents (simulated and real). Davidor's (1991) research on robot arm control bears some similarities to the present study, though apart from using GA's (as opposed to LCS's), it focuses on a very different sort of optimisation problem, and not directly concerned with the effects of the type, nature and extent of environmental information (via sensors and effectors) on learning. Also our emphasis is different since, *that* the robot arm would (eventually) be able to learn to carry out the task was not of primary issue. Of more immediate concern was determining the nature and the extent of environmental information that is necessary and sufficient for a robot arm to learn the behaviour. Ideally, of course, this should not be radically different from that provided to the AutonoMouse to learn the light seeking behaviour. On the other hand, given the totally different effectors and highly constrained space of potential movement of the arm a number of adjustments had to be made.

ALECSYS and related implementation issues and details

Findings of previous related research have clearly shown that ALECSYS (Dorigo, 1994), is particularly good at learning a set of fairly complicated behaviours by the AutonoMouse. This was the case for both computer simulations and real agent-environment interaction (Dorigo and Colombetti, 1994). One of the learning tasks that was particularly well supported by ALECSYS was that of detecting and approaching a light source (Dorigo and Schnepf, 1993; Colombetti and Dorigo, 1992). The rules of the trained LCS could carry out the task independent of the relative positions of the agent and the light and could even track a moving light source. In the results presented here, ALECSYS was implemented on a simulation of an industrial robot arm which was trained (in the same way as the zoomorphic agent of earlier studies) to detect and approach a randomly placed (but within the practical limit of the potential reach of the arm) light source. Of course, the transfer from a zoomorphic agent to a robot arm with a highly constrained range and extent of possible movements (which described a fairly eccentric non-symmetrical range of arm movement space as shown in Figure 1) deeply affected the nature of the solution, as will become clear in the results reported here. However, the range of the arm's movements closely resembled that of the real robot arm, which covered an arc of about 200 degrees (all the shaded area in the Figure 1).

While the use of ALECSYS to support adaptive (reinforcement) learning was relatively straightforward, determining other aspects such as the number and position of sensors or information about the resulting changes in the environment as a result of actions (via effectors) was a lot more complicated. In practice it is not at all clear how this might be done. Basically, the AutonoMouse had very simple sensory devices which could sense the direction and presence (or absence) of light source - the information that it got from the environment. It, however, got no direct feedback from its effectors (wheels) but often this information was implicit in the environmental input since if the direction of a stationary light source changed after an action then it must have been due to the agent turning in a particular direction. Initially it was decided to replicate as much of the zoomorphic agent's sensory perceptors as possible.

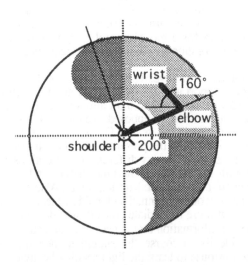

Figure 1. *The simulated mechanical arm.*

At the outset it was obvious that the robot arm would need to have visual capabilities to detect the presence and direction of the light source. This can be achieved simply by providing information about the environment in terms of visual sectors, that is, the presence of the light source is indicated as being in a specific part of the visual field. For example, the simulated visual sensor, with a four sector division of the 360 degree visual field (illustrated in Figure 2), indicates the presence of the light source in one of the four sectors; thus, a binary code indicates both the presence and the sector (and therefore the direction) of the light source simultaneously. During the initial stages it was decided to use two visual sensors located on the "wrist" and on the "elbow" of the arm. One reason for this was to maintain as much similarity as possible with the complexity of the task learned by the AutonoMouse. A much easier solution would have been to locate them both on a stationary part of the shoulder but this would have effectively rendered the learning problem trivial and it would have required very different sorts of sensors. Alternatively the reason for locating at least one sensor on the wrist was to avoid complicating the task too much, as without it the arm would have had to learn the position of its wrist relative to its sensor position *and* to manipulate it in order to direct it towards the light, which without any direction information about the distance between the target and the wrist would have rendered it a far more difficult task.

The issue about providing information on the angular extent of the elbow and shoulder joints was even more complicated. From a cursory analysis of possible arm positions relative to successful goal-seeking behaviour it is immediately obvious that the arm's different absolute degrees of angular extent at each joint can result in the same arm actions - that is, the arm can move in more than one way to "solve" a local problem. It seemed therefore that without some information about the relative position of these joints - referred to as proprioceptive information - it would be a lot more difficult for the robot arm to learn to solve the problem, and perhaps not as efficiently. Further, as already mentioned, the AutonoMouse did have implicit information about the change in the environment as a result of its preceding action. So it was decided to provide proprioceptive information about the angular position of the shoulder and elbow joints. This information however was minimal and coarsely defined. Each joint was divided into no more than four sectors and the arm position was encoded according to the sector in which the extended angle of the joint fell. However, as will be seen, this sort of information was not consistently useful for the experimental learning task.

After a number of experimental learning trials it became fairly clear that the above combination of the number and location of sensors, the sectional division of the visual field and the nature of the information about the angle of the effectors though seemingly optimal (based on AutonoMouse findings) was not ideal for the task that the robot arm had to learn to solve. As the results presented here illustrate, while in a number of trials the arm did learn to produce appropriate goal seeking behaviours, its performance was highly erratic across (and even at times, between) trials. Such partial or *fragile* learning reflects the fact that the learning system has not explored all the relevant search space and has therefore failed to learn a robust solution. However, running more learning cycles per trial did not radically improve matters.

It seemed that the real reason for the learning difficulty had a lot more to do with a fairly complicated interaction between the total space in which the arm could move and the nature of the sensory information (from the environment). Obvious aspects of this interaction were explored and provided the following pointers. Initially it seemed clear that for small number of arm positions vis-a-vis the target, there were limited opportunities for the learning; these scenarios did not (and could not, without any bias) occur with sufficient frequency to support adaptive reinforcement learning. More interestingly, it became clear that this problem would be minimised if the visual and proprioceptive information is reduced. Obviously this would follow since the search space would be reduced but there is an added reason for this outcome, which is an overall reduction in potentially ambiguous situations that need to be learned about. One example of such ambiguity is a consequence of the constraint on the arm movements such that there are situations in which the *same* kind of visual information about the target would require *different* optimal actions. Such problems are not, however, insurmountable since rules that are more sensitive to proprioceptive information (that have to be different for the two cases) can resolve this problem. However, it was not always possible to rely on the proprioceptive information in order to learn to resolve the ambiguity, since this information itself can lead to a different sort of potentially confusing situations during learning. For example, as a result of certain movements (changes in the elbow and shoulder angles), the visual information would remain unchanged from what is was previously, which effectively precluded any learning based on that stimuli. Overall, it is difficult to describe in detail or systematically the combined effect of these sorts of confusing, ambiguous and atypical situations that can occur during learning as there is much interaction between them and not all possible combinations have the same effect on learning, which can further vary between trials. This is part of the nature of open-ended adaptive learning and is a direct result of the complexity of the problem (or the large search space in which more than one solution is possible, that is, more than one different set of learned rules can in theory accomplish the task or solve problems such as ours).

However, on the basis of very general attempts at comprehending the reason for the relative lack of success in learning, it certainly became clear that not only was there no need for *all* the input information but that at times it was counterproductive as it served to confuse and complicate learning. Sensory-motor information from all the perceptors and effectors not only increased the possible solution search space which slowed down the rate of learning but also seemed to provide environmental information (that is, information based on immediately preceding action) that was potentially confusing. And on the basis of the above description of some of the problems, it could even be regarded as noise which hindered learning. Alternatively the combined information from all sensors could be also be regarded as too rich or detailed that required learning of fine discrimination between the separate environmental inputs to each visual and proprioceptive sensor. In general, in the field of self-learning autonomous agents and complex adaptive systems it would seem odd to regard environmental information as being too rich but in this instance it makes a great deal of sense as this outcome is a direct consequence of the *asymmetrically constrained* range of movements of the robot arm. However, we will return to this issue in the Discussion section of this paper.

In light of these sorts of problems encountered at the early stage we varied the number of visual sensors and tried out a greater variety of their sector divisions. Further we also carried out trials in which the angle position information was varied. However, neither all possible combinations and permutations were tried nor were the ones that were explored done so exhaustively. Apart from the limitations of time and resources, our increasing understanding of the unusual interaction between the robot movement constraints and the experimental task enabled us to be more selective and directed in our search for the minimally best solution. Hence less promising approaches were abandoned at a much earlier stage, though results reported are all based on extensive trials (further details in the next Section).

As will become clear after the presentation of the results, the best combination does not bear very close relation to the AutonoMouse, but this is by no means a negative finding for two reasons. First, at the non-implementational abstract level the two learning systems are solving very similar tasks and on the whole doing so with very similar sorts of information from the environment. The major differences, such as the number of sectors in the visual field and the proprioceptive information or input were necessary given the different nature and capabilities of the two learning (or control) systems. The second reason, and a far more interesting one, is that this study serves to illustrate how the *usefulness* of information from the environment depends on the nature of the task *and* the nature of the sensory-motor ability of the task solving agent. While this may sound trivially true, research findings presented here highlight the basic and fundamental nature of this relationship which is all too often seemingly overlooked as a result of a limited focus on zoomorphic agents coupled with anthropomorphic problem solving or task learning. Our experimental methodology and findings presented in the next two sections, apart from raising interesting theoretical issues, also have engineering and practical significance. We will touch upon both these aspects in the final, Discussion section.

Learning Trials' Method and Procedure

The results reported here are based on simulated trials, though work has now moved on to the training of a real robot arm. In order to constrain the level of complexity of the learning task, and to ensure that it did not depart too radically from that of the previous studies involving an autonomous agent, the simulation was carried out in a two-dimensional space.

ALECSYS is a tool for building hierarchies of learning classifier systems (LCSs). In ALECSYS an LCS is composed of a set of rules (its knowledge-base) which are used by a kind of parallel inferential engine to control the learning system of the robot. An LCS can be viewed as a set of three interacting systems (see Figure 2): the performance system, the rule-discovery system (genetic algorithm), and the credit apportionment system (bucket brigade algorithm). In ALECSYS all three are parallelized: A basic execution cycle of the sequential LCS can be conceptualised as an interaction between two data structures, the list of messages (ML) and the set of classifiers (CF) which are executed into two concurrent processes - MLprocess and CFprocess, which itself can be further split into separate CFprocesses. The GA can applied to each CFprocess, which is effectively a parallel GAprocess. This sort of parallelization results in gain in efficiency in learning (Dorigo 1992, 1994).

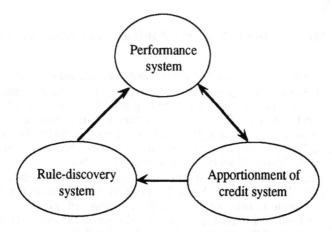

Figure 2. *Functional organization of LCS.*

In ALECSYS rules, (classifiers) are constrained to a fixed format with two conditions, connected by a logical AND operator, and one action. Conditions are strings on $\{0,1,\#\}^k$, where # is the don't care symbol, and actions are strings on $\{0,1\}^k$. The length of all strings, k, is determined by the length necessary for the description of the longest individual string for encoding the sensory input and the motor output. (Dorigo 1994; Dorigo and Colombetti 1994)

LCS's in ALECSYS learn by two algorithms, the bucket brigade (Holland, 1980) and the genetic algorithm (Holland, 1975); the latter is implemented in a slightly different way. In classic implementations of CSs, the genetic algorithm (i.e., reproduction, crossover and mutation) is called every N cycles with N a constant whose optimal value is experimentally determined. A better solution is to call the genetic algorithm when the bucket brigade has reached a steady state. In this way the genetic algorithm is called exactly when the strength of every classifier correctly reflects its usefulness to the system. The GA has also a modified mutation operator (mutespec) designed to reduce variance in the reinforcement received by default rules. Basically, this operator replaces "don't care" (#) with either 1 or 0 in rules. Hence, it effectively reduces the general nature of rules which otherwise fire in many different situations (as their conditions would match a large number of environmental inputs) with varying results. Thus, mutespec's application depends on a rule's strength variance relative to average variance of all classifiers (for further details see, Dorigo, 1993). The number of active rules can dynamically change during runtime (i.e., the classifier set cardinality can shrink) as rules of low utility are identified and discarded. When rule's strength drops to very low values its probability of winning a competition declines and even when it wins, its proposed action is unlikely to be appropriate. So in order to reduce computational time devoted to matching of rules with messages such low utility rules are not included in further learning. This enables an ALECSYS's LCS to perform a greater number of cycles (a cycle goes from one sensing action to the next one) than a standard LCS in the same time period leading to a quicker convergence to a steady state and therefore a higher frequency of genetic algorithm application, increasing the overall possibility of testing more rules in the same time (Dorigo, 1993).

We used a monolithic architecture, that is, we used a single LCS, split into three CFprocesses with 90 rules each. ALECSYS was set to its stimulus-response mode (that is, no internal messages were used). The GA was applied at steady-state with a crossover rate $p_c=0.5$, a mutation rate $p_m=0.25$, and the *mutespec* operator probability was set to 0.5.

The interface between the mechanical arm and the simulated environment

The mechanical arm was provided with the following sensors and effectors.
- *Visual sensors*. We experimented with many different combinations of visual sensors. These can be classified according to (i) the number and positioning of sensors on the arm, (ii) the number of sectors in which visual range is divided, as shown in Figure 3a, and (iii) whether the sensors divide the visual range into equal sectors or not, as shown in Figure 3b. We report and compare results of a relevant subset of all possible combinations .
- *Proprioceptive sensors* give information about the position of the two sections of the arm. The first sensor detects the position of the forearm, that is, whether the elbow angle is between 0° and 80° or within 81° and 160° (160° being the upper limit - as indicated in Fig.1). The second sensor provided similar information about the extent of the shoulder joint, that is, whether the position of the upper arm is between 0-90, 91-180 or 181-200 degrees (200 degrees being the upper limit).

- *Motor effectors.* The elbow and shoulder joints can *rotate right, rotate left* or *stay still.* These three movements are coded into two (one for each joint) 2-bit messages. For each joint bits 00 and 01 code taking no action, that is the first bit 0 indicates stay still, and conversely, 10 and 11 code movement to the right (10) and to the left (11). A proposed action which is not possible due to arm movement constraints is not performed.

Obviously the number of bits necessary for sensory messages varies according to the amount of information. For example, the information for the eight sector visual sensor (Fig. 3a) requires 3 bits while only 2 bits are necessary for a four sector visual sensor (Fig. 3b). The use of two visual sensors instead of one doubles the number of bits necessary to represent the input visual information. Similarly, proprioceptive information about shoulder and/or elbow angular position affects the size of the search space

We compare the use of a single visual sensor on the wrist with the use of two visual sensors, one on the wrist and the other on the elbow. For this two configurations we report experiments with sensors dividing the space into eight equal sectors, or four unequal sectors. The resulting combinations are compared with either information about both joint angles or the elbow alone. In a second set of experiments the best configuration (one visual sensor on the wrist with an eight sector division and one proprioceptive sensor on the elbow) was further investigated by varying the number of visual sectors - 4, 8 and 16 equal sectors.

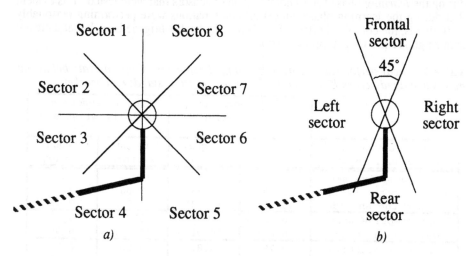

a)   *b)*

Figure 3. *The visual sensors.*
a) The visual field divided into eight equal sectors.
b) The visual field divided into four unequal sectors.

The reinforcement program

The reinforcement program (RP) compares at every time step the distance of the arm wrist from the light source with the distance as it was measured at the preceding time step. If this distant decrease the learning system is rewarded (+20), and otherwise punished (-20) if the distance increases or does not change. The reward and punishment values were experimentally determined to be appropriate for learning the task.

The training procedure

For each combination of number of visual and proprioceptive sensors we ran an experiment with 12 trials. In each trial the learning phase during which the arm is trained to learn the task of approaching the light source consisted of 20,000 cycles, followed by a test session of 5,000 cycles during which no further learning took place. To constrain the problem complexity, the target light source is placed in a small region of total space in which the arm could move. This is shown as the lightly shaded area in Figure 1. We collected data on *local* and *global* performance during the learning and testing phases. Local data refers to the number of (good) moves that resulted in the arm (wrist) being closer to the light source. Global data indicates the frequency of successful intercepts of the light by the arm (wrist "touching").

Results

Local and global performance data presented here are averages of performance over 12 trials. Each Table illustrates the relevant data for all experimental combinations of varying number and location of visual and proprioceptive sensors, and number of visual field sectors. Table 1 presents the local performance data, that is, the proportion of all moves that result in the wrist of the robot getting closer to the light target, during the *learning* phase for all combination of sensors that were tested. It is evident that during the learning phase almost all combinations were performing reasonably well; that is, the combined effect of GA's and the bucket brigade is such that there are enough good rules for appropriate action.

Table 1. *Local performance during learning: Proportion of appropriate behaviour moves averaged across 12 trials (20,000 cycles per trial; std. dev. in brackets).*

Type of division of the visual input:	Visual sensor only on the wrist		Visual sensors on both the wrist and the elbow	
	4 unequal sectors	8 equal sectors	4 unequal sectors	8 equal sectors
Proprioceptive sensors on both the elbow and the shoulder	0.82 (0.01)	0.84 (0.02)	0.86 (0.01)	0.78 (0.03)
Proprioceptive sensor only on the elbow	0.82 (0.01)	0.88 (0.01)	0.87 (0.02)	0.77 (0.06)
No proprioceptive sensors	0.78 (0.02)	0.87 (0.02)	0.90 (0.01)	0.84 (0.02)

However, during the test phase, when both the GA and the bucket brigade algorithm are switched off, performance shows a marked deterioration for more than one combination as illustrated in Table 2. In fact, apart from the combination of one proprioceptive sensor on the elbow together with one visual sensor in the wrist with 8 equal sectors (referred to as *8WE* hereafter, and marked in the appropriate cell in Table 2), all other combinations' local performance is poorer than during the training phase. Non-parametric statistical tests confirmed that the overall difference in mean test performance is significant (Kruskal-Wallis $p<10^{-4}$) as might be expected given the large observed variation. Further, 8WE's test performance when compared with all other combinations is highly significant (Mann-Whitney, $p<10^{-3}$, for all eleven pairwise comparisons).

Table 2. *Local performance during testing: Proportion of appropriate behaviour moves averaged across 12 trials (5,000 cycles per trial; std. dev. in brackets).*

Type of division of the visual input:	Visual sensor only on the wrist		Visual sensors on both the wrist and the elbow	
	4 unequal sectors	8 equal sectors	4 unequal sectors	8 equal sectors
Proprioceptive sensors on both the elbow and the shoulder	0.22 (0.27)	0.49 (0.23)	0.29 (0.16)	0.21 (0.18)
Proprioceptive sensor on the elbow	0.43 (0.28)	0.91 (0.02) 8WE	0.52 (0.28)	0.31 (0.22)
No proprioceptive sensors	0.51 (0.28)	0.68 (0.31)	0.67 (0.32)	0.40 (0.28)

These results, particularly those in combinations with two proprioceptive sensor and two visual sensors (the first row of Table 2), suggests partial learning of the task. At the end of 20,000 cycles appropriate rules had not been tested for long enough to determine optimal action during testing. One reason for this is that the system may only have learnt a small set of rules for appropriate action in certain situations but lacks rules for other situations - that is, rules covering some particular input combinations are missing because they have not been learnt (during the learning phase). Or it could be that while the good rules may be present they would be not selected for application since they have comparatively low strength; only those with high strength but inappropriate ones are tried. During learning this is not so obvious as operation of the genetic algorithm and the bucket brigade algorithm ensures to a certain extent that an appropriate rule would be generated or acquire sufficient strength. So one way of interpreting this difference is to treat the test indices as an indicator of the relative completeness of the learning.

This problem could be partially rectified if the bucket brigade algorithm is not switched off during testing, which therefore suggests that with longer learning phases good rules with relatively high strength would have a greater possibility of emerging. However, in our case increasing the learning phase (up to 50,000 cycles) did not result in an appreciable improvement during testing. Though it can be argued that even longer learning cycles, proportional to the corresponding extent of search space vis-a-vis the amount of input sensor information, may have led to improved test performance. While theoretically this proposition is valid some of the results presented here suggests the this relationship is not always so clear cut. For example, results in the first column of Table 2 are worse then those in the second column, even though the former combination have a smaller search space. We do not propose to analyse this sorts of observation in detail, since our immediate aim is to explain the complex nature of the learning task and the interaction between different sources of sensory information in this experimental task.

A similar trend is evident for the global test performance data. The learning phase global data is not presented since the general trend evident in Table 1 is replicated, that is, during the learning phase all combinations performed more or less equally well. The histogram in Figure 4 illustrates the average number of arm actions (cycles) per intercept, that is, the number of moves necessary for touching the target for each of the all 12 different combinations. Generally, the global test performance of various combinations resembles the trend observed for local performance (Table 2) which is also similarly statistically significant (Kruskal-Wallis $p<10^{-4}$), except that this data serves to highlight the significantly greater overall difference in performance between combination 8WE and the rest (for all eleven pairwise comparison Mann-Whitney test is significant, $p<10^{-3}$). Note that while the combination 8WE needs on average 14 moves to reach the target the next best combination (8 equal sector *without* any proprioceptive sensors) needs more than twice that, 38, moves (a significant difference that is not immediately obvious in the histogram)..

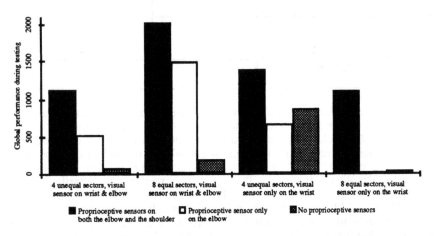

Figure 4. *Global performance index during test session: Proportion of rewarded behaviour moves averaged across 12 trials. 5,000 cycles per trial. Lower bars are better.*

Overall, it is worth noting how a reduction in amount of proprioceptive sensor information from the elbow and shoulder joints results in improved local and global performance (for both visual sensors conditions) with only combination 8WE bucking this trend. At first glance this inverse relationship seems counterintuitive though it makes sense given that more information about joint positions increases the size of the search space of the learning task. As the learning cycle length remained constant for all combinations this trend is to be expected. As already noted in a different context above, increasing the length of the learning phase (to 50,000 cycles) resulted in no significant improvement in local and global performance and certainly nothing that approached 8WE's performance based on a 20,000 cycle learning phase.

Though, there is another possible explanation that this trend reflects the drop in the number of ambiguous situations that the arm had to learn to resolve - in specific situations the same wrist position vis-a-vis the light position required a different action and having the information about the angle position of the joints added to the total complexity of the solution. However, without systematic analysis of the solution space (an unreasonable if not impossible requirement) it is difficult to be certain about this interpretation, partly because the test performance for all, except 8WE, combinations varied a good deal between trials; in a typical experiment of 12 trials it was not unusual for one trial to display high proportion of good behaviour while another to be totally abysmal (in the test phase local performance less than 0.01 were not infrequent). In this respect it is not quite clear why 8WE performs (locally and globally) so consistently well across all trials. It seems to suggest that while for most other combinations performance was negatively affected by a strong interaction between an increase in search space and potentially confusing information, that of 8WE had just the right balance of sensors *with respect* to the size of search space, the length of the learning phase and the sensors. However, a series of more detailed analyses of the rules in each trial would be necessary before we this conjecture can be confirmed.

Whatever the actual reason, it was quite clear that overall the more information about arm joints the less robust the learning, and that for the purpose of out experimental task the ideal combination seemed to be 8WE (elbow sensor combined with one visual sensor in the wrist with 8 sector field). So further exploratory experiments were confined to exploring the effect of changed in the number of sectors in the wrist visual sensor. The main reason for this was to check if increasing the sectors from 8 to 16 would increase the quality of learning. And for the sack of completeness results of trials with a sensor with 4 *equal* sectors are also presented here.
Table 3 illustrates results of local and global performance of the test phases of each experiments (of 12 trails of 20,000 cycle learning phase and 5,000 cycle test phase, except for the 16 sector condition in which the learning cycle was increased to 40,000).

For both sets of means the overall differences are highly significant (Kruskal-Wallis, $p < 10^{-2}$) which reflect the consistently better overall performance of 8WE combination. The pairwise comparison of difference in local and global performance between best combination, 8WE, and the other two conditions are also highly significant (for all four pairwise comparisons Mann-Whitney, $p < 10^{-2}$). While the performance of the 4 equal sector condition was an improvement on the previous 4 *unequal* sector condition (Table 2 and Figure 4) especially so in global performance (an improvement from 670 to 44 average moves per intercept), it still remained significantly worse than that for 8WE, at on average 14 intercepts per intercept. Increasing the number of sectors in the visual field to 16 does not result in any improvement over 8WE combination.

Table 3. *Local and global performance during testing averaged across 12 trials (5,000 cycles per trial; std. dev. in brackets).*

Type of division of the visual input:	Visual sensor only on the wrist		
	4 equal sectors	8 equal sectors	16 equal sectors
Proprioceptive sensor only on the elbow (local performance)	0.59 (0.34)	0.91 (0.02) 8WE	0.64 (0.33)
Proprioceptive sensor only on the elbow (global performance)	65 (92)	14 (1) 8WE	44 (52)

Obviously, increasing the number of sectors from 8 to 16 does not result in an improvement similar to doubling the number of sectors from 4 unequal to 8 equal even when the former is accompanied with a doubling of learning phase. On the basis of these finding it seems that increasing or decreasing the number of visual field sectors from 8 does not result in any major improvement in learning and therefore test performance. This further seems to support the idea that for this particular learning task by a robot arm (constrained in its range of movement) visual information about direction of light relative to the wrist position and information about the elbow angle position is sufficient (and perhaps even necessary). Hence, it was decided to implement the best combination, an eight sector visual field sensor on the wrist and a elbow proprioceptive sensor, on a real robot arm.

Discussion and Conclusion

The main finding seems to suggest that the experimental task is solved a lot more efficiently when sensory information is *reduced*. This is particularly so for proprioceptive sensor information. The obvious reason for this is that less information translates into shorter bit strings which effectively reduces the problem search space. So, if the length of learning cycle remains constant than performance should improve. Such redundancy in information increased the level of difficulty of the solution. The problem was made tougher because rules had to be far more sensitive to more subtle differences in information from the environment, which, as is evident for the purposes of our experimental task learning was not necessary. However, since the improvement in performance varied between trials, this can only be a partial explanation. In particular, it was quite clear that information about the shoulder joint was not helping matters at all. One of the reasons being that the angle position was not fine grained enough to be of much help. Most of the time knowledge about position of the upper arm in a quadrant was of little help in determining appropriate moves; it was simply noise that had to be learnt to be ignored. Further this sort of information actively led to an increase in ambiguity especially when coupled with information about the elbow joint. It gave rise to situations whereby the same information required different actions and *vice versa.*. And a third reason was that shoulder position information was particularly counter-productive when the arm was extend beyond 180° (i.e., in the third quadrant 181°-200°), as it contributed to the overall difficulties that the arm had in learning appropriate moves when in this position.

To a much lesser degree, information about the elbow joint was similarly a hindrance to learning the task because it too could lead to an increase in ambiguity but as is clear from the findings that in tandem with appropriate visual information it actually resulted in a very significant improvement in performance. Overall the confusion due to elbow position information is a lot less because unlike the shoulder joint it is not fixed. Being relative to the upper arm a task solution based on that information is not affected by inconsistency between elbow information and appropriate move, which is the case for shoulder information. Further in light of the fact that LCS rules would learn to correlate this information with a move towards the light (therefore information from the visual sensors) then the information is particularly useful. Tentatively this may partly explain the best performing combination. It is likely that respective information about the elbow position and the direction of light source relative to the wrist ideally supported learning of rules based on the consistent relationship underlying these two information sources. Of course, further research would be necessary in order to be able to confirm this conjecture. In particular it would be interesting to compare performance with an arm with a visual sensor on the elbow joint together with proprioceptive elbow information.

Of course, while some of the above raised ambiguity and noise and therefore an increase in the complexity of the task solution, none of it rendered the problem insoluble. Also in practice this effect was not consistent across trials; in some trials the robot arm learned to deal with the ambiguity and the test results were quite respectable, but when learning was incomplete then test performance was pretty bad. Usually such partial or fragile learning reflects the fact that the learning system has not explored all the relevant search space and has therefore failed to learn a robust solution. However, running more learning cycles per trials did not radically improve matters. Apart from the potential for hindrance of proprioceptive information, this inconsistency in learning was to a large extent due to the robot arm's unusual constraints on nature and range of movement. During the course of this work it became quite clear that for this experimental task (the robot arm) had to learn inherently contradictory solutions that no amount of learning of subtle distinction can overcome. Often, it was clear that the bad performance can be more or less explained by the arm getting stuck into a position where the direction of light source indicated movement that was physically unviable. This was essentially because the arm could not move in a complete circle. This meant that rules which were perfectly reasonable when the arm was towards the centre of its range were rendered useless at the extreme boundaries. And in such cases reducing the amount and source of sensory information would not have lead to an improvement in performance for *this particular* task.

The above discussion add up to no more than a partial explanation of some of the findings (not all presented here). This simply reflects the nature of these sort of studies of complex adaptive learning. There are far too many interacting variable to afford a clear picture of a precise effect of each of them on overall learning behaviour. Further, intra-experimental (sensor combinations) trial differences in performance makes it difficult to be confident about the general validity of some possible explanations. Nor, does any of the above completely account for success of the 8WE. Unlike any other combinations it does so consistently well (differences between learning and text, local and global performance were minimal). Our limited explanation of the poorer performance of other combinations provides only an implicit account. Only more detailed analyses of the learnt LCS rules and the simulated behaviour would give a precise idea of why a general robust solution of the experimental task by a highly constrained robot arm depends on a specific combination of sensors. At this stage we can given an implicit

Finally, it should be noted that while we have a pretty good solution for the simulated robot arm and task it is not equally clear that it will be an effective one for the real robot arm. For instance, the division of the visual sector field into eight sectors may prove to be too coarse grained to support actual touching/grasping of the target (light) by the wrist/palm of a robot arm. However, only actual implementation on a real robot enable us to determine such issues.

To conclude, thus, sensory information not only increased the possible solution search space but also seemed to provide environmental information that was potentially confusing or noisy as did not help in disambiguating the environment in the manner which would support successful learning of the light approaching task. It also became clear that not only was there no need for *all* the input information but that at times it was counterproductive as it served to confuse the learning system. This is an unexpected finding and highlights the relatively difficulty in comprehension and analysis of an evolutionary approach to adaptive learning. This is further affected by the interaction between the robot arm movement constraints and the task. Overall, the most interesting aspect of these findings is that they serve to illustrate the very complicated interaction between the problem task, the learning strategy, the nature (repertoire and limits of effectors) of the learning agent and the information from the environment. As we apply an evolutionary computational approach to adaptive learning in a wide variety of agents and tasks, we increasingly have to take these sort of interactions into account in the design and evaluation of learning tasks and testing the efficiency and robustness of learnt behaviour

Acknowledgements

This work has been partially supported by a ERCIM Research Fellowship funded by EC Human Capital and Mobility Programme to Mukesh J Patel, and by a NATO-CNR Advanced Fellowship to Marco Dorigo. We also thanks Marco Colombetti for the many stimulating discussions. The experiments were run by Massimo Papetti.

References

Booker L., D.E. Goldberg and J.H. Holland (1989). Classifier systems and genetic algorithms, Artificial Intelligence, 40, 1-3, 235–282.

Colombetti M. and M. Dorigo (1992). Learning to control an autonomous robot by distributed genetic algorithms. *Proceedings of From Animals To Animats, 2nd International Conference on Simulation of Adaptive Behaviour (SAB92)*, Honolulu, HI, MIT Press, 305–312.

Davidor Y. (1991). *Genetic Algorithms and Robotics: A heuristic strategy for optimization.* World Scientific: Singapore.

Dorigo M. (1992). Using transputer to increase speed and flexibility of genetics-based machine learning systems. North Holland, *Microprocessing and Microprogramming, Euromicro Journal*, 34, 147–152.

Dorigo M. (1993). Genetic and Non-Genetic Operators in ALECSYS. *Evolutionary Computation Journal*, 1, 2, 149–162.

Dorigo M. (1994). ALECSYS and the AutonoMouse: Learning to Control a Real Robot by Distributed Classifier Systems. *Machine Learning*, forthcoming.

Dorigo M. and M. Colombetti (1994). Robot Shaping: Developing Autonomous Agents through Learning. *Artificial Intelligence, in press.* Also available as *Tech. Report No.92-040*, International Computer Science Institute, Berkeley, CA.

Dorigo M. and U. Schnepf (1993). Genetics-based Machine Learning and Behaviour Based Robotics: A New Synthesis. *IEEE Transactions on Systems, Man, and Cybernetics*, 23, 1, 141-154.

Fogel D. B. (1994). *Evolutionary Computation: Toward a New Philosophy of Machine Intelligence.* IEEE Press, forthcoming.

Holland J.H. (1975). *Adaptation in natural and artificial systems.* The University of Michigan Press, Ann Arbor, Michigan.

Holland J.H. (1980). Adaptive algorithms for discovering and using general patterns in growing knowledge bases. *International Journal of Policy Analysis and Information Systems*, 4, 2, 217-240.

Meyer J-A. and S.W. Wilson (1991). *From Animals to Animats, Proceedings of the First International Conference on Simulation of Adaptive Behaviour*, Bradford Books, MIT Press.

Meyer J-A., H.L. Roitblat and S.W. Wilson (1993). *From Animals to Animats 2, Proceedings of the Second International Conference on Simulation of Adaptive Behaviour*, Bradford Books, MIT Press.

Patel M.J. (1994a). Concept formation: A complex adaptive approach. *Theoria* **20**

Patel M.J. (1994b). Situation assessment (S-A) and Adaptive Learning - Theoretical and Experimental Issues. *Proceedings of The Second International Roundtable on Abstract Intelligent Intelligence.* February, Rome, ENEA Headquarters.

Varela F.J. and P. Bourgine (1992). *Toward a Practice of Autonomous Systems, Proceedings of the First European Conference on Artificial Life*, Bradford Books, MIT Press.

Co-evolving Co-operative Populations of Rules in Learning Control Systems

Terence C. Fogarty

Faculty of Computer Studies and Mathematics
University of the West of England, Bristol, UK
tc_fogar@csd.uwe.ac.uk

Abstract. It is shown how co-evolving populations of individual rules can outperform evolving a population of complete sets of rules with the genetic algorithm in learning control systems. A rule-based control system is presented which uses only the genetic algorithm for learning individual control rules with immediate reinforcement after the firing of each rule. How this has been used for an industrial control problem is described as an example of its operation. The refinement of the system to deal with delayed reward is presented and its operation on the cart-pole balancing problem described. A comparison is made of the performance of the refined system using only selection and mutation to learn individual rules with that of the genetic algorithm to learn a complete set of rules. A comparison is also made of the performance of the refined system using only selection to learn individual rules with that of the bucket-brigade and other reinforcement algorithms on the same task.

1 Introduction

1.1 Background

Holland (1975) encapsulated, in the genetic algorithm, the elegant balance between discovery and reinforcement embodied in the process of the evolution of a natural system to provide us with a method for adapting an artificial system to achieve some specified goal. While Smith (1980) used the genetic algorithm to optimise a system specified as a complete set of rules Holland went on to use the genetic algorithm as the discovery component for individual rules within a classifier system (Holland and Reitman 1978). He introduced a second algorithm based on the paradigm of the competitive economy, the bucket-brigade (Holland 1985), as the reinforcement component. Much subsequent work on classifier systems has concentrated on the operation of the bucket brigade (Wilson 1987, Wilson and Goldberg 1989, Holland 1990, Riolo 1991) and its relation to other reinforcement learning methods such as Q-learning (Watkins 1989) and the DYNA architecture (Sutton 1991).

1.2 Motivation

In this paper one component of the genetic algorithm, selection, is isolated and it is shown how this can be used for the reinforcement of individual rules - selectionist reinforcement learning. This obviates the need for a separate reinforcement learning algorithm in rule-based control systems in cases where credit assignment is determined and clears the way for a comparison of genetic reinforcement with other reinforcement algorithms. It also prompts an analysis of the interaction between the mechanisms for reinforcement and discovery in genetic algorithms and classifier systems.

1.3 Outline

It is shown how co-evolving populations of individual rules can outperform evolving a population of complete sets of rules with the genetic algorithm in learning control systems. A rule-based control system is presented which uses only the genetic algorithm for learning individual control rules with immediate reinforcement after the firing of each rule. How this has been used for an industrial control problem is described as an example of its operation. The refinement of the system to deal with delayed reward is presented and its operation on the cart-pole balancing problem described. A comparison is made of the performance of the refined system using only selection and mutation to learn individual rules with that of the genetic algorithm to learn a complete set of rules. A comparison is also made of the performance of the refined system using only selection to learn individual rules with that of the bucket-brigade and other reinforcement algorithms on the same task.

2 Genetic Reinforcement Learning

2.1 A Rule-based Control System Using the Genetic Algorithm with Immediate Reinforcement

The system presented consists of a set of condition/action rules each of which has an associated weight, representing its cost or value, and a time, recording when it was created or produced. To begin with the system contains no rules. It receives inputs from the system to be controlled and all rules, if any, with conditions that match those inputs are activated. If the number of activated rules is below a certain size, let us say P, a new rule is created using the cover operator (Wilson 1985). This has conditions that match the inputs and an action randomly chosen from the set of allowable actions given those inputs.

If the number of activated rules is equal to P a new rule is produced from the set of activated rules with the genetic algorithm. One rule is chosen and copied as the basis for a new rule, using the weights of the activated rules as the basis of a probability distribution for the purpose of selection. With a high probability a second rule is chosen in the same way and that part of it from its beginning up to a randomly generated single point is used to replace the corresponding part of the copy of the first rule. The action of the resulting single rule is mutated at a randomly generated point with a low probability to produce the new rule. The action of the new rule, whether created with the cover operator or produced with the genetic algorithm, becomes the output of the system. The reward or punishment that is received as a result of the output action becomes the weight of the new rule - the weights of all existing rules in the system remain unaltered.

If the new rule was created with the cover operator it now becomes part of the rule-based control system. If it was produced with the genetic algorithm, how and when it becomes part of the system depends on the type of genetic algorithm used. With the standard generational genetic algorithm it must be saved until there are P saved rules with the same conditions whereupon all rules with those same conditions in the system can be replaced with the new set. An incremental genetic algorithm based on Holland and Reitman's (1987) original idea is easier to implement - the oldest activated rule is selected for deletion and this is immediately replaced with the new rule. Deletion based on weight has also been used. Finally new inputs are received and the process is repeated.

2.2 Relation to Other Systems

The main difference between the simple system described above and more complicated classifier systems is that the genetic algorithm is explicitly used for reinforcement as well as discovery. A new rule is created or produced at each interaction with the environment and its weight reflects the cost or value of its action to the system in the state defined by its conditions. Old rules are replaced when necessary so that the population of rules for a given set of conditions adapts to produce the best action for the indicated state of the system to be controlled. Actions which benefit the system in that state come to dominate the population with that set of conditions while those which do not wither away. Thus, a beneficial action for a given state can be randomly selected or discovered using crossover and mutation and then reinforced using selection. The value or cost of an action in a given state is given by the average strength of the corresponding rules while the probability of that action being taken in a given state is represented by the sum of the strengths of the corresponding rules relative to the sum of the strengths of rules with other actions for that state.

The system described knits together a number of previous ideas into a simple but powerful concept which has been demonstrated on an industrial control problem. The genetic algorithm is based largely on that outlined by Holland and Reitman (1978) except in its use. They generated two parent rules from the set controlling the same effector based on their predicted payoff and crossed them to produce a new rule which replaced one of the oldest in the population as a whole. However, they only used the genetic algorithm explicitly to do discovery - reinforcement was accomplished by altering the strength of existing rules to reflect their predicted payoff. The system described takes Booker's (1983) restricted mating to the extreme - rules with the same conditions form a distinct pre-defined species within the total population of the rule-based system that adapts to give the best action for a given state. Interaction with other species is external to the system demonstrating Brook's (1992) theory of co-evolving behaviours in robots. The strength, and weakness, of the system is that it is based completely on the paradigm of evolution - it is a simplified form of an alternative view of classifier systems which sees them as an ecology rather than an economy.

3 An Industrial Control Example

3.1 Rule-Based Optimisation of Combustion in Multiple Burner Installations

Fogarty (1988) developed, tested and used successfully a rule-based system for optimising combustion in multiple burner installations on a twelve burner zone of the 108 burner furnace of a continuous annealing line for rolled steel. When demonstrating the system on a double burner boiler in the steam generating plant of a tinplate finishing works it was found that the rule-base was not very efficient at dealing with the modulation of firing levels in response to changing levels of demand from the works. The rule-base was not built with this problem in mind since the multiple burner furnace on which it was developed only had one firing level and zones were turned on or off to satisfy requirements rather than having their firing levels modulated. The rules had to be modified manually over a three week period to work successfully (Fogarty 1989).

3.2 Learning Individual Rules with the Genetic Algorithm

A system based on the one described above has been used to control simulations (Fogarty 1990) of multiple burner installations and provides a good example of its operation. The inputs to the system are oxygen and carbon monoxide readings taken in the common flue of the burners each classified as: very low, low, o.k. or high, giving 16 possible states in the environment. The system has fixed actions when either of the readings is very high. There are 96 possible outputs from the system to the air-inlet valves controlling the air-fuel ratios of the burners. These are composed of 6 qualitatively different actions applied by any of 16 different amounts. The qualitatively different actions are: lean burner correction, rich burner correction, lean/rich burner correction, reduce air to all burners, increase air to all burners and no action. The first three of these are only applied to the current burner and attention is then switched to the next burner ready for the next action. The cost of an action is the energy loss calculated from the oxygen and carbon monoxide readings (Fogarty 1991) in the common flue after that action has been performed.

The maximum number of activated rules P is 30 with the probability of a second rule being chosen for single point crossover of 0.95 and a probability of mutation of 0.01. The rules are encoded with strings representing the conditions such as "low, high" when the oxygen reading is low and the carbon monoxide reading is high and bits representing the qualitatively different actions together with their associated amounts. The first three bits of the action are used for the qualitative part of the action with a bias of three different representations for doing nothing and the other bits are a binary encoding of the amount. The system has been run on ten different simulations of ten burner installations with two firing levels for 20,000 interactions each. In situations where the carbon monoxide reading is high there is a definite convergence of rules to the action of increasing air to all burners by about 10%. In situations where the carbon monoxide reading is very low there is a convergence of rules towards rich/lean or lean correction by various amounts or do nothing, depending upon the oxygen reading. The system is yet to be tested on a real multiple burner installation and it has some obvious limitations.

3.3 Limitations of the System

The first observation to make is that a definite action is not learned for every situation encountered. Some situations are not entered enough times in a run so that the action taken in that situation is still random at the end of the run. Other situations have populations suggesting two different actions at the end of a run which may be due to the deselection strategy of scaled proportional replacement of the worst rather than the oldest rule but is more likely to be due to the fact that there are hidden states not identified by the system. The main reason is that the state space has to be discretised by an expert rather than learnt by the system and an approach using point based classifiers with nearest neighbour matching has been proposed to overcome this (Fogarty and Huang 1992).

Secondly, the system can never learn an action that is qualitatively different from the ones it has at its disposal because the actions are sophisticated routines specified by an expert. They could be broken down into simpler actions from which new routines could be constructed but these would need to be able to make use of delayed rather than immediate reinforcement since some of them involve incurring a temporary loss in order to make a longer term gain.

4 Dealing With Delayed Reward

4.1 Extending the System to deal with Delayed Reinforcement

Holland and Reitman (1978) used an epochal credit allocation scheme in conjunction with their reinforcement learning algorithm to deal with delayed reinforcement and Grefenstette (1988) used a modified version of this in his profit sharing plan. The basic method is to keep a record of all rules that are activated after each external reward or punishment has been received until the next one is received and then update their strengths in proportion to the reward or punishment received. Thus the strengths of individual rules are altered to reflect their use to the system as a whole. In the system that uses only the genetic algorithm for reinforcement the strengths of individuals are never altered - reinforcement is done by selection of individuals to reproduce in a population because the genetic algorithm uses whole populations as the basis for reinforcement rather than any specific individuals in those populations.

To refine the system based on the genetic algorithm to deal with delayed reinforcement an epochal credit allocation scheme is also used, but in conjunction with the genetic algorithm rather than a second learning algorithm. The operation of the system is the same as before except in one respect: once a new rule is created with the cover operator or produced with the genetic algorithm and posts its action to the message list it goes into a queue until a weight has been assigned to it. When external reward or punishment comes into the system this is used to assign a weight to each rule waiting in the queue and it is only then that they are used to update the populations of the system. This method uses the genetic algorithm to do both reinforcement and discovery and uses no other method of reinforcement. Of course the question of credit assignment remains to be dealt with - are all rules in the queue assigned the same weight in proportion to the external reward or punishment or is this attenuated to the back or to the front of the queue in some way? This is largely problem dependent but we shall demonstrate that both schemes work on the cart-pole balancing problem.

5 Learning Control Rules for the Cart-pole Balancing Problem

5.1 The Cart-Pole Balancing Problem

The problem is as follows. A cart is free to move along a bounded track. A pole is hinged to the top of the cart and is free to move in the vertical plane defined by the track. Initially the cart is moved to the centre of the track with the pole more or less vertical. At any time step a force must be applied to the right or left of the cart, the object being to stop it hitting the ends of the track and to prevent the pole from falling into a position from which it cannot be raised. The only reinforcement is the number of time steps taken to fail, i.e. when the cart hits the end of the track or the pole falls over, or a success signal when it has been balanced for a desired number of time steps. After each failure the cart-pole system is re-initialised.

The system is modelled by a set of non-linear differential equations given in Anderson and Miller (1991). The state of the cart-pole system at any time step is represented by a point in 4-dimensional space of which the axes correspond to four state variables, i.e. the position of the cart on the track, the angle of the pole with the vertical, the velocity of the cart and the rate of change of the angle of the pole. A standard approach to solving this problem with a learning system is to partition the state-space into discrete states and learn whether to push the cart right or left in each of those states (Michie and Chambers 1968).

5.2 Controlling the Cart-Pole Simulation with Genetic Reinforcement Learning

Using a rule-based system to control a cart-pole simulation is simple. The input to the system at any time step is an integer indicating the state of the cart and pole as defined by the way in which the state space has been discretised. The output from the system at any time step is a 1 or a 0 indicating whether a force is to be applied to the right or left of the cart at that time step. Reward comes into the system when the cart hits the end of the track or the pole falls over, indicating failure, in the form of the number of time steps since the last failure. The simulation is run until the system is successful in balancing the cart-pole for a given number of time-steps or has failed for a certain number of trials.

The rules in the system consist of an integer condition to match the state of the cart-pole at any time and a bit action to indicate whether the cart is to be pushed to the right or left in that state. Since the genetic algorithm works only on the action part of the rule all the power of the genetic algorithm is not required in this case - it is impossible to crossover two strings each of which only contains one bit. Only selection and mutation are used and, as will be shown later, selection is the primary operator for reinforcement learning. Mutation has only been used in the first set of experiments described below to give a fair basis for comparison with using the genetic algorithm to learn the complete set of rules, it is not used in the comparison with other reinforcement learning algorithms.

5.3 The Operation of Selectionist Reinforcement in a Rule-based Control System on the Cart-Pole Simulation

The operation of the rule-based system on the cart-pole balancing problem is as follows. No rules exist in the system to begin with. The cart-pole simulation is initialised and its state is read by the rule-based system. A rule is created with its condition equal to the current state and its action randomly generated from the set {0,1}. This action is performed on the cart-pole simulation to take it to another state and the rule is put in a queue. This process is repeated until the cart-pole is no longer in balance on the track at which point reward equivalent to the amount of time the pole has been balanced comes into the system.

Each rule in the queue is then assigned a strength. In the first scheme used each rule is assigned a strength equal to the reward coming into the system for comparison with the approach of using the genetic algorithm to optimise the complete set of rules. In the second each rule is assigned a strength equal to the proportion of reward that accumulated after that rule fired, i.e. if i represents the position of the rule in the queue and r represents the external reward then rule i is assigned a strength of $r-i$, for comparison with the other reinforcement learning methods. The rules in the queue are now moved into the system starting with the first one in the queue until the queue is empty. When new rules are being moved from the queue into the rule-based system the number of rules with the same condition already in the system is checked and if it equals a given number P the oldest rule with that condition is deleted.

The cart-pole simulation is re-initialised and the process is repeated except in cases where the number of active rules (those matching the current state of the cart-pole) has reached the maximum allowable P. In this case one rule is selected using proportional selection over the strengths of the active (matching) rules and this rule is copied to make a new rule. The action part of the new rule can be mutated (flipped) with a very small probability, although this has no significant effect on the performance of the system given a sufficiently large P. The primary operator for reinforcement is selection and the probability of a particular action being selected in a given state is proportional to the strength of that action amongst the matching rules. This strength reflects the value of that action in the given state over the last P times that the system has been in that state. Apart from the rate of mutation, if mutation is used, the only parameter of the system is P the maximum size of the active population which can be optimised for a particular problem as shown in the following results.

6 Comparison With the Global Approach of Learning Complete Sets of Rules

6.1 The Global Approach

Odetayo and McGregor (1989) used the global approach of generating complete sets of rules with the genetic algorithm for the cart-pole problem. They divided the state space of the cart-pole into 54 discrete states and generated complete strings of 54 bits with the genetic algorithm. Each resulting set of rules was used to control the simulation from initialisation until failure and the number of time steps until failure was assigned as the value of that member of the population.

The simulation was initialised each time with the cart positioned randomly within 0.1 metres of the centre of the track, the angle of the pole positioned randomly within 6 degrees of the vertical and velocities set to zero. The objective was to balance the cart-pole for 10,000 time steps and the measure of performance was how many runs of the simulation it took to do this. A generational genetic algorithm was used with proportional selection, a population size of 300, one-point crossover, a string mutation probability of 0.01 (equivalent to a bit-wise mutation rate of 0.01/54) but with some other rather esoteric selection criteria to maintain diversity in the population. Over 50 runs the performance of the system varied between 729 and 3,808 iterations of the cart-pole simulation with a mean performance of 1,942 (Odetayo 1994).

6.2 Performance of the global approach

To replicate Odetayo and McGregor's results an incremental genetic algorithm with fitness proportional selection, replacement of the oldest, a population size of 300, one-point crossover and a string mutation rate of 0.01 was used on simulations of the cart-pole with the same 54 divisions of the state space. Over 50 runs the performance of this system varied between 420 and 30,000 (the maximum allowed) iterations with a mean of 3,392 and a standard deviation of 5,947. Using fitness proportional selection scaled to the worst member in the population the performance varied between 385 and 24,602 with a mean of 3,280 and a standard deviation of 5,092. Odetayo and McGregor's results are not significantly different from these.

In order to get a fair benchmark, for comparison with selectionist reinforcement learning, 50 experiments were conducted using first proportional selection and then scaled proportional selection with each of a range of population sizes for the global approach. The results of these are shown in table 1.

It is only with very large population sizes that the standard deviation in the performance of the global approach using scaled proportional selection starts to come down on the cart-pole problem as show in table 2.

From these results it would seem that population size is not a significant factor below about 5,000 and that the performance of the global approach on the cart-pole problem is better than random search.

Pop.	Prop. mean	Prop. s.d.	Scaled mean	Scaled s.d.
100	4450	6107	4004	4256
200	5585	8416	4715	5351
300	3392	5947	3280	5092
400	3140	4453	3829	5849
500	3581	4968	3164	3859
600	4317	7058	3064	3448
700	4790	6995	5568	8204
800	4011	5056	4068	6348
900	3082	4292	4225	5916
1000	4119	5488	3572	6158

Table 1: Performance of the global approach

Pop.	mean	s.d.
2000	3363	2002
3000	4674	3944
4000	4935	1123
5000	5813	922

Table 2: Performance of the global approach with large populations

6.3 Comparison of Selectionist Reinforcement with the Global Approach

For comparison with the global approach, a rule-based system using selectionist reinforcement to learn individual rules, an incremental genetic algorithm with fitness proportional selection, replacement of the oldest, a maximum size P of the active rule set of 300 and a bit mutation rate of 0.01/54 was used on simulations of the cart-pole with the same 54 divisions of the state space. Rules in the queue were assigned an exact copy of reward coming into the system at the end of each epoch, using the first credit assignment scheme described, before being entered into the population of the rule-based system.

All corresponding parameters were thus set the same in the two systems, the only difference being that in the global approach the genetic algorithm is working on a population of complete sets of rules whereas in the local approach it is working on populations of individual rules. During a single run of the simulation the global approach assigns a single action to each of the states. However many times the simulation enters a particular state during that run the same action is used in that state. When the simulation stops the same weight is implicitly assigned to the action for each state regardless of how many times it was used, if at all. In the local approach, different actions may be used in a single state at different times of the run and although these will all be assigned the same weight, using the chosen credit assignment scheme, states that were not entered will have their rule sets unaltered at the end of a run.

Over 50 runs the performance of the local approach of learning individual rules varied between 1,529 and 3,089 iterations with a mean of 2,022 and a standard deviation of just 342. Using fitness proportional selection scaled to the worst member in the population the performance varied between 1,366 and 2,856 with a mean of 1,669 and a standard deviation of just 256. These results are significantly better than the global approach both as described above and implemented by Odetayo and McGregor. The full set of results for varying P from 100 to 1000 and using both proportional and scaled proportional selection are given in table 3.

P	Prop. mean	Prop. s.d.	Scaled mean	Scaled s.d.
100	2934	3254	1293	1237
200	1800	2007	1258	324
300	2022	342	1699	256
400	2466	312	2156	302
500	3086	486	2723	347
600	3524	549	3142	300
700	3984	300	3598	217
800	4697	497	4064	296
900	5158	489	4471	260
1000	5599	329	4943	333

Table 3: Performance of selectionist reinforcement with mutation

With population sizes of between 100 and 600 selectionist reinforcement learning significantly outperforms the global approach on the cart-pole problem.

7 Comparison With Other Reinforcement Learning Approaches

7.1 Other Reinforcement Learning Approaches to the Cart-Pole Problem

Twardowski (1993) experimented with four reinforcement learning algorithms in a rule-based control system for the cart-pole problem. He divided the state space of the cart-pole into 162 discrete states and had only two rules for each of these states, one for each possible action in each state, whose weights were altered by the reinforcement learning algorithms. These rules were assigned an initial arbitrary weight and placed in the rule-based control system at the beginning of an experiment and no rules were created or deleted during a run - the genetic algorithm was not used at all. The simulation was initialised each time with the cart positioned at the dead centre of the track, the pole positioned exactly vertically and velocities also set to zero. The objective was to balance the cart-pole for 100,000 time steps and the measure of performance was how many runs of the simulation it took to do this.

The four reinforcement learning algorithms used were: backward averaging (similar to Holland and Reitman's (1978) original reinforcement algorithm but based on Barto, Sutton and Anderson's (1983) adaptive heuristic critic and Liepens et al's (1989) geometric attenuation), Holland's (1985) bucket-brigade, Liepens et al's (1991) hybrid bucket brigade-backward averaging scheme and Watkins'(1989) Q-Learning. Twardowski experimented with various values for the learning parameter and the probability of exploration in the algorithms on the cart-pole problem. Backward averaging never consistently reached the basic criteria of 100,000 time steps but the other three algorithms did. The mean performance of the bucket brigade averaged over 15 runs varied between about 1,250 and 5,000 depending upon the settings for learning and exploration parameters. The mean for the hybrid of bucket brigade and backward averaging was 250 to 3,000 and the mean for Q-Learning was 1,000 to 4,000. Best results were obtained when setting the exploration probability to zero. In this case Q-Learning is equivalent to the bucket brigade with a mean performance of 273 but the hybrid scheme performs the best with a mean of 228.

7.2 Comparison of Performance of Selectionist Reinforcement with Other Reinforcement Learning Approaches in a Rule-based Control System

For comparison with the other approaches to reinforcement learning a rule-based system, using: selectionist reinforcement learning, an incremental genetic algorithm with scaled fitness proportional selection, no mutation, replacement of the oldest and a maximum size P of the active rule set varied between 100 and 600, was used on simulations of the cart-pole with the same 162 divisions of the state space and the same initial conditions. Rules in the queue were assigned the difference between their position in the queue and the reward coming into the system at the end of each epoch, using the second credit assignment scheme described, before being entered into the population of the system. This scheme gives each rule a weight corresponding to how long the cart remains on the track with the pole balanced after that rule has effected its action until the simulation fails.

The mean performance of the system averaged over 15 runs varied between 1000 and 3000 depending on the value of P which is as good as that of the other reinforcement learning algorithms with exploration probability greater than zero. The full set of results for varying P from 100 to 600 are given in table 4.

The best performance was obtained with a P of 180 which gave a mean over 15 runs of 1015 with a standard deviation of 138. P must obviously be of a minimum size to ensure that each action in each area of the state space is given adequate initial trials to survive because once lost an action will not be rediscovered if no mutation is used. P is the only parameter to tune when using selectionist reinforcement learning although there is obviously some lee-way in the choice of the precise method of selection (Goldberg and Deb 1991) and deletion (Fogarty 1993).

P	mean	s.d.
100	2342	3831
150	2060	3116
200	1194	264
250	1317	133
300	1583	189
350	1851	115
400	2025	212
450	2278	207
500	2575	323
550	2753	326
600	2997	197

Table 4: Performance of selectionist reinforcement

7.3 Conclusion

It has been shown how the genetic algorithm can be used to do reinforcement in a rule-based control system and that whether reward or punishment is immediate or delayed there is no need to use an algorithm other than the genetic algorithm for reinforcement in a rule-based system given a reasonable credit assignment scheme. This has been demonstrated in the case of delayed reward on a simulated cart-pole problem. Selection was isolated as the sole genetic reinforcement operator from crossover and mutation, the discovery operators, which need not be used. Selectionist reinforcement in a rule-based control system was shown to outperform the global approach and to have comparable performance to other methods of reinforcement, using a probability of exploration greater than zero, in rule-based control systems on the cart-pole problem.

References

Anderson C W and Miller,III W T (1991) Challenging Control Problems. In Miller,III W T, Sutton R S and Werbos P J (eds) Neural Networks for Control, p.475-510.

Barto,A.G., Sutton,R.S. and Anderson,C.W.(1983) Neron like elements that can solve difficult learning control problems. IEEE Transactions on Systems, Man and Cybernetics, SMC-13(5), p.834-846.

Booker,L.B.(1985) Intelligent behaviour as an adaptation to the task environment. PhD Dissertation. The University of Michigan.

Booker,L.B.(1992) Viewing classifier systems as an integrated architecture. Presented at the First International Conference on Learning Classifier Systems, Houston, Texas.

Brooks,R.A.(1992) Artificial life and real robots. In Varela F J and Bourgine P (eds) Towards a Prctice of Autonomous Systems - Proceedings of the First European Conference on Artificial Life, p.3-10.

Compiani,M., Montanari,D. & Serra,R (1990) Learning and bucket brigade dynamics in classifier systems. In S.Forrest (Ed) Emergent Computation. Amsterdam: North Holland, p.202-212.

De Jong,K.(1988) Learning with genetic algorithms: an overview. Machine Learning, vol.3, p.121-137.

Fogarty,T.C.(1988) Rule-based optimisation of combustion in multiple burner furnaces and boiler plants. Engineering Applications of Artificial Intelligence, vol:1, iss:3, p.203-9.

Fogarty,T.C.(1989) Adapting a rule-base for optimising combustion on a double burner boiler. Second International Conference on Software Engineering for Real Time Systems (IEE Conf. publ. no.309), p.106-110.

Fogarty,T.C.(1990) Simulating multiple burner combustion for rule-based control. Systems Science, vol.16, no.2, p.23-38.

Fogarty,T.C.(1991) Putting energy efficiency into the control loop. I Mech E Technology Transfer in Energy Efficiency Session of Eurotech Direct 91, p.39-41.

Fogarty,T.C.(1993) Reproduction, ranking, replacement and noisy evaluations: experimental results. Proceedings of the Fifth International Conference on Genetic Algorithms, edited by Forrest,S., Morgan Kaufman.

Fogarty,T.C. & Huang,R.(1992) Systems control with the genetic algorithm and nearest neighbour classification. CC-AI, vol:9, nos 2 & 3, p.225- 236.

Goldberg,D.E. and Deb,K.(1991) A comparative analysis of selection schemes used in genetic algorithms. In Rawlins,G.J.E. (ed) Foundations of Genetic Algorithms, p.69-93, San Mateo, CA, Morgan Kaufmann.

Grefenstette,J.J.(1988) Credit assignment in rule discovery systems based on the genetic algorithm. Machine Learning, vol 3, p.225-245.

Holland,J.H.(1975) Adaptation in natural and artificial systems. Ann Arbor: University of Michigan Press.

Holland,J.H.(1985) Properties of the bucket brigade algorithm. Proceedings of the First International Conference on Genetic Algorithms and their Applications (p.1-7) Hillsdale, New Jersey: Lawrence Erlbaum Associates.

Holland,J.H.(1986) Escaping Brittleness: the possibilities of general- purpose learning algorithms applied to parallel rule-based systems. In R.S.Michalski, J.G.Carbonell & T.M.Mitchel (Eds.), Machine Learning, an artificial Intelligence Approach. Volume II. Los Altos, California: Morgan Kaufmann.

Holland,J.H.(1990) Concerning the emergence of tag-mediated lookahead in classifier systems. Physica D, vol.41, p.188-201.

Holland,J.H. and Reitman,J.S.(1978) cognitive systems based on adaptive algorithms. In D.A.Waterman and F.Hayes-Roth (Eds), Pattern-directed Inference Systems. New York: Academic Press.

Liepens,G.E., Hilliard,M.R., Palmer,M. and Rangarajan,G.(1989) Alternatives for classifier system credit assignment. Proceedings of the Eleventh Int. Joint Conference on A.I. p.756-761, Los Altos, CA, Morgan Kaufmann.

Liepens,G.E., Hilliard,M.R., Palmer,M. and Rangarajan,G.(1991) Credit assignment and discovery in classifier systems. Int. Journal of Intelligent Systems, vol 6, p.55-69.

Michie D and Chambers R A (1968) BOXES: an Experiment in Adaptive Control. In Dale E and Michie (eds) Machine Intelligence 2, p.137-152.

Odetayo M O and McGregor D R (1989) Genetic Algorithm for Inducing Control Rules for A Dynamic System. In Proceedings of the 3rd International Conference on Genetic Algorithms, p.177-182.

Odetayo M O (1994) Personal Communication.

Riolo,R.L.(1991) Lookahead planning and latent learning in a classifier system. In From Animals to Animats: Proceedings of the First International Conference on Simulation of Adaptive Behaviour, p.316-26

Smith,S.F. & Greene,D.P.(1992) Cooperative diversity using coverage as a constraint. Presented at the First International Conference on Learning Classifier Systems, Houston, Texas.

Smith,S.(1980) A learning system based on genetic algorithms. PhD Dissertation. University of Pittsburgh.

Sutton,R.(1991) Reinforcement learning architecture for animats. In From Animals to Animats: Proceedings of the First International Conference on Simulation of Adaptive Behaviour, p.188-296. Cambridge, MA: MIT Press.

Twardowski,K.(1993) Credit assignment for pole balancing with learning classifier systems. Proceedings of the Fifth International Conference on Genetic Algorithms, p.238-245.

Watkins,J.C.H.(1989) Learning with delayed rewards. PhD Dissertation, Kings College, London.

Wilson,S.W.(1985) Knowledge growth in an artificial animal. Proceedings of the First International Conference on Genetic Algorithms and their Applications, p.16-23.

Wilson,S.W.(1987) Classifier systems and the animat problem. Machine Learning, vol.2, p.199-228.

Wilson,S.W. & Goldberg,D.E.(1989) A critical review of classifier systems. In Proceedings of the Third International Conference on Genetic Algorithms, p.244-255.

Learning Anticipatory Behaviour Using a Delayed Action Classifier System

Brian Carse

Faculty of Engineering
University of the West of England, Bristol
Coldharbour Lane, Bristol BS16 1QY, UK
email: b_carse@csd.uwe.ac.uk
tel. +44-272-656261 x2705

Abstract. To manifest anticipatory behaviour that goes beyond simple stimulus-response, classifier systems must evolve internal reasoning processes based on couplings via internal messages. A major challenge that has been encountered in engendering internal reasoning processes in classifier systems has been the discovery and maintenance of long classifier chains. This paper proposes a modified version of the traditional classifier system, called the delayed action classifier system (DACS), devised specifically for learning of anticipatory or predictive behaviour. DACS operates by delaying the action (i.e. posting of messages) of appropriately tagged, matched classifiers by a number of execution cycles which is encoded on the classifier. Since classifier delays are encoded on the classifier genome, a GA is able to explore simultaneously the spaces of actions and delays. Results of experiments comparing DACS to a traditional classifier system in terms of the dynamics of classifier reinforcement and system performance using the bucket brigade are presented and examined. Experiments comparing DACS with a traditional classifier system, which appear encouraging, for a simple prediction problem are described and considered. Areas for further work using the delayed-action classifier notion are suggested and briefly discussed.

1 Introduction

In many applications a learning system is required to perform in an environment possessing complex temporal characteristics. Such applications include process control, robotics, and distributed systems control. In such environments, the learning system must be endowed with the capability of developing anticipatory behaviour based on accurate predictions and timely actions.

Classifier systems [1] provide a promising approach to learning in such environments. The classifier system is a parallel rule-based technique which attempts to avoid the brittleness encountered with traditional high-level symbolic methods such as production systems [2]. The strength of the classifier system approach lies in the use of a simple, abstract representation combined with general learning mechanisms enabling the system to dynamically build its own internal models through associations between rules [3]. This simple representational scheme (usually bit strings) provides basic building blocks in the form of schemas from which good rules can be discovered by application of the genetic algorithm [4]. In a similar fashion, but at a higher level, individual rules form the building blocks of rule associations representing higher-level concepts and behaviours.

To learn behaviour that goes beyond simple stimulus-response, a classifier system must evolve internal reasoning processes based on chains of classifiers coupled by internal messages [5]. However, serious difficulties arise here in the discovery and maintenance of long chains of classifiers [6]. These are discussed in section 2 below. Further, to learn behaviour with a temporal element, the classifier system must have some notion of the passage of time. For example, the classifier system may be required to learn temporal rules such as "If the environment is in state A at time T, and action X is taken, then the environment will be in state B at time $T+t$". One way in which a traditional classifier system might discover such behaviour is by evolving long coupled chains, using the classifier execution cycle as a basic clock 'tick'. However, the aforementioned problems of discovery and stability of long classifier chains arise here. Pass-through symbols (i.e. 'don't cares' in the action part of a classifier) can also be used to form the basis for simple memory and therefore predictive behaviour, but then again this mechanism requires the discovery and maintenance of long classifier chains.

This paper proposes a modified version of the traditional classifier system, called the delayed action classifier system (DACS) which may be of use in the learning of such temporal and predictive behaviour. The paper is organised as follows. Section 2 outlines related work, particularly focusing on the problem of discovery and maintenance of long chains of classifiers. Section 3 describes the delayed action classifier system. Section 4 presents and discusses the results of preliminary experiments using DACS. Finally, section 5 concludes and suggests possible future work.

2 Related Work and Problems with Long Chains of Classifiers

The difficulties of discovery and maintenance of long chains of classifiers have received much attention. These are briefly discussed below together with related work which has attempted, with a significant degree of success, to resolve them. For a more detailed discussion see [7].

2.1 Discovery of Classifier Chains

In many cases, all classifiers in a chain must be simultaneously present in the classifier population for any of them to receive the benefit of environmental payoff. Perhaps expecting the GA to discover a complete chain of rules is expecting too much. Complementing the GA with rule discovery strategies which encourage the building of classifier chains, for example Holland's 'triggered chaining' operator [8], has been shown to produce better results than using the GA alone [6]. The triggered chaining operator takes a classifier with high environmental payoff, X, and crosses it with a classifier from the previous time step, Y. Two additional classifiers are produced by the operator - the first one fires on Y's condition producing an internal message and the second fires on this internal message producing X's message. The hypothesis that the new linkage is a good one is tested in subsequent trials. Robertson and Riolo's experiments [6] in learning letter sequences demonstrate the effectiveness of incorporating this technique into the rule discovery strategy.

2.2 Maintenance of Classifier Chains

Long chains of classifiers are prone to disruption under the action of the genetic algorithm. A conventional 'Michigan-style' classifier system has all classifiers in the population compete to reproduce based on fitness alone. Clearly in this situation, classifiers in the same chain which are cooperating with each other to effect good actions and receive payoff, later go on to compete with each other under the action of the GA [7,9]. High fitness classifiers at the end of a chain, and their offspring, may well displace earlier 'stage-setting' classifiers in the population. Wilson and Goldberg [7] propose a classifier system which clusters classifiers into 'corporations'. Classifiers belonging to the same corporation do not compete with each other under the action of the GA and corporations form and break up under the action of a modified crossover operator. This is effectively a fusion of the Pitt and Michigan approaches. A successful implementation based on this approach is Shu and Schaeffer's [10] hierarchical classifier system in which classifiers are grouped into 'families' which form the basic units of selection by the GA.

Another problem with long chains, particularly using the bucket-brigade is the allocation of credit to early stage-setting classifiers in the chain. Using the bucket brigade early classifiers are reinforced only after many passes through the chain. Wilson [11] and Riolo [12] estimated the number of passes through a chain required to fully reinforce the first classifier in that chain. For certain bucket-brigade parameters (used in the experiments we describe later), Riolo estimates the number of passes required through a chain of length N for the first classifier in the chain to achieve a strength equal to 90% of its asymptotic strength to be approximately 22 + 11.9N. A number of candidate solutions to this problem have been proposed. Holland's bridging classifiers [13] are designed to feed environmental payoff to the first classifier in a chain regardless of its length. Riolo [12] experimentally verified the efficacy of bridging classifiers in reinforcement of stage-setting classifiers. However, the problem still remains that the chain (and the attendant bridging classifier) must be discovered in the first place. Wilson [11] suggested the use of a hierarchical credit assignment method in which the modularity of the hierarchy keeps chains short and thus easily reinforceable. This technique uses modules (chains) of classifiers as well as individual classifiers as the units for credit assignment. Grefenstette [14] replaced the bucket-brigade altogether with an epochal credit assignment scheme, the 'profit-sharing plan' in which classifiers which have been active since the last receipt of environmental reward are paid on the next, and demonstrated its effectiveness experimentally.

3 A Delayed Action Classifier System (DACS)

DACS operates by delaying the action (i.e. the posting of messages) of, appropriately marked, matched classifiers by a number of classifier execution cycles. Markers located in the non-coding loci of the action part (i.e. the otherwise non-interpreted part of the message) are used by DACS to interpret whether or not a particular classifier is a delayed-action classifier and, if so, the delay that is to be applied to its action (Figure 1 shows one possible message interpretation scheme). For example, using the interpretation scheme shown in Figure 1, a classifier with action part 110001011 would be interpreted as a delayed-action classifier producing an environment output message delayed by three classifier execution cycles. Such a classifier performs the same action as a linearly-coupled (i.e. non-branching) chain of four classifiers in a traditional classifier system. However, the single delayed action classifier is much more likely to be discovered quickly, at least by genetic search alone, than the equivalent four-classifier chain.

```
Bit Position
┌──────┬──────┬──────┬──────┬──────┬──────┬──────┬──────┬──────┐
│  0   │  1   │  2   │  3   │  4   │  5   │  6   │  7   │  8   │
├──────┴──────┴──────┴──────┴──────┴──────┴──────┴──────┴──────┤
│                                                              │
│  Tag Bits    Non-Coded Loci    Delay   Delay Bits            │
│  00=EnvIP                       0=OFF   000 = Delay 0         │
│  11=EnvOP                       1=ON    001 = Delay 1         │
│  01,                                    .......              │
│  10=IntMsg                              111 = Delay 7         │
│                                                              │
└──────────────────────────────────────────────────────────────┘
```

Fig. 1. Message Interpretation Scheme Used in DACS

DACS maintains two message lists - an active-message list which corresponds to the message list of a traditional classifier system, and a delayed-message list containing the actions of delayed-action classifiers. Messages from matched delayed-action classifiers are stored on the delayed-message list, which serves as a form of short-term memory, until they are due to be posted to the active-message list. The classifier(s) posting a delayed message are also remembered for later credit assignment. A variable associated with each delayed message, TIMETOGO, is initialised to the delay encoded within the message at the time of posting to the delayed-message list and decremented on each classifier execution cycle. When TIMETOGO reaches zero, the message is posted to the active message list. In our initial implementation, DACS uses the bucket-brigade credit assignment method.

Payoffs (whether from the environment or other classifiers) to a delayed-action classifier are paid directly and in full in the same way as to non-delayed classifiers. The significance of this is that a delayed-action classifier which does the same job as a long sequence of linearly-coupled classifiers receives direct payoff. Discounted payoff (based on delay) permits delayed action classifiers to coexist with traditional classifier chains (see section 4 for details). A slight modification to the standard bucket-brigade algorithm, as described in [6], is used which assigns environment payoffs only to those classifiers posting the effector-activating messages. The rule discovery mechanism used in DACS is the genetic algorithm. Since classifier delays are encoded on the classifier genome, the GA is able to explore simultaneously the spaces of actions and delays.

4 Experiments with DACS

This section describes and discusses preliminary experiments which have been conducted using DACS. The first experiment pertains to credit assignment using the bucket-brigade algorithm, in particular to how quickly classifier strengths approach their fixed-point (asymptotic) values. Early stage-setting classifiers in chains containing delayed-action classifiers achieved 90% of asymptotic strength faster than in an equivalent chain of simple classifiers. The experiments and results for this are presented in section 4.1. In a second set of experiments we investigated the potentially deleterious effect that the presence of delayed action classifiers in a population has on reinforcement of competing traditional classifier chains. Details are presented in section 4.2. In a third experiment the GA was turned on and DACS was applied to a simple learning problem - the letter prediction task. This task was chosen because it is representative of the sort of prediction problem for which DACS is intended, and also because it has been widely studied [6,15]. In our experiments, DACS consistently outperformed a simple classifier system in solving this problem. Results and a discussion of this experiment are presented in section 4.3.

4.1 Chains of Classifiers

The purpose of this experiment was to compare how quickly classifier strengths approached their asymptotic values for classifier chains with and without delayed action classifiers. The classifier sets chosen are based on those used in [12], 'Sequences with multiple message sources'. A difference between our environment here and Riolo's finite-state world, is that we are only interested in a correct output message being generated a certain time delay after the initial stimulus. Each classifier is of the form *Condition1/Condition2/Action*. The two classifier chains used are shown in Figure 2 (in this figure *InX* denotes an environment input message, *MX* an internal message and *OutX* an environment output message; the notation z^{-n} implies a delay of n classifier execution cycles). Both sets of classifiers effectively implement the following composite temporal rule:

IF   (EnvironmentInput = In0 at step 1) AND
    (EnvironmentInput = In1 at step 4) AND
    (EnvironmentInput = In2 at step 7)
THEN
    (EnvironmentOutput = Out0 at step 9)

Parameters were set as follows: initial classifier strengths = 50, environmental reward = 100, bid constant = 0.1. Note that in this experiment there are no competitions between classifiers and the GA is not applied.

<u>Traditional Classifiers</u>	<u>Delayed Action Classifiers</u>
C1 : In0 AND In0 = > M0	C1 : In0 AND In0 = > z^{-2}M0
C2 : M0 AND M0 = > M1	C2 : In1 AND M0 = > z^{-2}M1
C3 : M1 AND M1 = > M2	C3 : In2 AND M1 = > M2
C4 : In1 AND M2 = > M3	C4 : M2 AND M2 = > M3
C5 : M3 AND M3 = > M4	C5 : M3 AND M3 = > Out0
C6 : M4 AND M4 = > M5	
C7 : In2 AND M5 = > M6	
C8 : M6 AND M6 = > M7	
C9 : M7 AND M7 = > Out0	

Fig. 2. Classifier Chains Used in Experiment 4.1

Figure 3 plots the strengths of classifiers against the number of runs through the chains. The asymptotic strengths of the last three classifiers in each chain (not shown in Figure 3) are all 1000 which is equal to the environmental reward (100) divided by the bid constant (0.1). The asymptotic strengths of traditional classifier C4 and delayed action classifier C2 are 500, since traditional classifier C6 and delayed action classifier C2 must share half of their bucket-brigade payoffs with the environment. Similarly, the asymptotic strengths of traditional classifier C1 and delayed action classifier C1 are 250, since traditional classifier C3 and delayed action classifier C1 must also share half of their bucket-brigade payoffs with the environment.

The first classifier in the chain of traditional classifiers achieves 90% of its asymptotic strength after around 130 runs through the chain (1300 classifier cycles in total) whereas the first classifier in the chain including delayed action classifiers achieves 90% of its asymptotic strength after only 80 runs through the chain (800 classifier cycles). Clearly the delayed-action mechanism is providing faster reinforcement to stage-setting classifiers and the improvement in reinforcement rate becomes greater with increased delay.

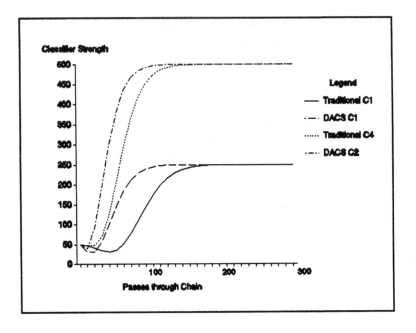

Fig. 3. Classifier Strengths in Experiment 4.1

4.2 Effect of DACs on Traditional Classifier Chains

In this section we describe experiments which show that under certain conditions, the presence of a delayed action classifier in the population can substantially slow down reinforcement of a competing, higher environmental payoff traditional classifier chain. Again we borrow from Riolo's [12] test-beds, this time from the section on "Choosing between simple sequences of classifiers". Since the dynamic behaviour of the classifier set employed, described below, is not deterministic (there is a random element in conflict resolution between competing classifiers), all results presented in this section are the average of ten independent runs, each using a different initial random seed.

In the first experiment described in this section, two competing paths each consisting of a linear chain of ten classifiers are used. The first chain involves classifiers CS #1 to CS #10, the second chain classifiers CS #11 to CS #20. For each classifier chain, classifier CS #N posts a message which matches the condition part of CS #(N+1). At the start of each pass, one chain is selected randomly based on the strengths of the first rule in each chain. This simulates the competition between two matched classifiers competing to post their messages using roulette-wheel selection. The first chain receives an environmental payoff of 400 and the second competing one a payoff of 100.

Initial strengths of all classifiers were set to 1000 with a bid constant of 0.1 used by the bucket-brigade. Figure 4a shows the results of running through 400 passes (4000 classifier execution cycle steps). With the parameters chosen, the strength of the first classifier (CS #11) in the second chain remains constant at its initial value since this is also its asymptotic value. The strength of the first classifier (CS #1) in the first chain increases to 90% of its asymptotic value after about 2000 or so cycle steps. The marginal payoff per 20 passes from the system is also shown. The maximal marginal payoff is 8000 (a reward of 400 for each of 20 passes). The results observed are in close agreement to those obtained by Riolo [12].

In a second experiment we ran the same set of classifiers but this time with an initial strength of 200, far away from the asymptotic strength of either classifier chain. The results are shown in Figure 4b. The first classifier (CS #1) in the first chain achieves its asymptotic strength faster due to decreased competition from the second chain, and the rate of increase in marginal payoff is correspondingly larger.

In a third experiment we replaced the second (low payoff) chain with a single delayed action classifier. The effect of this, shown in Figure 4c, is quite devastating in terms of the marginal payoff achieved by the system in earlier passes. The delayed action classifier, since it receives direct payoff, quickly achieves its asymptotic strength (1000), effectively starving the better first chain of reinforcement. The first chain eventually recovers and marginal payoff increases but clearly the presence of the delayed action classifier is detrimental to system learning dynamics. If a population containing delayed action classifiers is expected also to maintain chains of simple classifiers, then some discount must be applied to the reward (or payoff) received by delayed action classifiers and this discount should be an (increasing) function of delay.

In the fourth and final experiment described in this section, we used the same classifier set as just described, but with the bid constant (used for payoffs) and environmental reward for the delayed action classifier reduced by a factor of 0.1 (this discount factor was arrived at simply by taking the inverse of the delayed action classifier delay). Choosing the same factor for both ensures the asymptotic strength of the classifier does not change - it merely affects the system dynamics. The classifier bid made for roulette-wheel selection is not changed. The results of this experiment are shown in Figure 4d. Clearly this has improved the situation somewhat, with the system achieving earlier high marginal payoff. The effect of using discounted bid constants and rewards effectively removes any advantage of DACS over a traditional classifier system in terms of reinforcement. Also, further work is required on dynamics of the bucket-brigade to determine a better discount function to be applied to bid constants and environmental rewards for delayed action classifiers.

However, the advantage in terms of rule discovery is still present and we address this aspect in the experiment described in the next section.

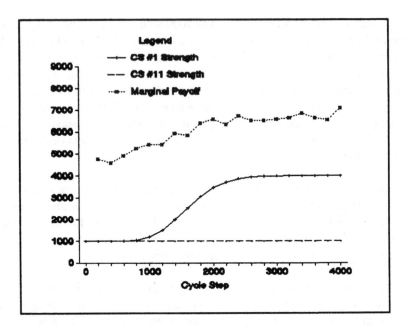

Fig. 4a. Marginal Performance and Strengths of Classifiers CS#1 and CS#11 for two Competing 10-chain Sequences. Initial Classifier Strengths = 1000.

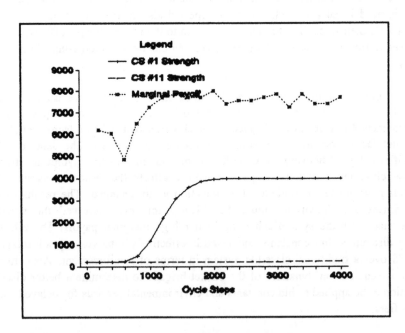

Fig. 4b. Marginal Performance and Strengths of Classifiers CS#1 and CS#11 for two Competing 10-chain Sequences. Initial Classifier Strengths = 200.

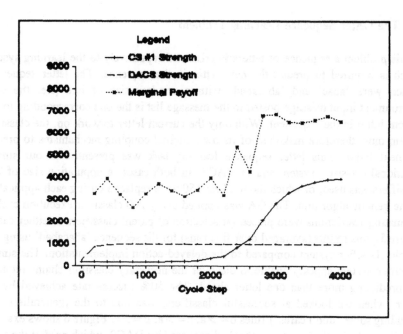

Fig. 4c. Marginal Performance and Strengths of First Classifier in a 10-chain Sequence and a Competing Delayed Action Classifier. Initial Classifier Strengths = 200. No Discount Applied to DAC Reward/Bid Constant.

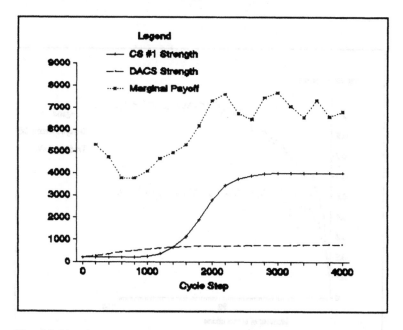

Fig. 4d. Marginal Performance and Strengths of First Classifier in a 10-chain Sequence and a Competing Delayed Action Classifier. Initial Classifier Strengths = 200. Discount Factor of 0.1 Applied to DAC Reward/Bid Constant.

4.3 The Letter Sequence Learning Problem

In this problem a sequence of letters is presented repetitively to the learning system which is required to predict the next letter in the sequence. The letter sequences chosen were 'abac' and 'abacabad' with a window size of one (i.e. the only environment input message posted to the message list is the one corresponding to the current letter in the sequence). With only the current letter to work on, the classifier system must therefore make use of its own internal coupling mechanisms to predict the next letter. This letter sequence learning task was presented to our simple traditional classifier system and to DACS. In both cases, a population size of 200 classifiers was used, of which the weakest 150 were replaced during each application of the genetic algorithm. The GA was applied every 200 classifier execution cycles. No mating restrictions were placed on selection of parent classifiers in either case. Figure 5 plots results (averaged over five runs) for the sequence 'abacabad' using our simple classifier system compared to the delayed-action implementation. The simple classifier system appears unable to discover the necessary classifier chains required for predicting more than one letter ahead. The 50% success rate achieved by the latter, when we looked at successful classifiers, was due to the (generalised i.e. including some "don't cares") rules b=>a, c=>a, d=>a. Figure 6 shows example classifier sets for the sequence 'abac', discovered by DACS, which predict the letter sequence correctly (the interpretation scheme described in Section 3 is used here with loci 2 through 4 used as follows: 000 is 'a', 001 is 'b', 010 is 'c' etc.).

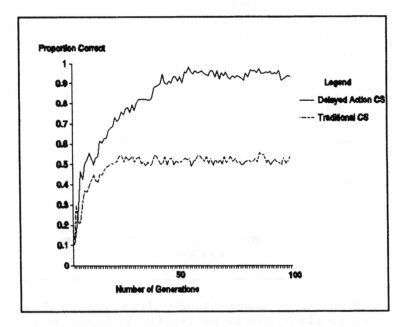

Fig. 5. Proportion of Letters Guessed Correctly in Experiment 4.3

Solution Set A

Classifier	LHS	RHS	Interpretation
C1	0000000#0	110001111	$a => z^{-7}a$
C2	0#00100#0	110011111	$b => z^{-7}b$
C3	0001#0#00	110101011	$c => z^{-3}c$

Comments: DACS has simply learned the period of repetition of the sequence.

Solution Set B

Classifier	LHS	RHS	Interpretation
C1	000010000	110101001	$b => z^{-1}c$
C2	0#00##000	110010001	a OR b => b
C3	000010000	110000010	b => a
C4	000100000	110000100	c => a

Comments: This classifier set represents a 'dynamical' solution described in [15]. Conflicts between C1 and C2 are resolved in C1's favour due to its higher acquired strength.

Solution Set C

Classifier	LHS	RHS	Interpretation
C1	000100000	110001100	$c => z^{-4}a$
C2	#00010000	110011111	$b => z^{-7}b$
C3	#00000000	110101100	$a => z^{-4}c$
C4	000100000	110001110	$c => z^{-6}a$

Comments: This classifier set is also a dynamical solution. Conflict between delayed action classifiers C2 and C3 are resolved in C2's favour due to its higher acquired strength.

Fig. 6. Three Solution Classifier Sets Discovered by DACS for the Sequence 'abac'

5 Conclusions and Future Work

This paper has outlined the motivation behind the delayed-action modification to learning classifier systems, namely the learning of predictive behaviour, and has described one possible implementation of such a scheme. Experimental results have been presented which demonstrate that this modification may be of use in learning temporal or predictive behaviour. Although these initial results are encouraging, further work needs to be done on the fundamental idea and its application to more complex domains. The experiments presented demonstrate that the presence of delayed action classifiers in the population can have deleterious effects on traditional classifier reinforcement methods using the bucket-brigade. To overcome this problem, discounted bid constants and rewards can be used, but this removes the advantage of delayed action classifiers in terms of reinforcement. The discount used in the experiments presented here was obtained heuristically and further work on the dynamics of the bucket-brigade is needed to determine optimal discount values. An interesting application of the delayed-action classifier idea which we are currently investigating is learning of real-time behaviour. The results presented here (and in many other studies of learning classifier systems) have assumed that environment state changes are somehow synchronised with classifier execution cycles. Clearly this is inappropriate for real-time systems since the classifier execution cycle time is variable and depends on many factors (above all, it is not deterministic). We are currently investigating ways in which the delayed-action classifier notion may be employed in such environments. One possibility is the design and implementation of a "real-time" classifier system, involving timed interrupts and background and foreground processes. Additional modifications of the basic DACS execution cycle may be worth investigating. One possibility is to have a delayed action classifier persist on the main classifier message list for a number of time steps rather than to have it posted to that list a number of time steps after its activation, which is the approach we have used here.

References

1.   Holland J.H.: Escaping brittleness: The possibilities of general purpose machine learning algorithms applied to parallel rule-based systems. In R.S. Michalski, J.G. Carbonell, and T.M Mitchell (eds.), Machine learning: An artificial intelligence approach, vol. 2, Los Altos, California: Kaufmann (1988)

2.   Davis R. and King J.: An overview of production systems. In E.W. Elcock and D. Michie (eds.), Machine Intelligence, 8. Chichester: Ellis Horwood (1977)

3.   Booker L.B., Goldberg D.E. and Holland J.H.: Classifier systems and genetic algorithms. Artificial Intelligence 40 pp235-282. Elsevier Science Publishers B.V.,North-Holland (1989)

4.   Holland J.H.: Adaptation in natural and artificial systems. University of Michigan Press, Ann Arbor, MI (1975)

5.   Forrest, S. and Miller J.H.: Emergent behaviour in classifier systems. In S. Forrest (ed), Emergent Computation. Elsevier Science Publishers B.V., MIT/North-Holland (1991)

6.   Robertson G.G. and Riolo R.L.: A tale of two classifier systems. In D.E. Goldberg and J.H. Holland (eds), Machine Learning 3:pp139-159, Kluwer Academic Publishers (1988)

7.   Wilson S.W. and Goldberg D.E.: A critical review of classifier systems. Proceedings of the 3rd International Conference on Genetic Algorithms, pp244-255 (1989)

8.   Holland J.H., Holyoak K.J., Nisbett R.E. and Thagard P.R.: Induction: processes of inference, learning, and discovery, MIT Press (1987)

9.   Grefenstette J.J.: Multi-level credit assignment in a genetic learning system. Proceedings of the 2nd International Conference on Genetic Algorithms. pp202-209. Erlbaum, Hillsdale, NJ (1987)

10.   Shu L. and Schaeffer J.: HCS: Adding hierarchies to classifier systems. In R.K. Belew and L.B. Booker (eds) Proceedings of the 4th International Conference on Genetic Algorithms, pp339-345 (1991)

11.   Wilson S.W.: Hierarchical credit allocation in a classifier system. Proceedings of the 10th International Joint Conference on Artificial Intelligence. pp217-220. Los Altos, CA: Morgan Kaufmann (1987)

12.   Riolo R.: Bucket brigade performance I. Long sequences of classifiers. In J.J. Grefenstette (ed) Genetic Algorithms and their Applications, Proceedings of the Second International Conference on Genetic Algorithms. Erlbaum, Hillsdale, NJ (1987)

13.   Holland J.H.: Properties of the bucket-brigade algorithm. Proceedings of the 1st International Conference on Genetic Algorithms and Their Applications. pp1-7. Erlbaum, NJ (1985)

14.   Grefenstette J.J.: Credit assignment in rule discovery systems based on genetic algorithms. Machine Learning, 3: pp225-245 (1988)

15.   Compiani M., Montanari D. and Serra R.: Learning and bucket brigade dynamics in classifier systems. In S.Forrest(ed), Emergent Computation. Elsevier Science Publishers B.V., MIT/North-Holland (1991)

Applying A Restricted Mating Policy To Determine State Space Niches Using Immediate and Delayed Reinforcement

Chris Melhuish, Intelligent Autonomous Systems Lab
Faculty of Engineering, University of the West of England
Bristol BS16 1QY, UK. E-mail cr_melhu@csd.uwe.ac.uk

Terence C. Fogarty, Faculty of Computer Studies and Mathematics
University of the West of England, Bristol BS16 1QY, UK

Abstract. Approaches for rule based control often rely heavily on the pre-classification of the state space for their success. In the pre-determined regions individual or groups of rules may be learned. Clearly, the success of such strategies depends on the quality of the partitioning of the state space. When no such a priori partitioning is available it is a significantly more difficult task to learn an appropriate division of the state space as well as the associated rules. Yet another layer of potential difficulty is the nature of the reinforcement applied to the rules since it is not always possible to generate an immediate reinforcement signal to supply judgement on the efficacy of activated rules. One approach to combine the joint goals of determining partitioning of the state space and discovery of associated appropriate rules is to use a genetic algorithm employing a restricted mating policy to generate rule clusters which dominate regions of the state space thereby effecting the required partioning. Such rule clusters are termed niches. A niching algorithm, which includes a 'creche' facility to protect 'inexperienced' classifiers, and the results of determining a simple 2D state space using an immediate reward scheme are presented. Details of how the algorithm may modified to incorporate a delayed reinforcement scheme on a real-world beam balancing control problem are reported.

1. Introduction

It has been shown that, with a priori knowledge, it is possible to partition a state space such that a system can subsequently learn an appropriate action in each of the partitioned regions [Michie 68],[Odetayo 89],[Fogarty 89]. It may be required to control systems which are difficult, if not impossible, to model and therefore to determine in advance an appropriate division of the state space. For such a class of problem a control system which not only learned the actions but also the state space partitioning would be advantageous.[Huang 93], [Goodwin 93], [Woodcock 91], [Moore 91], [Renders 92]. This paper sets out to explore how a simple real numbered classifier system can be used to partition a state space into regions or niches by converging to stable sub-populations of classifiers with the same action. Davidor [Davidor 91] gives a brief history of the mechanisms that have been suggested to answer the control of convergence issue. The system described here relies on reward sharing [Booker 82] which encourages niche formation in the genetic search and the continued creation of offspring requiring a restricted mating policy to minimize the proliferation of inappropriate classifiers.

In this paper for the problem of determining a 2D simulated test state space and a real-world state space, each individual in the population is a classifier which describes a rectangular region of the state space as well as the action to be executed. Restricted mating is achieved by determining firstly, which individuals have a range which covers the point in the state space under inspection and secondly, applying the conventional procedures of selection of parent rules, crossover and mutation to the selected sub-set of the pool of rules which make up the niche. In the experiments described a parent is chosen by a stochastic selection process and the second parent, again chosen stochastically, must then have the same action as the first parent if mating is to proceed. In order to reduce the disruption to a niche by allowing 'young' classifiers, which have not been used sufficiently and therefore lack credibility, to be given too much 'weight' in the selection process[Valenzuela-Rendon 91], the niche population is considered in two classes, namely; mature classifiers - those 'experienced' classifiers which have been matched, since they cover the point in the

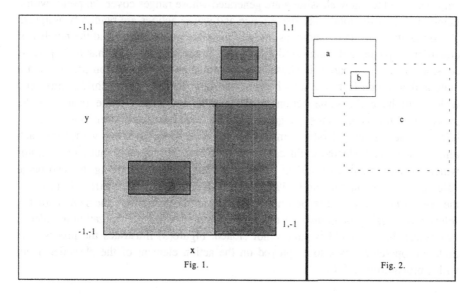

Fig. 1.

Fig. 2.

state space under inspection, and had their strengths altered more than a set number of times and immature classifiers which are considered 'inexperienced' since they have not been matched and had their strengths altered on sufficient occasions.

2 Converging On a 2D State Space using Immediate Reward

A simple experiment was conducted in which a 2D state space was divided into 6 regions, in which one of two actions was appropriate. The task set was for a dynamically sized population of classifiers with initial random range settings to converge to a set of sub-populations covering the 6 state space regions using an immediate reward scheme. Figure (1) shows the state space regions and the appropriate action in each region. Each classifier has a structure in which the condition part consists of four floating point values which describe the range of the classifier ie its 2D coverage, an action to be performed (0 or 1) and an associated strength as well as a number of housekeeping elements. In this experiment there is no bucket brigade type of credit assignment and no internal message list. Credit assignment is accomplished by taking a random position in the state space and rewarding all those classifiers which cover that point and have the same action as the underlying model and punishing those classifiers which cover the point but whose action does not agree with the underlying model.

3 The Immediate Reward Algorithm

When a random point is selected those classifiers which cover it are grouped into a match set *[M]*. If the match set is considered to be viable the mechanism of parent selection and mating will proceed but if it has less than eight classifiers, set empirically, then new classifiers are generated whose ranges cover the point with a range of 5% of the total state space range and whose actions are randomly assigned. The population size is dynamic as initially there are no classifiers and the number of classifiers expands and contracts during a run. In the case where a match set *[M]* is considered viable, a parent is chosen using a roulette wheel selection process and a mate is then found from the set which has the same action *[A]* . A child is produced which will have the same action as the parents but whose range results in the crossover of the parent ranges. In general, for the two possible children, one child will have a wider range and one a smaller range. The 'from' and 'to' values for each dimension of the selected child can then be altered by a 'creep' mutation operator which has a probability of 0.01 of being employed. If used it moving the end range value by a small amount (up to 5%) of the range length of the classifier. In this way the area of the classifier can be finely adjusted. In the case where the same parent is selected there is a probability of 1 that both end points will suffer mutation in order to guarrantee that the child is simply not cloned. Figure(3) illustrates the process. A mutation operator was also employed on the action element of the classifier again with a probability of 0.01.

4 Restricted Mating Algorithm

The mechanism to be explored is that of selecting classifiers which best agree with the underlying state space model and due to the influence of the genetic algorithm create better classifiers which cover more of the state space correctly. The algorithm employed is:

Begin
  For a given number of trials
    Select a random position in the state space
    Determine the match set *[M]*
    If *c[M]* is insuffcient Then
      Create new classifiers
    Else
        Apply reward and punishments to *[A]* and *[A′]* respectively
        Attempt to Generate a child from the Action set *[A]*
        If a child is produced Then
            Cull sufficient of the weakest mature classifiers in *[M]* to give
            smallest number of mature classifiers which can support immature
            Insert the child into the population.
        Endif
      Endif
    Next Trial
End

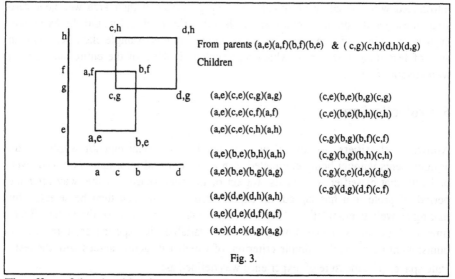

Fig. 3.

The effect of the algorithm is to promote the 'merging' together of higher strength, ie more appropriate, classifiers to evolve new classifiers whose ranges match more closely the regions of the underlying state space. Figure(7a) illustrates the result of running the algorithm for several hundred trials and shows that the regions have been identified but with a coarse resolution. The coarseness of the resolution is affected by the ranges of the classifiers set at initialization with a tradeoff between resolution and computation time.

5 Fitness

Since larger area classifiers have the opportunity to receive more reward than their smaller counterparts, a large area classifier which overlaps a small region where its action is incorrect may still have a higher strength than a smaller area completely correct classifier which snugly fits a region of the state space. Figure (2) shows a region of the state space which is overlapped by two classifiers, a and b. In the case of b, this classifier correctly matches the action in the region but the larger classifier a correctly matches the action specified in the region outside c. In this case the majority of the area bounded by a is correct and since the reward given to a classifier in the simulation is directly proportional to the area of the classifier because of the random selection of each point in the state space, then classifier a will accrue a greater strength than b. Parental selection using strength alone is inappropriate as it will favour larger classifiers with a larger incorrect region over smaller but more correct classifiers. A strategy which overcame the problem was to replace strength with a measure of how well the classifier had fared over a number of reward/punishment experiences and to prevent classifiers being eliminiated until they had accumulated a minimum number of experiences. For these experiments a mechanism was introduced which disallowed classifiers being generated if their area was near zero since these would simply accumulate in the classifier pool and would be extremely unlikey to gain sufficient experiences to mature. In this example classifiers with an area of less then 0.001 units, which represented 0.025% of the entire state space, were disqualified.

6 Protectionism

Allowing classifiers to experience a number of punishment/rewards credit assignmnents without being eliminated can provide useful information on how well or badly that classifier covers its region of the state space. In this way after the period of protection has lapsed, an experienced classifier can then be selected for mating as well as elimination along with other 'mature' members of the niche. Every time a classifier is matched with the state variables it experiences a reward or punishment based on the simple criterion of whether its action agrees with the test state space. The measure of experience was defined as:

$$\beta = \frac{n(C)}{n(C) + n(I)}$$

where $n(C)$ is the number of matches resulting in the correct action, that is, in agreement with with the underlying state space model and $n(I)$ the number of incorrect matches. The use of such a fitness parameter gives a normalized indication of reliability for comparison between classifiers but will not, by itself, force selection for larger classifiers.

However, if both parents have $\beta = 1$, then the simple rule: -------- [β Rule]
   If both parents are experienced and both have $\beta = 1$ Then
      If child area is larger than at least one of the parents Then
         Child is considered valid
      Else
         Child is considered invalid
      Endif
   Endif

should promote the growth of high integrity classifiers. If a child of high β parents survives and is shown also to have $\beta = 1$ then it will correctly cover a larger area of the state space than at least one of its parent and will therefore increase the selective pressure to delete from the niche other classifiers with a value of $\beta < 1$.

7 Increasing Selection Pressure For Larger Classifiers

In order to decrease the overall number of classifiers in the gene pool, which would reduce the computational overhead, the restricted mating algorithm was then modified to increase selection pressure in favour of larger classifiers and set limits to the number of mature and immature classifiers which could be maintained in a niche. In this way selection pressure for niches which cover larger areas of the state space would be promoted. Ideally, the state space partitions should then coincide with the niches with the end goal being to cover the state space with the minimum number of rectangular niches.

This was accomplished by modifying both the parent selection and culling operations. In the first case, the number of mature and immature classifiers in a niche was restricted. A niche could support a maximum of, typically 4, immature and, typically 8, mature classifiers respectively. A new 'protected' immature classifier could only be generated if the number of immature individuals is less than the set 'creche' size and, of course, if mature classifiers are present in the niche. In the second case, once mating has occurred then the number of mature individuals is culled in order, once more, to reduce the number of classifiers back to the fixed value. By ordering the classifier pool not only by β but also additionally employing a secondary ordering based on strength of the classifier, the culling operation will select for deletion those primarily with lower β and secondarily those with lower strengths consequently leaving the larger high integrity classifiers in the classifier pool.

8 Experimental Results

The β Rule does not materially affect the number of overall trials required to correctly cover over 90% of the state space as can be seen in Figure (4) where graph (a) shows the result of running the algorithm without the β rule and (b) with. However, each trial demands less computation since there are fewer classifiers to be processed as shown in Figure (5) which shows the affect of the primary and

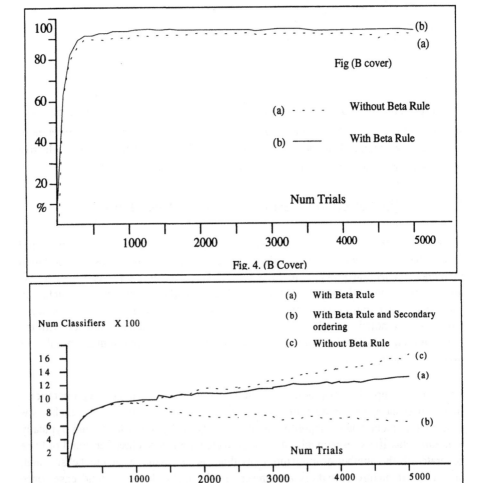

Fig. 4. (B Cover)

Fig. 5.

secondary ordering of the classifier pool in relation to β and strength respectively.

The figure shows that without employing the β rule the number of classifiers increases (c) but using the rule with secondary ordering (b) does significantly reduce the classifier population. The solution appears stable since if graph (b) is extended to 50,000 trials then it is found that the number of classifiers in the pool remains around 700 from about trial 2500 onwards.

The above techniques used in the context of a restricted mating policy have been shown to maintain an equilibrium between the conflicting demands of exploration and exploitation but clearly the 'stable' number of classifiers P_S is still larger than the optimum which in the above case would be simply:

$$P_S = nR.(nM + nI)$$

where nR is the number of distinct regions of the state space, nM the number of mature classifiers in a niche and nI the number of immature classifiers in a niche. For the above test case the state space could be optimally partitioned into 12 regions and with each 'perfect' niche containing 8 mature and 4 immature classifiers an optimal classifier pool of 144 classifiers is suggested. From observation of the converged state space at the end of a run the number of classifiers actually used is considerably more than the optimal number. It appears that the β rule may be working against convergence later in the run in that its action will generate classifiers which will tend to 'overgrow' the region and therefore accumulate in the classifier pool.

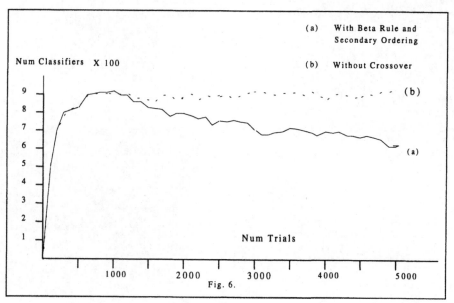

Fig. 6.

9 Crossover Vs Creep Mutation

It is not unreasonable to consider that an algorithm based purely on expansion of good classifiers using the creep mutation mechanism could partition a state space and not have to employ any mating and crossover strategy. Figure (6) shows the results from one study. Graph (b) shows how the population of classifiers behaves over 5000 trials using the simple strategy of stochastic selection of a parent from the niche then cloning a child followed by the creep mutation operator being applied to each of the four x and y end range values with a probability of 0.25. This strategy is outperformed by the algorithm described above with no action mutation being employed as shown by graph (b) which provides justification for the use of a mating and crossover strategy.

10 Region Bias

Some tests were conducted on partioning non-rectangular regions. Figure (7b) shows the successful results of running the algorithm on an underlying non rectangular, sinusoidally partitioned state space.

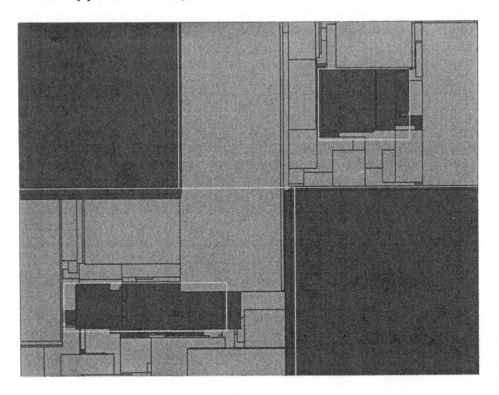

Fig. 7a

11 Applying the Algorithm to a Real Time Sytem using Delayed Reward

11.1 The System

This section deals with the attempt to apply the algorithm to a delayed reward real world scenario. The hover beam provided a simple test bed for the algorithm on a

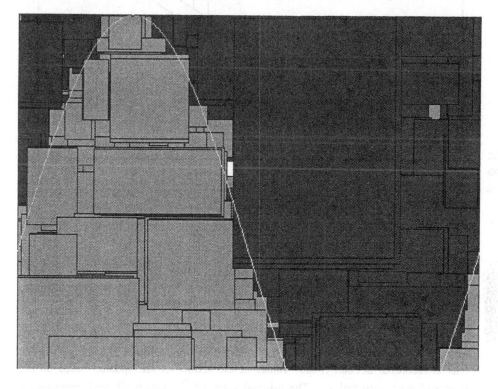

Fig. 7b

real system. Like the cartpole the beam is essentially an unstable non linear system but is less complex and easy to reset. Figure (8) shows the experimental configuration of a beam 1 metre long with an electric motor driving a propellor at one end and a counter weight at the other. The beam angle and angular velocity could be established by reading a potentiometer fixed to the pivot shaft. Two motor speeds were chosen to represent on and off, one causing the beam to lift and the other to sink. Sampling and motor action selection was carried out at 10Hz and the goal was for the beam to enter and remain in a region confined to $30°$ above and below the horizontal. Reward, which is proportional to the time spent in the 'safe' region, is attributed to the system either when the beam moves out of the region or the trial has timed out.

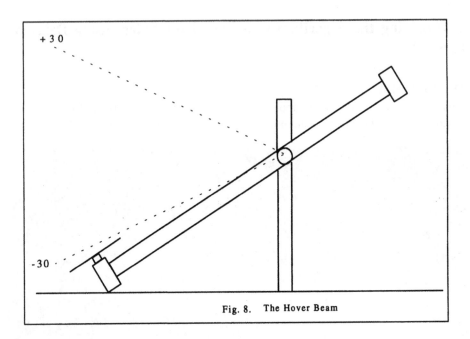

Fig. 8. The Hover Beam

11.2 Algorithm Details

The algorithm was applied to a real hoverbeam system with a niche size of eight supporting two immature classifiers. For a real time system using delayed reward a number of adjustments were made to algorithm described above.

The algorithm is run in three phases. Firstly, as before, actions are selected from the niche classifiers which match the state variables but now a log of the actions in each trial is recorded. Secondly, the values of the action set are updated based on the length of time until failure. The original value of β, which allowed comparisons between mature and immature classifiers, was simply the ratio of the correct to the sum of correct and incorrect actions taken. Clearly, this is inappropriate for a delayed reward scheme. The following cumulative function was used which would give a value between 0 and 1. [Fogarty 94]

$$\beta = (\beta + \frac{T_f}{T\max}) \cdot \frac{1}{N_e}$$

where T_f is the length of time until failure after the action has been execute $T\max$

is the length of the required sequence and N_e is the number of times the action has been taken.

Thirdly, every 35 trials, the genetic algorithm is applied to every niche visited in the sequence if the number of immature classifiers is less than the 'creche' size. In an attempt to reduce the real time computational overhead the number of classifiers in the total population was reduced on each occasion that the genetic algorithm was used. This was simply implemented by ranking the classifiers according to their value and allowing the top 750 to survive while culling any mature classifier ranked lower in the population.

11.3 Results

After 890 trials a sequence of greater than 200 seconds was found. Learning was then switched off and actions simply selected as the fittest classifier in the niche. Learning was then switched on again until trial 1000 and then, once more, learning was switched off. The table below shows how the system performed, ie how long the beam remained in the safe region, with learning off between trials 891 to 900 and 1001 to 1010.

+ Trial Number	After 890 Trials Sequence Length	After 1000 Trials Sequence Length
1	2010	2010
2	1892	2010
3	1904	2010
4	18	2010
5	2010	2010
6	1136	2010
7	41	39
8	2010	2010
9	211	174
10	352	2010
	mean 1158	mean 1629

12 Conclusions

Algorithms have been described which allow classifiers to converge on an arbitrary state space if an immediate reward is given to each subset of classifiers which match the state variables. A number of strategies have been employed to reduce the number of classifiers generated and increased the selection pressure for larger area classifiers:

- The use of the β rule early on in a run.
- Setting the minimum area of a classifier
- Allowing a niche population to support a fixed number of immature classifiers
- Not allowing mating of the mature classifiers in a niche until the number of immature classifiers fell below a preset 'creche' size.
- Use of the primary and secondary ordering of the mature niche population to cull low β and low strength classifiers.
- Crossover is a useful strategy.

However, there are a number of drawbacks. In particular, a large number of trials are required for even a simple problem and this will become increasingly worse as the dimensionality of the state space increases. Further work is required to define optimal values for the empirically determined values of niche size, ratio of mature to immature classifiers, minimum number of experiences as well as considering other reinforcement strategies.

References

Booker L.B. 1982. Intelligent Behaviour as an Adaptation to the Task Environment.Doctoral Thesis, University of Michigan. Dissertation Abstracts International, 43(2),469B

Davidor Y. 1991. A Naturally OccurringNiche and Specied Phenomenon. In Proc. 4th. Int. Conf. on Genetic Algorithms pp 257-263 Ed: Belew R. and Booker L.

Fogarty T.C. 1989. Learning New Rules and Adapting Old Ones with the Genetic Algorithm. In Artificial Intelligence In Manufacturing. Ed: G.Rzevski. Springer Verlag pp275-290

Fogarty T.C. 1994. Selectionist Reinforcement in Learning Classifier Systems. BTC-CSM Technical Report University of the West of England, Bristol.
Huang R. 1993. Evolving Prototype Rules for Process Control. Ph.D. thesis. University of the West of England, Bristol.

Michie D. and Chambers R.A. 1968. BOXES: An Experiment in Adaptive Control. In Machine Intelligence 2 .

Moore A. 1991. Variable Resolution dynamic Programming: Efficiently Learning Action Maps in Multivariate Real-valued State-spaces. In Machine Learning: Proc. 8th Int. Workshop. Ed: Birnbaum L. & Collins G. Pub: Morgan Kaufmann.

Odetayo M.O. and McGregor D.R. 1989. Genetic Algorithm for Inducing Control Rules for a Dynamic System. Proc 3rd Int. Conf. on Genetic Algorithms pp177-182 .

Renders J-M., Nordvik J-P. and Bersini H. 1992. Genetic Algorithms for Process Control: A Survey. In Proc. of International Symposium on Artificial Intelligence In Real-Time Control.

Rosen B. and Goodwin J. 1993. Dynamic Flight Control with Adaptive Coarse Coding. In Proc. 2nd Int. Conf. on Simulation of Adaptive Behaviour . MIT Press.

Twardowski K. 1993. Credit Assignment for Pole Balancing with Learning Classifier Systems. In Proc. 5th. Int. Conf. on Genetic Algorithms.

Valenzuela-Rendon M. 1991. The Fuzzy Classifier System: A classifier System for Continuously Varying Variables. In Proc. 4th. Int. Conf. on Genetic Algorithms pp346-353 .

Woodcock N., Hallam N. and Picton P. 1991. Fuzzy Boxes as an Alternative to Neural Networks for Difficult Control Problems. Proc. 6th Int. Conf on AI in Engineering .

A Comparison between two Architectures for Searching and Learning in Maze Problems

A. G. Pipe, B. Carse

Faculty of Engineering
University of the West of England
Coldharbour Lane, Frenchay, Bristol BS16 1QY

email ag_pipe@csd.uwe.ac.uk, b_carse@csd.uwe.ac.uk

Abstract

We present two architectures, each designed to search 2-Dimensional mazes in order to locate a "goal" position, both of which perform on-line learning as the search proceeds. The first architecture is a form of Adaptive Heuristic Critic which uses a Genetic Algorithm to determine the Action Policy and a Radial Basis Function Neural Network to store the acquired knowledge of the Critic. The second is a stimulus-response Classifier System (CS) which uses a Genetic Algorithm, applied "Michigan" style, for rule generation and the "Bucket Brigade" algorithm for rule reinforcement. Experiments conducted using agents based upon each architectural model lead us to a comparison of performance, and some observations on the nature and relative levels of abstraction in the acquired knowledge.

1   Introduction

In attempting to solve problems which involve goal oriented movements through an unknown or uncertain environment the need for a good balance between exploration of the environment and exploitation of the knowledge gained from that exploration has been well established [eg. Thrun 1992]. Such problems are typified by, though not restricted to, mobile robots navigating through 2-dimensional maze-like environments. Much work has been done, embracing a wide range of disciplines. Recently the techniques of Temporal Differences learning [Sutton 1991], Neural Networks [Lin 1993], Genetic Algorithms and "Holland" style Classifier Systems [Roberts 1993] have all been brought to bear with promising results, some researchers using more than one technique to build hybrid systems.

We wish to focus here on a comparison between an agent based upon the Classifier System (CS) [Booker et al 1989] and one based upon our own hybrid Adaptive Heuristic Critic (AHC) architecture [Pipe 1 et al 1994]. Below we first describe the maze problem we used as a test platform, then each architecture. We then go on to present some results from experiments which are pertinent to the aims of this paper.

We compare the performance of agents based upon each architecture. However we do not focus on absolute numbers of positions evaluated to reach the goal, clearly small changes in parameters of each architecture, or the maze, could favour one or the other of them in this respect. Rather we are interested in the innate strengths and weaknesses of each. Finally we make some brief observation about the levels of abstraction in the acquired knowledge.

We chose a relatively small maze with simple static obstacles so as to ease our comparison. As we expected our CS agent did not perform well on larger state spaces since, unlike the AHC agent, it possesses no local generalisation capabilities.

2 The Maze Environment

Our maze is set on a 32 X 32 square grid. The edges of the grid form four permanent obstacles enclosing the environment. Other obstacles may be formed from any straight vertical or horizontal line. The agent may be set down on any square within the maze. The goal is made up of a group of four adjacent squares, forming a larger square target two maze positions on each side. This goal can likewise be centred on any position. The agent is deemed to have found the goal if it lands on any one of these four positions.

In both of the cases described below the agent is capable of movement in a single straight line motion from any square in the maze to any other square. The environmental model must therefore be capable of detecting a collision part way through a movement and returning the position of that collision if required.

3 A Classifier System (CS) for a Maze Solving Agent

A good example of the use of a Classifier System (CS) in this context is the work of Gary Roberts [Roberts 1993] who has reconstructed Wilson's Animat (WA) [Wilson 1985, Roberts 1989, Roberts 1991], and his experimental environment "WOODS7" in order to compare the performance of the traditional "Bucket Brigade" reinforcement learning algorithm with Q-learning. In both of Roberts' architectures the CS creates rules (and chains of rules) for movement. A Genetic Algorithm (GA) operates on the rule set, displacing weak rules with new ones evolved from the strong rules. The reinforcement learning algorithm increases the strengths of good rules after interaction with the environment. We do not use one of Roberts' architectures directly here, but make some changes to aid our comparison.

In the following description we will assume that the reader is familiar with the general architectural features of the "Holland" style Classifier System, the "Bucket Brigade" reinforcement learning algorithm and the "Michigan" approach to applying a Genetic Algorithm to the classifiers. If this is not the case then the reader is guided to [Booker et al 1989] as a starting point.

3.1 Classifier Structure

Each classifier is 20 elements long. The left-hand side consists of x- & y-coordinates which are 5 ternary elements each. Each element may be 0, 1 or # to be matched with the input message which is the current position of the agent in the maze (# matches either a 0 or 1 in the input message). The right-hand side consists of a pair of 5-bit x- & y-coordinates, ie. a new absolute position rather than a relative movement. There are 1000 classifiers in the rule set. Initially their strengths are all set to an equal midrange value.

3.2 Classifier Execution Cycle

The following cycle of action and reinforcement is repeated 5000 times. If the agent finds the goal during this time it is placed back at the start position with all classifiers and their strengths unchanged from their goal state levels. At the end of this phase the Genetic algorithm is executed. Each run, consisting of 5000 movements and one application of the Genetic Algorithm, may be repeated as many times as required.

Action Policy

The cycle begins by selecting one from the set of active classifiers matched by the current (x,y)-input message. Each classifier with a left-hand side matching the input message generates a bid to fire based upon its strength and specificity (ie. a larger number of #s reduces the bid). These bid strengths are then used in a "noisy auction" [Goldberg 1989] to determine the single classifier right-hand side to be used in attempting to make the next move. If no matched classifier is found then a "cover operator" [Wilson 1985] generates a new classifier with matching left-hand side and a random right-hand side, replacing the weakest classifier in the population.

The move produced by either the noisy auction or cover operator is tried out in the environment on a move-and-return basis. If a collision with a wall or obstacle occurs then a second "Lamarckian operator" generates another classifier with matching left-hand side but random right-hand side, replacing the one that resulted in collision. This is also tried out in the environment as before. The process continues until a collision free movement is identified. This move is then executed.

Reinforcement

After each move classifier strengths are modified using bucket brigade reinforcement as follows. The classifier responsible for the move receives any environmental reward/punishment which may be attributable. This could be reaching the goal state or a small reward/punishment for moving closer to or further from the goal. The classifier then makes a payoff to the classifier responsible for the previous movement. To prevent cyclic behaviours from becoming established a "firing tax" is paid. Finally all classifiers pay a "survival tax" to prevent "freeloading".

3.3 The Genetic Algorithm

In the "Michigan" approach the Genetic Algorithm operates on the CS at the level of individual classifiers. In our architecture we replace the weakest 200 classifiers using the following operators. "Roulette wheel" selection is used to pick two parents based on strength. One point crossover and mutation are applied to generate a pair of child classifiers. To encourage exploration on the next run all classifiers have their strengths set back to the initialised value rather than inheriting the strengths of their parents.

4 Hybrid Adaptive Heuristic Critic (AHC) Agent

Our second architecture is a hybrid system based around a Genetic Algorithm (GA) and a Radial Basis Function (RBF) Neural Network [eg. Poggio & Girosi 1989, Sanner & Slotine 1991]. It learns a continuous valued internal model of the environment through interaction with it. It employs the GA to search & optimise each movement directly. The GA fitness function is supplied by the RBF Neural Network which acts as an Adaptive Heuristic Critic (AHC) [Sutton 1984, Werbos 1992, Barto et al 1989] . Over successive trials the V-function is learned, a mapping between real numbered positions in the maze and the value of being at those positions.

The underlying architecture is based on some excellent work conducted by Long-Ji Lin [Lin 1993], but with some important modifications. Firstly Lin used a simple action policy based upon stochastic selection from a small list of possible action at each step. In maze problems, for example, the list consisted of movements to the squares adjacent to the current position. As previously stated we have opened this up to allow the action policy to include any straight line movement to any other square in the maze. Clearly this makes the search policy much more complex. We replaced Lin's "stochastic action selector" with a Genetic Algorithm.

Secondly Lin used Multilayer Perceptron (MLP) Neural Networks to store acquired information regarding the value attributable to being at one of the squares in the maze (Adaptive Heuristic Critic), or the value of making a movement from one square to one of the adjacent squares (Q-learning). Here we chose to use the local learning Radial Basis Function (RBF) Neural Network rather than the global learning MLP since this class of Neural Network has certain advantages for real time control applications. A straightforward advantage is that learning times can be orders of magnitude faster with no local minima to get stuck in. More importantly however the characteristic of "local generalisation" which this Neural Network type possesses means that changing the V-function mapping at one point in the input space has a tendency to modify the local region to a gradually diminishing degree as distance from this point increases. This property is very useful here in extracting as much knowledge as possible from each movement through the maze. It translates roughly as, for example, "if a position is a good one then the immediate region around it will also be good".

We chose gaussian basis functions for our Neural Network, with an equal spacing between centres. We will not go into the details of the choice of functions radius but refer the interested reader to [Sanner & Slotine 1991] for the guidelines which we followed.

4.1 AHC Execution Cycle

The processes described below are repeated from each new position in the maze until the goal is attained. This completes one trial which can be repeated any number of times, knowledge of the maze being passed on through the weight vector of the Neural Network.

Action Policy

The Genetic Algorithm (GA) is used directly as the main exploring part of the agent. At each "movement time step" the GA is restarted with a random population and with the entire maze at its disposal searches for the "best next movement", any straight line traversal of the maze from the current position. Lamarckian interactions with the environment occur during the search. The fitness function used to rate GA population members is supplied by the RBF Neural Network, which performs a mapping between a given continuous real valued xy-coordinate input and a measure of "value" for that position as its output, ie. the V-function as described above.

Reinforcement

The Action Policy continues through the normal process of multiple GA generations until a maximum number of generations has been reached. The highest rated population member is then used to make a "non-return" movement to the next position in the maze. After this has happened a Temporal Difference algorithm is executed on the RBF network to change the V-function's shape. Changes are made in the regions of the maze surrounding the movements made so far in the current trial by distributing a discounted reward or punishment back to them. The amount of this reward/punishment is simply derived from the value of the V-function at the new position.

For example if the V-function value at the new position is 55.5 and the value at the previous position was 30.5 then some proportion of the difference (25.0) is added to the previous position's value. According to a simple variant of the standard TD algorithm [Sutton 1991], this process is "daisy-chained" back through time until a "horizon" of backward time steps is reached. If the goal has been attained then the V-function mapping is increased by a FREWARD factor at this position before the final TD learning pass for that trial. If the mapping at the goal position has reached a preset MAXfitness value then this factor is no longer applied.

5 Experiments & Results

Many experiments were undertaken, only a representative set of results is given below, sufficient to support the following discussion. We have used diagrams to illustrate selected runs of agents based on one or the other architecture to assist comprehension. The following parameters not already mentioned in the above descriptions applied during all experiments unless otherwise stated.

5.1 Maze Parameters

Position coordinates are numbered from 0 to 31 on each axis, on all diagrams the x-axis increments from left to right and the y-axis from top to bottom. (1,1) is the agent's starting position (ie. the top left corner in figure 1) and (29,29), (29,30), (30,29), (30,30) are the goal positions. The start and end coordinates of the straight line obstacles are as below.

(6,0)	(6,25)	(12,6)	(12,31)
(18,25)	(31,25)	(18,11)	(18,25)

5.2 Classifier System Parameters

GA Population size = 1000, Crossover probability = 0.5, mutation probability = 0.1
BB BIDRATIO = 0.1, BIDNOISE = 10, FIRETAX = 0.01, SURVIVETAX = 0.0001
RWD FINALREWARD = 100

5.3 Adaptive Heuristic Critic Parameters

GA Population size = 20, number of generations = 10, Crossover probability = 0.9, mutation probability = 0.01
RBF 16 Basis Function centres on each axis, ie. 2 maze squares between each centre. Pre-trained to mapping output of 50.0 for any input.
TD TDRATE = 0.5, HORIZON = 10, MAXfitness of goal = 100.0
RWD FREWARD = 10.0, PUNISH = 5.0

5.4 The First run to the Goal Position

In these experiments we wished to simulate an environment in which the agent has no *a priori* knowledge about either the obstacles or the position of the goal. In the case of the CS agent this meant that no reward was obtainable in the maze except at the goal position. In the case of the AHC agent it meant pre-training the RBF Neural Network with a flat V-function landscape, again receiving reward only at the goal.

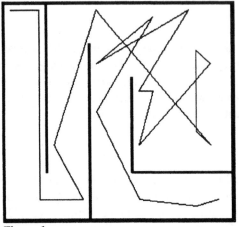

Figure 1

Figure 1 illustrates a typical set of movements made in finding the goal on the first run for the AHC agent, which took 22 moves. Only the "non-return" moves from each position are shown, ie. the 200 (population of 20, 10 generations) "move-and-return" explorations made by the GA at each position are not illustrated for obvious reasons of clarity. The agent therefore required a total of 4400 evaluations to solve the maze on the first run.

The Action Policy of the CS agent results in quite different activity at this level. There are no "move-and-return" evaluations as found in the AHC agent. The first move selected by "noisy" strength auction which does not cause a collision takes the agent to a new position in the maze. Clearly there are a large number of such movements early in the search, covering the maze in a mass of lines when displayed diagrammatically. By contrast the AHC agent investigates the maze thoroughly from the current position then makes a move.

The total number of moves required by the CS agent to find the goal for the first time typically varied by two orders of magnitude, from 47 to 6638 in two sample runs we conducted. This is not surprising considering that the agent is essentially conducting a random search. By contrast the AHC agent is **not** conducting random search. In fact its search has certain deterministic aspects to it with run times to the goal varying by a much smaller factor (we never witnessed a first trial taking more than 100 moves). There are two main factors at work here.

Firstly the PUNISH factor incorporated into the Temporal Difference (TD) learning depresses the V-function landscape behind the agent as it proceeds through the maze since there is no reward available except when the goal is reached. By coincidence the characteristics of this particular maze are beneficial to the AHC agent here. It starts at the top left corner of the maze and drives itself through to the bottom right. Since the goal is in this corner surrounded by relatively close walls accidental discovery is quite likely. The second factor concerns the initial training of the RBF Neural Network. It is impossible to pretrain the Network to a flat landscape with zero error. The landscape has shallow peaks repeated at the RBF centre separations. The GA search will operate on any gradient information available, now matter how small, so it is actually conducting its initial search on a grid determined by the RBF centre density, in this case one every other square of the maze. Since the goal covers four squares the agent behaves as if it was searching a 16 X 16 maze for a single square goal the first time through. After this, and on bigger mazes investigated elsewhere

[Pipe 1 et al 1994], the second factor becomes inconsequential. However they help to explain the initial behaviour of the AHC agent here.

5.5   The Stabilised "Best" Route

After solving the maze once the AHC agent knows the position of the goal because it is the highest V-function mapping in the environment. This puts it into a similar situation to that described in [Pipe 1 et al 1994], ie. it knows where the goal is but not the best way of getting to it. The analogous situation for the CS architecture is that after it has found the goal it only has to re-discover a move which matches the left hand side of the "goal classifier", which hopefully contains a number of #s, to finish. However the AHC agent's knowledge about the goal is known globally from anywhere in the maze, clearly this is not the case for the CS agent used here.

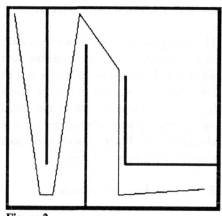

Figure 2

After the first 22 move trial described above the AHC agent took between 8 and 43 moves in the subsequent 6 trials, mostly spent investigating some small area not previously searched. By the 8th trial the route was stabilised, taking 7 direct moves. Figure 2 shows this route.

In most cases the CS agent found a similar stabilised route. Figure 3 shows one such, achieved after 154 trials of the maze and a total of 9843 moves. The agent took just 47 moves in the first trial, then subsequently between 2097 and 9, settling at the illustrated 9 moves. Sometimes the CS agent failed to converge on a stable route over many thousands of moves. We are not clear about this behaviour yet, more work is required.

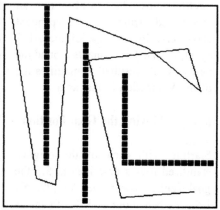

Figure 3

It should be observed that the truly optimal route is unlikely to be found by either of these "blind-leading-blind" architectures. However, for this maze problem at least, the sub-optimal path is not usually too far short for either architecture.

5.6 Moving an Obstacle

Once the route was stabilised we introduced a "short cut" by opening up a hole at the bottom of the second vertical obstacle at coordinates (12,28) & (12,29), then tested it on each agent. The AHC agent immediately took advantage of this, changing on the next trial to a 6 move solution.

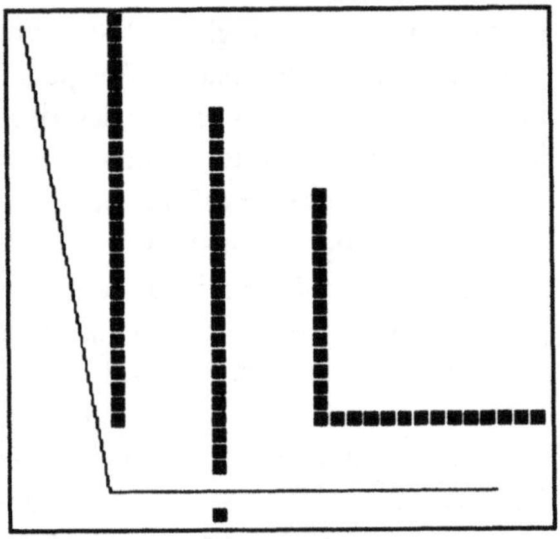

The CS agent was able to accommodate such changes in the environment whenever the GA was run at the end of 5000 moves since all 1000 classifiers have their strengths reset to the initial value, thus giving alternative rules a chance to fire. Figure 4 shows one such change to a 2 move solution.

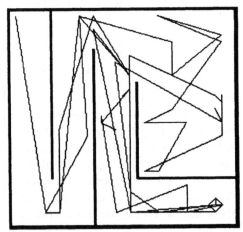

Figure 4

5.7 Adding an Obstacle

Both agents responded to the introduction of new obstacles into the established route in similar ways. Dependent on the position of the new obstacle the established route was disrupted to a lesser or greater degree followed by a re-convergence on a new route.

In the case of the CS agent this essentially required that some local modifications be made to an established string of interdependent classifiers. In the case of the AHC agent localised movements around the new obstacle resulted in modification of the V-function in that area.

5.8 Moving the Goal Position

Starting with each of the original stabilised routes described in 5.5 the goal centre was moved to (29,15). Like the behaviour of 5.7 both agents displayed a similar response. There is

Figure 5

an initial "reluctance" to leave the former goal area, followed by re-evaluation of the maze, and then reinforcement of a new path in subsequent trials. Figure 5 shows the initial response of the AHC agent.

5.9　Moving the Start Position

Again, with the original established rule set of the CS agent and V-function mapping of the AHC agent we moved the start position to (29,1). The AHC agent immediately used the gradient information for this region of the maze, evaluated during previous trials, to complete the maze in 4 moves, as shown in figure 6.

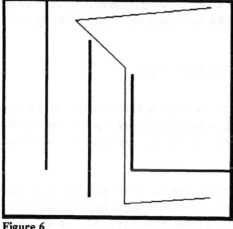

Figure 6

The converged CS agent was much more severely disrupted by this. Because the set of classifiers had converged on the linked chain beginning from the original start position it had no knowledge of other areas of the maze. Because it also possesses no concept of locality it simply reverted to random search until it either landed by chance on a position mapped by a classifier in the original chain or just resolved the entire maze again from scratch. This difference in behaviour is perhaps the single largest disparity between the two algorithms.

6　Discussion

Clearly both architectures utilise the GA's known strengths in global and local optimisation to explore their environment. Whilst the AHC based agent uses it at a low level to directly generate the "optimal" movement from a given position, the CS based agent uses it less often in a more "hands-off" way to generate whole new sets of movements after periods of reinforcement. Clearly also both architectures use their own form of TD reinforcement learning as the main method of exploiting the acquired knowledge from the search. Whilst the AHC based agent reinforces the absolute value of being at positions in the maze the CS based agent reinforces the relative value of movements from one position to another. This derivative relationship is reminiscent of that found between AHC and Q-learning.

In comparing the performance of agents based upon the two architectures described here we conclude that, in our opinion, any differences in their initial search and convergence on a stabilised route to the goal are within the limits of the various parameters which each architecture possesses or minor modifications which could be made. However once converged on an established route there are some significant differences in their responses to changing environments. The AHC based agent is able to respond to a "short-cut" immediately because it continuously re-evaluates the maze at each movement, whereas the CS based agent does this at regular intervals after the

GA executes. However this is a "costly" feature for the AHC agent when considering static environments. Clearly the AHC agent could be made more like the CS agent in this respect by oscillating between its natural searching mode and a "memory recall" mode once converged.

More importantly perhaps the CS agent is also unable to use acquired knowledge to respond to a repositioning of the start position. Once a stable route is established it "forgets" all other knowledge it may have acquired during the search. By contrast the AHC agent retains a V-function mapping for the whole maze, which is clearly useful under these circumstances.

7   Concluding Remarks & Further Work

Clearly there are many areas of similarity in both the architectures and performance of the AHC and CS based agents. Perhaps the biggest disparity in performance results from the differences in the knowledge contained in the converged system in each case.

The CS based agent only possesses knowledge of the "best" path, when the start position is changed, it must start over again. The AHC agent by contrast retains a mapping of the entire maze environment that has been investigated on successive trials, it can therefore respond flexibly under such conditions. The CS architecture might be improved in this respect by one or more of the following modifications. It might be possible to improve the CS agent's abilities in rediscovering the established path to the goal if partial matching [Booker 1982] of the input message were used. We could employ techniques designed to prevent GA convergence, eg. niching. Alternatively the state space could be divided into sub-regions, each of which being covered by a separate Classifier System.

We chose a maze of this size because it lies somewhere near the boundary of feasibility for each architecture. If we scale down the problem it becomes too coarsely discretized for the Neural Network approach to be worthwhile. When we consider scaling up the CS based agent in its present form is at a severe disadvantage. In other work we have found that the AHC architecture scales well to very large state spaces [Pipe 1 et al 1994] and to continuous spaces where the concept of a discrete state space ceases to exist [Pipe 2 et al 1994]. This is principally because of the local generalisation feature of the RBF Neural Network. However a Fuzzy [Parodi & Bonelli 1993] or partial matching [Booker 1982] approach might help solve these problems for the CS agent.

Neither architecture in its present form is capable of learning non-situated general rules such as "if next to a wall move sideways". However the CS architecture at least looks as though it might be **capable** of such an abstraction in this application if the form of the classifier were restructured appropriately.

References

Barto A. G., Sutton R. S., Watkins C. J. C. H., 1989, 'Learning and Sequential Decision Making', COINS Technical Report 89-95

Booker L. B., 1982, 'Intelligent Behaviour as an Adaptation to the Task Environment' PhD dissertation, University of Michigan

Booker L. B., Goldberg D. E., Holland J. H., 1989, 'Classifier Systems and Genetic Algorithms', Artificial Intelligence 40, pp.235-282

Goldberg, D. E., 1989, 'Genetic Algorithms in Search, Optimization and Machine Learning', Addison Wesley

Lin L., PhD thesis, 1993, 'Reinforcement Learning for Robots using Neural Networks', School of Comp. Science, Carnegie Mellon University Pittsburgh, USA

Parodi A., Bonelli P., 1993, 'A New Approach to Fuzzy Classifier Systems', Proceedings of the 5th International Conference on Genetic Algorithms, pp.223-230

Pipe A. G. 1, Fogarty T. C., Winfield A., 1994, 'A Hybrid Architecture for Learning Continuous Environmental Models in Maze Problems', to appear in Procs. of 3rd International Conference on Simulation of Adaptive Behaviour, Brighton

Pipe A. G. 2, Fogarty T. C., Winfield A., 1994, 'Hybrid Adaptive Heuristic Architectures for Learning in Mazes with Continuous Search Spaces', to appear in Procs. of 3rd Parallel Problem Solving from Nature, Jerusalem

Poggio T., Girosi F., 1989, 'A Theory of Networks for Approximation and Learning', MIT Cambridge, MA, AI lab. Memo 1140

Roberts G., 1989, 'A rational reconstruction of Wilson's Animat and Holland's CS-1', Procs. of 3rd Int. Conf. on Genetic Algorithms, pp.317-321, Ed. Schaffer J. D., Morgan Kaufmann

Roberts G., 1991, 'Classifier Systems for Situated Autonomous Learning', PhD thesis, Edinburgh University

Roberts G., 1993, 'Dynamic Planning for Classifier Systems', Proceedings of the 5th International Conference on Genetic Algorithms, pp.231-237

Sanner R. M., Slotine J. E., 1991, 'Gaussian Networks for Direct Adaptive Control', Nonlinear Systems Lab., MIT, Cambridge, USA, Tech. Rep. NSL-910503

Sutton R. S., 1984, PhD thesis 'Temporal Credit Assignment in Reinforcement Learning', University of Massachusetts, Dept. of computer and Info. Science

Sutton R. S., 1991, 'Reinforcement Learning Architectures for Animats', From Animals to Animats, pp288-296, Editors Meyer, J., Wilson, S., MIT Press

Thrun S. B., 1992, 'The Role of Exploration in Learning', Handbook of Intelligent Control: Neural, Fuzzy, & Adaptive Approaches, Van Nostrand Reinhold, Ed. White D., Sofge D.

Werbos P. J., 1992, 'Approximate Dynamic Programming for Real-Time Control and Neural Modelling', Handbook of Intelligent Control: Neural, Fuzzy, and Adaptive Approaches, Van Nostrand Reinhold, Ed. White D., Sofge D.

Wilson S. W., 1985, 'Knowledge growth in an artificial animal', Procs. of Int. Conf. on Genetic Algorithms and their Applications, pp.16-23, Editor Grefenstette J. J.

Fast Practical Evolutionary Timetabling

Dave Corne, Peter Ross, Hsiao-Lan Fang

Department of Artificial Intelligence, University of Edinburgh, 80 South Bridge,
Edinburgh EH1 1HN, U.K.

Abstract. We describe the General Examination/Lecture Timetabling Problem (GELTP), which covers a very broad range of real problems faced continually in educational institutions, and we describe how Evolutionary Algorithms (EAs) can be employed to effectively address arbitrary instances of the GELTP. Some benchmark GELTPs are described, including real and randomly generated problems. Results are presented for several of these benchmarks, and several research and implementation issues concerning EAs in timetabling are discussed.

1 Introduction

A number of researchers have applied evolutionary algorithms (EAs) to timetabling problems [2, 1, 7, 3, 8, 9]. Work so far has however tended to be isolated, applying a range of techniques to disconnected problems with little cross-comparison. The intent of this paper is to fully set out the nature of the problems addressed in EA timetabling research, and to present a series of results on some real and randomly generated problems which form part of a benchmark set we are collecting, using the (so far) most successful variant of the 'direct' approach. These benchmarks will hopefully spawn further, focussed work in the EA timetabling arena.

Section 2 describes the general form of the timetabling problem addressed by researchers using EAs and/or other techniques. Section 3 then notes how an EA may be set up to address an arbitrary instance of such a problem. In Sect. 4, we describe various test problems, both real and random, and gives tables of results on these problems. Section 5 then discusses a variety of aspects of the general approach which are worth mentioning, and some summary and conclusions appear in Sect. 6.

2 Evolutionary Timetabling

A large class of timetabling problems can be described as follows. There is a finite set of events $E = \{e_1, e_2, \ldots, e_v\}$ (for example, exams, seminars, project meetings), a finite set of potential start-times for these events $T = \{t_1, t_2, \ldots, t_s\}$, a finite set of places in which the events can occur $P = \{p_1, p_2, \ldots, p_n\}$, and a finite set of *agents* which have some distinguished role to play in particular events (eg: lecturers, tutors, invigilators, ...) $A = \{a_1, a_2, \ldots, a_m\}$. Each event e_i can be regarded as an ordered pair $e_i = (e_i^l, e_i^s)$, where e_i^l is the length of event e_i (eg:

in minutes), and e_i^s is its size (eg: if e_i is an examination, e_i^s might be the number of students attending that examination). Further, each place can be regarded as an ordered pair $p_i = (p_i^e, p_i^s)$, where p_i^e is the *event* capacity of the place (the number of different events that can occur concurrently in this place), and p_i^s is its size (eg: the total number of students it can hold). The event capacity matters: two exams can take place in the same room simultaneously, but two lectures cannot. There is also an $n \times n$ matrix D of travel-times between each pair of places.

An *assignment* is an ordered 4-tuple (a, b, c, d), where $a \in E$, $b \in T$, $c \in P$, and $d \in A$. An assignment has the straightforward general interpretation: "event a starts at time b in place c, and with agent d". If, for example, the problem was one of lecture timetabling then a more natural interpretation would be: "lecture a starts at time b in room c, and is taught by lecturer d".

Given E, T, P, A, and the matrix D, the GELTP involves producing a timetable which meets a large set of constraints C. A timetable is simply a collection of v assignments, one per event. How easy it is to produce a useful timetable in reasonable time depends crucially on the kind of constraints involved. In the rest of this section, we discuss what C may contain.

2.1 GELTP Constraints

Different instances of GELTPs are distinguished by the constraints and objectives involved, which typically make many (or even all) of the s^{vmn} possible timetables poor or unacceptable. Each constraint may be hard (must be satisfied) or soft (should be satisfied if possible). Many conventional timetabling algorithms address this distinction inadequately: if they cannot solve a given problem they relax one or more constraints and restart, thus trying to *solve a different problem*. The kinds of constraints that normally arise can be conveniently classified as follows.

Unary Constraints Unary constraints involve just one event. Examples include: "The science exam must take place on a Tuesday", or "The Plenary talk must be in the main function suite". They naturally fall into two classes:

Exclusions : An event must not take place in a given room, must not start at a given time, or cannot be assigned to a certain agent.

Specifications : An event must take place at a given time, in a given place, or must be assigned to a given agent.

Binary constraints A binary constraint involves restrictions on the assignments to a pair of events. These also fall conveniently into two classes:

Edge Constraints : These are the most common of all, arising because of the simple fact that people cannot be in two places (or doing two different things) at once. A general example is: "event x and event y must not overlap in time". The term 'edge' arises from a commonly employed abstraction of simple timetabling problems as graph colouring problems [12].

Juxtaposition Constraints : This is a wide class of constraints in which the ordering and/or time gap between two events is restricted in some way. Examples include: "event x must finish at least 30 minutes before event y starts", and "event x and event y must start at the same time".

Edge constraints are of course subsumed by juxtaposition constraints, but we single out the former because of their importance and ubiquity. Edge constraints appear in virtually all timetabling problems, and in some problems they may be the *only* constraints involved.

Capacity Constraints Capacity constraints specify that some function of the given set of events occurring simultaneously at a certain place must not exceed a given maximum. Eg: in lecturing timetabling we must usually specify that a room can hold just one lecture at a time. In exam timetabling this capacity may be higher for many rooms, but in both cases we need also to consider the total student capacity of a room. We may allow up to six examinations at once in a given hall, but only as long as that hall's maximum capacity of 200 candidates is not exceeded.

Event-Spread Constraints Timetablers are usually concerned with the way that events are spread out in time. In exam timetabling, for example, there may be an overall constraint of the form "A candidate should not be expected to sit more than four exams in two days". In lecture timetabling, we may require that multiple lectures on the same topic should be spread out as evenly as possible (using some problem-specific definition of what that means) during the week. Event-spread constraints can turn a timetabling problem from one that can be solved easily by more familiar graph-colouring methods into one which requires general optimisation procedures and for which we can at best hope for a near optimal solution.

Agent Constraints Agent constraints can involve restrictions on the total time assigned for an agent in the timetable, and restrictions and specifications on the events that each individual agent can be involved in. In addition to those already discussed (exclusions and specifications) we typically also need to deal with constraints involving agnet's preferences (for teaching certain courses, for example), and constraints involving teaching loads.

3 Applying EAs to the GELTP

In applying an EA to a problem, central considerations are the choice of a chromosome representation and the design of the fitness function. In this section we describe the approach we have found most successful so far.

3.1 Representation

For the GELTP, a chromosome is a vector of symbols of total length $3v$ (where v is the number of events), divided into contiguous chunks each containing three genes. The three alleles in the ith chunk, where $1 <= i <= v$, represent the time, place, and agent assignments of event i. Naturally, the set of possible alleles at time, place, and agent genes are respectively identified with the sets T, P, and A. The simple example chromosome "abcdef" represents a timetable in which event e_1 starts at time a in place b, involving agent c, and event e_2 starts at time d in place e, involving agent f.

Very many timetables thus represented will involve edge-constraint violations ('clashes'). Eg, the chromosome "abcabcabc......" is well-formed, even though it puts every event in the same place at the same time and involving the same agent. The job of the EA is to gradually remove such violations of constraints during the artificial evolutionary process.

3.2 The Evaluation function

It is important to be able to differentiate the relative quality of different timetables. An apparently satisfactory solution is widely used in the EA literature: it is simply to have fitness inversely proportional to the number of constraints violated in a timetable with each instance of a violated constraint weighted according to how important or not it is to satisfy it.

Let C_j be the set of constraints of type j (for example, event-spread constraints). The specific type classification used can be tailored to suit the problem. Each violated member of C_j attracts a specific penalty w_j. For each $c \in C_j$, let $v(c,t) = 1$ if c is violated in timetable t, and $v(c,t) = 0$ if c is satisfied. A simple fitness function for a GELTP is thus:

$$f(t) = 1/(1 + \sum_{\text{types}_j} w_j \sum_{c \in C_j} v(c,t)) \tag{1}$$

Assuming that all the penalties are positive, this function is 1 if and only if all the constraints hold, otherwise it is less than 1. The general idea is that an appropriate choice of values for the penalty terms should lead both to reasonable tradeoffs between different kinds of constraint violations, and (by virtue of defining the shape of the fitness landscape) effective guidance of the EA towards highly fit feasible timetables.

The approach works best if the constraint set C is fine grained. For example, if C contained only the 'single' constraint: "the timetable has no clashes", then the fitness landscape would contain a few spikes in an otherwise flat space, which is quite intractable to any form of search. C is hence best composed of low order constraints, each of which involves only one or two events. For example, the important constraint "No lectures which share common students should clash" appears in C as a large collection of separate constraints each involving a distinct pair of lectures which (are expected to) share common students or teachers. The set of such edge constraints are usually the largest single block of constraints in a timetabling problem. Commonly, applications will be faced with data in a form like: "student Jones sits exams Maths, Physics, Chemistry, ..."; this is then easily transformed to collections of binary constraints between events. In this case, each distinct pair of exams taken by the same student constitutes an edge constraint; typically, we may also automatically create an event-spread constraint between the same pair.

Given such a collection, a question arises as to how to treat them: they may either be treated as a uniform collection of distinct edge constraints of identical importance, or each edge may be weighted according to how many students share these events. In this way, and typically throughout this 'problem transformation' process, the constraint satisfaction problem (CSP) of finding a timetable which satisfies all the constraints can be transformed into any of several different constrained optimisation problems (COPs)[4]; each such COP will share at least one global optimum with the CSP (and preferably all of them) but will be otherwise different. In general, it seems important to make the landscape of this COP as meaningful as possible. For example, in an exam timetabling problem in which we treat each distinct edge constraint the same and each distinct event-spread constraint the same (but typically with event-spread constraints being less important edge constraints), we may find that the best we can do is find a collection of answers which violate a single event-spread constraint. These may be markedly different however; although each answer has just one pair of edge-constrained events timetabled too closely, some may involve only one student, while others may involve several. By weighting edge-constrained pairs of events (in this case according to the number of students sharing the given exams), the COP becomes more meaningful in that it is able to distinguish between such cases. This is particularly important in cases where the CSP has no solution, or where its solution is difficult to find; one or other of which is quite common norm in exam and lecture timetabling problems. In such cases, the COPs addressed by the EA (or some other method) may have different global optima, and so it becomes particularly important to use as 'meaningful' a COP as possible. In an example similar to that just discussed, for example, the less meaningful COP may have an optimum which violates just one event-spread constraint (which involves 50 students suffering consecutive exams, say), while a global optimum of the more meaningful version may involve just 2 students suffering consecutive exams, although having violations of two distinct event-spread constraints.

3.3 Speed

An important general consideration is that calculation of fitness be fast. Fortunately, this is usually true for most of the constraints we need to consider in the GELTP, which mainly comprise unary and binary constraints. More to the point, however, when using the direct representation it is particularly easy to set up 'delta evaluation', whereby to evaluate a timetable we need only consider the changes between it and an already-evaluated reasonably similar one such as one of its parents.

[9] discusses the use of delta evaluation further, noting that it is slightly more than just an obvious speedup measure. In particular, [9] notes that speed comparisons made in terms of 'number of evaluations', as commonly done, may often be overturned when delta evaluation is in use. That is, EA configuration X might regularly find results in fewer evaluations than EA configuration Y (and hence be faster when *full evaluation* is in use, but Y may be found to be faster than X when delta evaluation is employed. The reason for this is just that evaluations when using X in conjunction with delta evaluation generally take longer than with Y, because, for example, X employs a highly diversifying recombination operator which means that delta evaluation has to consider several changes each time (though will typically still be faster than full evaluation). It is important to note here that delta evaluation leads to the notion of *evaluation equivalents* EEs, which we employ later on. This simply records the time taken for a run in terms of the number of *full evaluations* that would have been done in the same time. We measure this, for example, by dividing the total number of constraint checks made during a delta evaluation run by the (constant) number that would be made during a full evaluation. This measure hence allows a machine independent measure of the time taken by an EA/timetabling run employing delta evaluation. An alternative is simply to record the total number of constraint checks made, but EE's provide a more accessible measure and give a more reasonable indication of total time taken.

3.4 Penalty Settings

Clearly, we can choose penalty terms for different constraints according to our particular idea of how we would trade off the advantages and disadvantages of different solutions to the problem in hand. Penalties must be set with care, however. If the ratio between two penalties (say, ordering constraints *vs* event-spread constraints) is too high, then search will quickly concentrate on a region of the space low in violations of the more penalised constraint, but perhaps missing a less dense region in which better tradeoffs could be found. If too low, then the capacity for the search to trade off between different objectives is lost. Evidently, optimal penalty settings depend on many things, primarily involving the density of regions of the fitness landscape in relation to each constraint, as well as the subjective relative disadvantages of different constraint violations in the problem at hand. Alternative possibilities include the approach in [10] (in the context of multiple objective facility layout problems), in which penalty settings are revised dynamically in accordance with the gradually discovered nature of the constrained fitness landscape. Also, a principled method for constructing scalar functions for multi-objective problems is discussed in the context of EA optimisation in [6]. In this method, called MAUA (Multi-Attribute Utility Analysis), extensive questioning of an expert decision maker (in our case, an experienced timetable constructor) on example cases (pairs of distinct timetables) would lead to the construction of a nonlinear function M of the vector of summed constraint violations, designed so as to best match the judgement of the expert in that the ordering on timetables imposed by M optimally matches that imposed by the expert. The effort involved in performing MAUA, however, is unlikely to be rewarded with a function M which is significantly closer to the expert decision-maker's judgement than an essentially *ad-hoc* but intuitive linear weighted penalty function and it is not clear that this is desirable. MAUA involves no attempt to structure M such that the fitness landscape is more helpful to the search process.

It suffices to say here that extensive experience so far suggests that for a wide range of problems we can settle for a simple linear penalty-weighted sum of violations with an intuitive choice of penalty settings.

3.5 Violation Directed Mutation Operators

In [9], a family of Violation Directed Mutation VDM operators for timetabling problems are examined, and it is found that a certain subclass of variations on VDM are particularly powerful for use on a range of realistic problems. Similar operators are studied in [4] for graph colouring and other constraint satisfaction problems. Here we adopt the use of one particular VDM variant, called (rand, tn10), as standard. The action of this operator is roughly as follows (fuller description appears in [9]): an application of (rand, tn10) to a timetable amounts to randomly choosing an event (gene), and then selecting a new allele (timeslot) for it via tournament selection with a tournament size of 10; an allele's 'fitness' for this purpose is a measure of the degree to which it will reduce the degree of constraint violation involving the chosen event. This 'allele choice' operation involves some computational expense, but since the substantial part of it involves numbers of constraint checks, the time it takes can be meaningfully absorbed into our EE measure.

4 Some Benchmark Timetabling Problems

We first look at five real examination timetabling problems, and later consider 32 randomly generated highly over-constrained test problems. These are not fully general, in the sense of having a full repertoire of place and agent constraints as well as several kinds of timeslot constraint, but only consider edge, event-spread, exclusion, and, in the case of two of the real problems, timeslot capacity constraints (arising from a limit on the number of seats available in examination halls at any one time). Realistic fully general benchmarks will appear anon, but for now it seems reasonable to provide more simply defined problems (and hence more accessible for comparative performance research) which are nevertheless realistically difficult and/or common GELTP variants..

4.1 Some Real Examination Timetabling Problems

Three of these arise from MSc examinations at the EDAI[1], and two from Kingston University, London. Each of the EDAI problems, named in turn: edai-ett-91,

---

[1] University of Edinburgh Department of Artificial Intelligence

edai-ett-92, and edai-ett-93, involve a four slots per day timetable structure, and involve a number of edge constraints and exclusions. In each case, there is an event-spread constraint as follows: if any pair of edge-constrained events are timetabled to appear on the same day, then they must have at least one full slot between them. That is, they must occupy the first and third, first and fourth, or second and fourth slots. edai-ett-91 has 314 edge constraints, and no exclusions, and must be timetabled over six days (hence 24 slots). edai-ett-92 has 431 edges, no exclusions, and must occupy seven days, while edai-ett-93 has 414 edges, 480 exclusions, and must occupy nine days.

The Kingston University problems respectively represent the first and second semester exams faced by Kingston University students in 1994. In the first semester problem, ku-ett-94-1, 97 exams have to be arranged over 5 days with with 3 slots per day. There are 399 edge constraints, and hence 399 individual event spread constraints. The event-spread constraint in this case is that if an edge constrained pair of exams both occur on the same day, they must occupy the first and third slot. ku-ett-94-2 is the second semester problem; 128 exams must be arranged over an 8 day period, with 3 slots per day. The event spread objective this time is to avoid a student facing more than one exam per day. Hence, edge-constrained pairs of events should not occupy slots on the same day. There are 536 edge constraints, and hence 536 event spread constraints too. An additional constraint faced by both of the Kingston University problems is that a maximum of just 470 candidates can be seated in any timeslot. Hence, associated with each event is a weight representing the number of students taking the appropriate exam. These weights are then used by the fitness function (as well as by the VDM operator) to penalise (avoid) violations of this capacity constraint.

4.2 Default EA Configuration

The EA configuration used in all experiments was as follows. A reproduction cycle consisted of a breeding step (in which one new chromosome was produced) followed by an insertion step, in which this new chromosome replaced the currently least fit individual (but only if the new individual was fitter). With probability 0.2, a breeding step involved the selection of one parent and the simple gene-wise mutation of it with a probability of 0.02 of randomly reassigning the allele of each gene in turn. With probability 0.8, a breeding step involved the selection of one parent, and the application of the VDM operator (rand, tn10. Tournament selection was used with a tournament size of 6, and the population size was always 1,000.

Using the default configuration, we examine the reliability of this EA on five real timetabling problems. The default configuration was applied in 100 separate trials to each of the five problems detailed above. We record, in each case, the number of such trials which found an optimum (the 'number of perfect trials' column; on each of these problems, optima violating no constraints exist), the least, average, and most evaluations taken to reach an optimum for those trials which did, and the the least, average, and most evaluation equivalents taken to reach an optimum for those trials which did. These results appear in Table 1. Each trial on an **edai** problem was run for 25,000 evaluations, while trials on the **ku** problems ran for 40,000 evaluations each.

Table 1. Performance on five real lecture timetabling problems

Problem	No. Perfect Trials	Evaluations			Eval. Equiv's		
		Lowest	Mean	Highest	Lowest	Mean	Highest
edai-ett-91	84	4595	7693	24933	4793	8084	26241
edai-ett-92	50	7701	14087	21703	5000	13771	21179
edai-ett-93	98	6611	9592	14818	4717	8501	13166
ku-ett-94-1	93	10273	14890	20243	4227	6181	8420
ku-ett-94-2	57	12594	19241	28808	3731	6109	9103

It may first be pointed out that these results show great general potential for EAs on timetabling problems. All trials are relatively fast; for example. Speed on the problems above ranged 250 to 400 evaluations per second[2] Even in the two cases where the EA found an optimum only 50% or so of the time, this means that a small number of trials would be more or less guaranteed to stumble on a perfect timetable soon enough. In the case of the **edai-ett-91** and **edai-ett-92**, the timetables actually used for these problems were independently produced by the relevant course administrators. As detailed in [3], these were very poor in relation to average EA-produced timetables on the same problems, and certainly far from the perfect timetables that the EA used here can typically find. From the **edai-ett-93** case on, the course administrators have been (thankfully) using an EA to do their timetabling work, and hence we do not have independently produced efforts for comparison. In the case of the **ku** problems, the course administrators at Kingston University tried to manually produce timetables for these problems but were having great difficulty, owing in particular to the recent modularisation of the underlying course; they also note that satisfying the capacity constraint in each case was particularly troublesome. Kingston University's timetablers notified us of their problems halfway through their troubled attempts, and sent us the relevant data, whereupon the EA managed to find perfect results regularly in each case.

The main intent in presenting these results is to provide benchmarks for future comparisons with other techniques. As detailed at the end of the paper, all problems we use are freely available. Comparative performance results on the above problems will focus on the *reliability* figure. That is, in looking for improvements on our timetabling EA, we are looking for a single method which will improve reliability in finding the optimum over all of the above set (and others). without any major speed sacrifice. Alternative valid targets include achieving

---

[2] On a Sun Sparcstation 2, using a not-necessarily optimised C program.

similar reliability faster, or perhaps less (but >50%, say) reliability but *much* faster.

4.3 Some Random Over-Constrained Timetabling Problems

A further kind of target, as is commonly the case with benchmark job shop scheduling problems, for example [11], is find increasingly lower bounds on the total penalty values for a suite of problems. Here we describe some timetabling problems involving edge and event-spread constraints which are constrained enough for this purpose, and present our best results so far using the EA described, and running to a limit of 200,000 evaluations in each of 10 trials for each problem. Each of these problems involves the same temporal structure and event-spread constraint as the edai-ett problems; that is, there are four timeslots per day, and for each edge constrained pair of events there is also an edge constraint which specifies that there should be at least one entire slot between them these events if they appear on the same day.

Each problem is named bench-ett(X,Y,Z,W), where:

- X is simply an identifier of the set of random edge constraints involved.
- Y is the number of events to be timetabled.
- Z is the number of days allowed; hence there will be $4 \times Z$ slots altogether.
- W is the number of edge constraints. Each edge constraint is a pair of events, (e_1, e_2), constrained not to appear in the same timeslot. Associated with each constraint is the edai-ett style event-spread constraint, Further, associated with each edge constraint c in the set of edge constraints C is a weight c_w. This is an integer between 1 and 100 inclusive.

In each case, the EA applied a penalty P to a candidate timetable t as follows, in which $v(c,t)$ for $c \in C$ is 1 (0) if edge constraint c is violated (satisfied) in timetable t, while $e(c,t)$ for is 1 only if the event-spread constraint is violated but the corresponding edge constraint is *not* violated.

$$P(t) = \Sigma_{c \in C} 2v(c,t)c_w + e(c,t)c_w \qquad (2)$$

Figures in the second column of Table 2 are hence the minimal value for P we have so far found on these problems. The third and fourth columns respectively give the weighted penalty for edge constraints violated by the best timetable found, and the weighted penalty for event-spread constraints violated in the same timetable.

The problems in Table 2 are extremely highly constrained; almost certainly more so than any real timetabling problem. Nevertheless, comparative performance of techniques (EA-based or not) on this set (and similar sets) of problems will be useful because these are indeed timetabling-style problems, similar in the nature of the constraints involved than a very common class of real exam timetabling problems. Hence, better performance on these benchmarks will almost certainly reflect potential for better performance on real timetabling problems. Also, since these are very highly constrained COPs, it seems likely that

Table 2. Performance on Very Highly Constrained Random Timetabling Problems

Problem	Smallest Penalty	Edge Penalty	Event-Spread Penalty
bench-ett(1,50,5,1000)	1116	212	692
bench-ett(1,50,6,1000)	613	81	451
bench-ett(1,50,7,1000)	248	17	214
bench-ett(1,50,8,1000)	79	2	75
bench-ett(2,50,5,1000)	1011	174	663
bench-ett(2,50,6,1000)	477	57	363
bench-ett(2,50,7,1000)	151	15	121
bench-ett(2,50,8,1000)	84	7	70
bench-ett(3,50,5,1000)	1129	180	769
bench-ett(3,50,6,1000)	550	65	420
bench-ett(3,50,7,1000)	205	24	157
bench-ett(3,50,8,1000)	61	3	55
bench-ett(1,100,10,4000)	1359	159	1041
bench-ett(1,100,11,4000)	825	125	575
bench-ett(1,100,12,4000)	479	49	381
bench-ett(1,100,13,4000)	312	29	254
bench-ett(2,100,10,4000)	1492	265	962
bench-ett(2,100,11,4000)	1025	158	709
bench-ett(2,100,12,4000)	620	60	500
bench-ett(2,100,13,4000)	346	44	258
bench-ett(3,100,10,4000)	1493	331	831
bench-ett(3,100,11,4000)	874	170	534
bench-ett(3,100,12,4000)	400	46	308
bench-ett(3,100,13,4000)	232	35	162
bench-ett(1,500,1,5000)	122502	30619	61264
bench-ett(1,500,2,5000)	23576	3948	15680
bench-ett(1,500,3,5000)	4580	411	3758
bench-ett(1,500,4,5000)	453	48	357

there is great scope for *continual* improvement on the 'best so far' figures given above. That is, techniques better than the EA we have used so far may be found which achieve 100% reliability on the five real timetabling problems discussed earlier, and hence be indistinguishable on those problems. However, such techniques are likely to be separated in terms of their comparative performance on these highly constrained benchmarks, while still leaving room for further improvement. Lastly, being COPs with (almost certainly) no perfect solutions possible in each case, these problems reflect the difficulties involved in many real modular lecture timetabling problems; in such a problem, for example, a typical approach may be to attempt to allow *any* combination of courses as feasible in the timetable, to give students greater flexibility in their module choices. Since this goal is almost always infeasible, such a problem becomes a COP, with weightings on pairs of module choices reflecting the desirability of allowing these as viable combinations for a student.

5 Prospects for EAs in Timetabling

Indications so far point to the approach we have described as highly promising for general timetabling needs, at least of the kind usually found in educational institutions. However, several matters need pointing out with respect to its more general application. Firstly, as may be discerned, and as we have found, it is not always immediately apparent how best to describe the problem itself in terms of the kinds of constraints processed. Eg, to incorporate the overall event-spread constraint "no student should sit more than four exams in two successive days", one choice would be to have the objective function (partially) calculate each student's individual timetable, and directly penalise every instance of a student suffering the appropriate constraint violation. A potentially far less computationally expensive choice, however, might be simply to penalise all consecutive edge-constrained events which are less than, say, 2 hours apart in the overall timetable; though not addressing the constraint directly, this certainly applies the artificial evolutionary pressure in the right direction. These are just two of several possibilities for addressing this constraint. Generally, choices occur at many stages in the process of translating the constraints of a real GELTP into a penalty function, and much work is needed to discern the best way to do this. The difficulty may not be too great, however. At least *some* way of handling any given constraint will be naturally apparent, and, in our experience, it is unlikely that different choices will bring significantly different results, unless the problem instance is very large and hence solution speed is a major factor.

Another important aspect is comparative performance with other techniques. There is reason to suggest that the EA approach described will succeed more as a *general timetabling tool* than some other methods. In many cases, however, some other technique may be significantly preferable. Eg, a large, continually faced problem which changes little (in terms of the kinds of constraints involved) from instance to instance may be better off treated with some specifically designed algorithm. This may itself be based on an EA, involve an alternative EA-based approach, or be based on best-first search with a specifically designed heuristic, for example. In general, much work is required to discern what promises to be the best method for different kinds of GELTP.

Also, several aspects of the approach itself warrant much further study. Among these are the design of the objective function itself, as briefly discussed above and also in Sect. 3.4, the use of alternative operators, and various aspects of the underlying EA configuration.

Finally, timetablers (ie; human course organisers, for example) often have needs which are not directly met within the described approach. There may be some desire to generate several different possible timetables, for example, or there may be need for extensive preprocessing and altering of the constraints (which may initially contain several inconsistencies). Use of the EA described (via iterative applications, for example) may be useful as a tool in each case, but such considerations evidently require more useful refinements or extensions, if not other methods entirely. A planned extension to a future version of the EA described here, for example, is to have an ATMS-based constraint-checking front end. Also, experiments are under way which involve the generation of multiple distinct solutions in a single run, making use of EA variants specifically designed for multimodal optimisation.

6 Summary and Conclusions

We have described and noted various matters in the application of EAs to general educational-institution based timetabling problems. EAs seem to have great potential in this arena, and we illustrate this via presentation of various results. First, we show that a particular EA described here (but more fully in [9]), finds, quickly and reliably, perfect timetables for each of five real examination timetabling problems. The particular reliability results are offered as benchmarks against which to examine alternative techniques. Second, a collection of very highly constrained random timetabling-style problems are described, and our best results on these are given. Various research issues and considerations are then noted.

Finally, the test problems addressed may be freely obtained (and explained) from the authors, and/or via the FTP site ftp.dai.ed.ac.uk .

Acknowledgements

Thanks to Bob Fisher and Peter Sutcliffe for providing problem data. Thanks also to the UK Science and Engineering Research Council for support of Dave Corne via a grant with reference GR/J44513, and to the China Steel Corporation, Taiwan, R.O.C., for support of Hsiao-Lan Fang.

References

1. Abramson D., Abela, J. : A Parallel Genetic Algorithm for Solving the School Timetabling Problem. IJCAI workshop on Parallel Processing in AI, Sydney, August 1991
2. Colorni, A., Dorigo, M., Maniezzo, V.: Genetic Algorithms and Highly Constrained Problems: The Time-Table Case. Parallel Problem Solving from Nature I, Goos and Hartmanis (eds.) Springer-Verlag, 1990, pages 55–59
3. Corne, D., Fang H-L., Mellish, C.: Solving the Module Exam Scheduling Problem with Genetic Algorithms. Proceedings of the Sixth International Conference in Industrial and Engineering Applications of Artificial Intelligence and Expert Systems, Chung, Lovegrove and Ali (eds.), 1993, pages 370–373.
4. Eiben, A.E., Raue, P.E., Ruttkay, Z Heuristic Genetic Algorithms for Constrained Problems. Working papers of the Dutch AI Conference, 1993, Twente, pages 341–353.
5. Corne, D., Ross, P., and Fang, H-L.: Fast Practical Evolutionary Timetabling. Proceedings of the AISB Workshop on Evolutionary Computation, Springer Verlag, 1994, to appear.

6. Horn, J., and Nafpliotis, N.: Multiobjective Optimisation Using The Niched Pareto Genetic Algorithm. Illinois Genetic Algorithms Laboratory (IlliGAL) Technical Report No. 93005, July 1993.

7. Ling, S-E.: Intergating Genetic Algorithms with a Prolog Assignment Problem as a Hybrid Solution for a Polytechnic Timetable Problem. Parallel Problem Solving from Nature 2, Elsevier Science Publisher B.V., Manner and Manderick (eds.), 1992, pages 321–329.

8. Paechter, B. Optimising a Presentation Timetable Using Evolutionary Algorithms. Proceedings of the AISB Workshop on Evolutionary Computation, Springer Verlag, 1994, to appear.

9. Ross, P., Corne, D., and Fang, H-L.: Improving Evolutionary Timetabling with Delta Evaluation and Directed Mutation.: Proceedings of PPSN III, Jerusalem, October 1994, Springer Verlag, to appear.

10. Smith, A.E., and Tate, D. M.: Genetic Optimisation Using a Penalty Function. Proceedings of the Fifth International Conference on Genetic Algorithms, San Mateo: Morgan Kauffman, S. Forrest (ed), 1993, pages 499–503.

11. Taillard, E.: Benchmarks for basic scheduling problems. European Journal of operations research, Volume 64, 1993, pages 278–285.

12. Wilson, R. J.: Introduction to Graph Theory. Longman, London, 1979.

Optimising a Presentation Timetable Using Evolutionary Algorithms

Ben Paechter

Department of Computer Studies
Napier University
219 Colinton Road
Edinburgh EH14 1DJ
email:benp@uk.ac.napier.dcs

Abstract

This paper describes a solution to the problem of scheduling student presentations which uses evolutionary algorithms. The solution uses a permutation based approach with each candidate schedule being coded for by a genotype containing six chromosomes. Five systems (chromosome representation and genetic operators) are described and their suitability assessed for this application. Three of the systems use direct representations of permutations, the other two use indirect representations. Experimental results with different fitness equations, operator rates and population and replacement strategies are also given. All the systems are shown to be good at solving the problem if the algorithm parameters are correct. The best parameters for each system are given along with those parameters that do particularly badly.

1. Introduction

Each year the Napier University Computing degree course conducts a 'Vertical Project' for a week during which teams built up across academic years work together on a project. The climax of the week is an afternoon in which the groups give presentations and are assessed by their peers. Since there are twenty four groups and only time for six presentations there must be four parallel sessions. Each group has to make one presentation and assess four others. Each group must have a break before their presentation. The peer assessments are relative scores, and to ensure fairness, each presentation should be marked against a maximum number of other presentations. In other words, if Group A is marked by Groups B, C, D, and E then the union of the presentations seen by these groups must be as large as possible. This means that the groups must move around the sessions. The problem then is to produce a schedule which maximises the amount of cross-marking and minimises the amount of moving around between venues. The number of possible schedules is of the order of 10^{47}, and so exhaustive search is not possible. A search by evolutionary means was therefore attempted.

2. Problem Analysis

Each group can be allotted a timeslot and venue to give their presentation and a timeslot to have a break just before their presentation (or at the end of the day if they are presenting in the first timeslot). We then have to assign assessing groups for each of the venues at each of the timeslots. This amounts to putting four assessing groups into each of the four venues at each of the timeslots. These can be assigned by using a permutation of the groups for each timeslot. The permutation determining the order in which the groups are assigned to venues. Six permutations of the groups are required - one for each timeslot.

3. Genetic Representation

The six permutations required to produce each schedule can be represented by six chromosomes. The representation can either be direct, with the chromosomes being the permutations, or indirect, with the chromosomes giving instructions on how to build the permutation. Two systems using an indirect representation and three systems using a direct representation were implemented and they are described below.

3. 1 Systems Using Indirect Representation

With these representations the chromosome tells us how to build the permutation. The first system is called "Naive Repair"; the second is called "Displacing Ordinal".

The Naive Repair (NR) System

In this system a permutation p of n elements is encoded by a chromosome c:

$$c[i], 0 \leq c[i] \leq n-1, i=0, \dots , n-1.$$

The permutation p is produced from c by working through the chromosome, inserting the chromosome values into the permutation. If at any time the chromosome value already exists in the permutation then it is changed to be the next value which does not exist in the permutation. The algorithm is:

```
for i=0 to n-1 do
        if(i>0)
                        while c[i] ∈p[0..i]
                                c[i]:=(c[i]+1) mod n
                        end-while
                end-if
        p[i]:=c[i]
end-for
```

It should be noted that this system repairs the chromosome to make it equal to the permutation it produces. This not only saves processing time in future evaluations but also reduces the effect on the permutation of small changes in the chromosome, making the system less brittle. If the chromosomes are not updated then values which appear late in the chromosome can be highly dependent on those which precede them. This is particularly true if the distance between a chromosome value and the value inserted into the permutation is great (as it certainly would be for the later values in the chromosome). Writing the permutation back into the chromosome effectively reduces these long distances to zero, thereby reducing the problem. The scheme is illustrated in Figure 1.

The Displacing Ordinal (DO) System

In this system a permutation p of n elements is encoded by a chromosome c:

$$c[i], 0 \leq c[i] \leq i, i=0, \dots ,n\text{-}1.$$

The permutation p is produced from c in the following way:

```
for i=0 to n-1 do
    if(i>0)
                p[i] := p[c[i]]
    end-if
    p[c[i]] := i
end-for
```

As the permutation p is built, each element of c dictates where in the permutation to put the next item. The item that is already in that position is moved to the end of the permutation. This system is described in [1] and is similar to a system presented in [2].

Crossover and Mutation in the Indirect Systems

The crossover used for both of these systems is classical single point crossover of the array of integers. The crossover rate is defined in the following way: when any two individuals are selected to be added to the population each of the six chromosome pairs is considered separately. The crossover rate is the percentage chance that any one pair will undergo crossover. Hence it is possible that some chromosomes in the children will have been crossed over and others will not.

Mutation is achieved by randomly resetting an element of the chromosome to a permissible value. The mutation rate is defined in the following way: for each of the six chromosomes of the new individual the mutation rate is the percentage chance that the chromosome will be mutated. If mutation occurs then a single value is chosen from the chromosome to be reset (this may have a large effect on the resultant permutation).

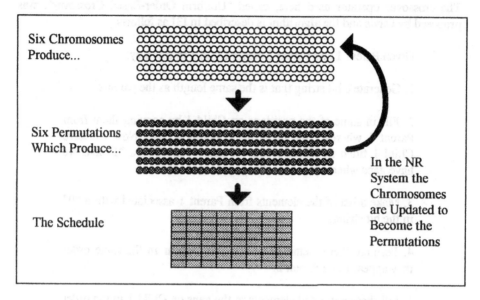

Six Chromosomes Produce...

Six Permutations Which Produce...

The Schedule

In the NR System the Chromosomes are Updated to Become the Permutations

Figure 1: Schedule Production in the Indirect Systems

3.2 Systems Using Direct Representations

With the direct representations, rather than the chromosomes telling us how to build a permutation, the chromosomes are themselves the permutation. It is important to define crossover and mutation operators which leave the chromosome as a valid permutation. The three systems implemented here use the same mutation operator but different crossover operators. The systems are referred to as "Davis", "Oliver" and "Goldberg" after the authors who first proposed the crossover operators used. The crossover rate for the direct representations is defined in the same way as for the indirect representations. The scheme is illustrated in Figure 2.

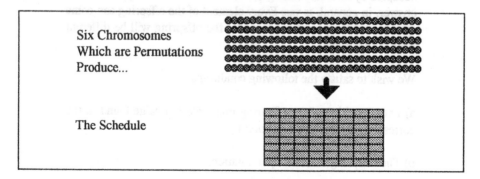

Six Chromosomes Which are Permutations Produce...

The Schedule

Figure 2: Schedule Production in the Direct Systems

Crossover in the Davis System

The crossover operator used here, called "Uniform Order-Based Crossover", was proposed by Davis and the algorithm is described in [3] as follows:

Given Parent 1 and Parent 2 create Child 1 in this way:

1. Generate a bit string that is the same length as the parents.

2. Fill in some of the positions on Child 1 by copying them from Parent 1 wherever the bit string contains a "1". (Now we have Child 1 filled in wherever the bit string contained a "1" and we have gaps wherever the bit string contained a "0".)

3. Make a list of the elements from Parent 1 associated with a "0" in the bit string.

4. Permute these elements so that they appear in the same order they appear in on Parent 2.

5. Fill these permuted elements in the gaps on Child 1 in the order generated in 4.

To make Child 2, carry out a similar process.

Crossover in the Oliver System

The crossover operator used here, called "Cycle Crossover" was proposed by Oliver et al., and is described in [4] as follows:

The cycle crossover is an answer to the question: Can we create an offspring different from the parents where every position is occupied by a corresponding element from one of the parents? The answer, in general, is yes. Every element of the offspring can come from one of the parents and usually the offspring will be different from both parents.

We wish to satisfy the following conditions:

a) Every position of the offspring must retain a value found in the corresponding position of a parent.

b) The offspring must be a permutation.

Consider... these parents:

Parent 1   *h k c e f d b l a i g j*

Parent 2   *a b c d e f g h i j k l*

Consider position one, condition a) says the offspring must contain either *h* or *a*. Suppose we choose *h*. Using condition b), *a* cannot be chosen from parent 2 since that position is now occupied by *h*. So *a* must also be chosen from parent 1.

Since *a* in parent 1 is above *i* in parent 2 we must also choose *i* from parent 1. Continuing the argument, having chosen *h* from parent 1 we must also choose *a*, *i*, *j*, and *l*.

The positions of these elements are said to form a cycle, choosing any of the positions from parent 1 or 2 forces the choice of the rest if the above conditions are to hold.

Crossover in the Goldberg System

This system uses a crossover operator called "Partially Match Crossover" (PMX), which was first proposed by Goldberg et al. in [5] and is described by Goldberg in [6] as follows:

Under PMX, two strings (permutations and their associated alleles) are aligned, and two crossing sites are picked uniformly at random along the strings. These two points define a matching section that is used to effect a cross through position-by-position exchange operations.

To see this consider two strings:

$$A = 9\,8\,4 \mid 5\,6\,7 \mid 1\,3\,2\,10$$

$$B = 8\,7\,1 \mid 2\,3\,10 \mid 9\,5\,4\,6$$

PMX proceeds by positionwise exchanges. First, mapping string B to string A, the 5 and the 2, the 3 and the 6, and the 10 and the 7 exchange places. Following PMX we are left with two offspring, A' and B':

$$A' = 9\,8\,4 \mid 2\,3\,10 \mid 1\,6\,5\,7$$

$$B' = 8\,10\,1 \mid 5\,6\,\ 7 \mid 9\,2\,4\,3$$

where each string contains ordering information partially determined by each of its parents.

Mutation in the Direct Representations

The mutation operator for the direct representation has to be one which leaves the chromosome as a permutation. The one used here Davis calls "Scramble Sublist Mutation" and is described in [3]. A portion of the chromosome is selected by randomly choosing a start and end point and that portion of the chromosome is permuted. For example:

$$a\,b\,c\,d \mid e\,f\,g\,h\,i\,j \mid k\,l$$

might become:

$$a\,b\,c\,d \mid f\,h\,i\,j\,e\,g \mid k\,l$$

Davis reports that a useful variation "when chromosomes are quite long involves limiting the length of the section of the chromosome that can be scrambled". This has not yet been attempted for this application.

The mutation rate for this operator is defined in the following way: for each of the six chromosomes of the new individual the mutation rate is the percentage chance that the chromosome will be mutated.

4. Evaluation

The candidate schedules are evaluated by assigning penalty points according to the amount of movement between venues each group has to make, and according to the distance away from the optimum amount of cross-marking.

In a perfect schedule each group would be cross-marked against twelve other groups. The average amount of cross-marking is calculated and subtracted from twelve. The result is squared and multiplied by one hundred to give the number of penalty points for non-optimal cross-marking.

In a perfect schedule no groups would have to move between venues. The average number of times a group moves between venues (while not on a break) is calculated. The result is squared and multiplied by four hundred to give the number of penalty points for groups moving between venues.

The optimum penalty for this system is unknown; however the minimum penalty achieved over all the test runs described below was 555 penalty points. A test was also carried out to determine the penalties likely from random permutations. 10,000,000 random permutation sets were generated; the lowest number of penalty points found was 1338. A manually produced back-up solution scored 1752 penalty points. A schedule in which each group moved once only and was cross-marked against ten other groups would score 800 penalty points and would be considered better than adequate for practical purposes.

5. Experiments

Experiments were carried out with each of the five systems described above. The systems were tested with two operator rate policies, two population sizes, four replacement strategies and two fitness equations, giving a total of 32 tests for each system. In each test the system was allowed to run for 60,000 evaluations and the result collected was the number of penalty points for the best chromosome after 30,000 evaluations. Each test was run 10 times and the results averaged. Having determined good parameter sets for each system, the tests with these parameter sets were run a further 40 times in order to obtain an accurate comparison between the systems. The details of the experimental variations are described below:

5.1 Operator Rates

In the first variant the operator rates were fixed at 60% for crossover and 2% for mutation. In the second variant the operator rates were allowed to change linearly between evaluation 1 and evaluation 30,000. The crossover rate started at 80% and decreased to 20% and the mutation rate started at 0% and increased to 20%. This method of decreasing crossover rate and increasing mutation rate as the evolution progresses was used for a timetabling problem by Corne et al. in [7]. The intention is that a large amount of crossover at the beginning will allow fairly good solutions to be found without too much interference from mutation. Later in the evolution, crossover is less likely to produce better individuals, but high mutation rates allow good individuals to be fine tuned through a local random walk.

5.2 Population Sizes

Population sizes of 30 and 100 individuals were used.

5.3 Replacement Strategies

The four replacement strategies varied in the number of individuals replaced in the population at one time. In the first case the whole population was replaced (true generational replacement); in the second case all but the best of the population was replaced (elitism); in the third case half the population was replaced; and in the last case only two individuals were replaced (steady state population). In each case the individuals removed from the population were those which were least fit. The parents of the new member of the population were chosen by classical roulette wheel selection.

5.4 Fitness Equations

The fitness of an individual was calculated from the penalty points in one of two ways. The first method used the following equation:

$$\text{fitness} = 1 / (1 + \text{penalty})$$

The second method linearly normalised the fitnesses. The n chromosomes were ordered according to the penalty points they incurred and then integer fitnesses were assigned ranging from 0 for the chromosome with the largest number of penalty points to n-1 for the chromosome with the smallest penalty.

6. Test Results

In general the results were good and showed that this problem can be solved in reasonable time using an evolutionary algorithm. The full test results are available from the author, but a summary of the interesting points follows:

Each of the systems had a variation of parameters which produced good results. A typical penalty point descent can be seen in Figure 3. The parameters of the best run for each system, according to the mean penalty after 30,000 evaluations, and the associated penalties are shown in Table 1. The relative final mean penalties of the systems can be seen in Figure 4.

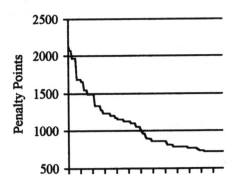

Figure 3: Typical Penalty Point Descent Over 30,000 Evaluations

As can be seen, the parameter sets that the systems respond well to all included a population size of 100, and a variable operator rate. The NR, DO and Goldberg systems worked best with linear normalisation of fitnesses while the other two worked better without this. A replacement size of 50 was best for all systems apart from the Goldberg system which worked better with a replacement size of 2.

System	Population Size	Replace-ment Size	Operator Rate	Linear Normal-isation	Mean Penalty After 30,000 Evaluations	Standard Error
NR	100	50	Variable	Yes	660	7.6
DO	100	50	Variable	Yes	670	9.2
Davis	100	50	Variable	No	646	7.5
Oliver	100	50	Variable	No	644	7.2
Goldberg	100	2	Variable	Yes	639	8.1

Table 1: Parameters with Best Mean Final Penalties for each System

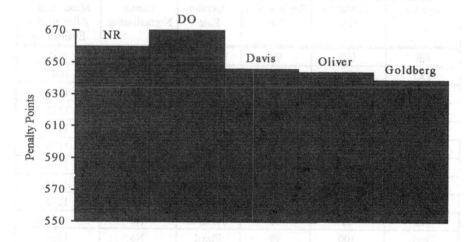

Figure 4. Mean Penalty Points After 30,000 Evaluations

Using the criterion of best average penalty after 30,000 evaluations the direct systems generally did better than the indirect systems, with no significant difference between the members of each type.

Of the 160 experiments 24 did particularly badly with at least one of their ten test runs failing to get a penalty of 1000 or less within 60,000 evaluations. These were also the 24 experiments with the worst average penalties after 30,000 evaluations, The parameters and penalty points after 30,000 evaluations are listed in Table 2.

It is interesting to note that 19 of the 24 experiments with poor results used true generational replacement, with the other 5 using a single elite individual. Replacing half or less of the population at one time meant that all test runs got results under 1000 points in under 60,000 evaluations. In attempting to establish the reason for this it is useful to look at the replacement policy used. This is a harsh system with the least fit individuals always being replaced. Replacing fewer individuals at a time gives those with poorer fitnesses a greater chance of surviving, so reducing the likelihood of premature convergence.

Just 3 of the 24 experiments that did badly used linear normalisation of fitness. This is surprising since the parameters to the linear normalisation (giving fitnesses from 0 to n-1) are very aggressive, and one would expect this to make premature convergence more likely. One reason might be that the major reason for premature convergence in these experiments is the harsh replacement policies and whilst linear normalisation will only increase likelihood of this, the increased selection pressure makes it more likely that the penalty of the best individual will get below 1000 before the convergence takes place.

System	Population Size	Replacement Size	Operator Rate	Linear Normalisation	Mean Penalty After 30,000 Evaluations
NR	100	99	Variable	No	960
DO	30	30	Fixed	No	1044
Oliver	30	30	Fixed	No	1102
NR	30	30	Variable	No	1125
Davis	100	99	Variable	No	1174
DO	100	100	Variable	No	1179
Davis	100	100	Variable	Yes	1181
NR	100	100	Variable	No	1232
DO	30	30	Variable	No	1235
NR	30	30	Variable	No	1255
Davis	100	99	Fixed	No	1264
Oliver	100	100	Variable	No	1274
Oliver	30	30	Variable	No	1278
Goldberg	100	100	Fixed	Yes	1470
Goldberg	100	100	Variable	Yes	1481
Davis	30	30	Fixed	No	1513
Goldberg	100	99	Fixed	No	1518
Goldberg	100	99	Variable	No	1521
Goldberg	30	30	Fixed	No	1538
Davis	30	30	Variable	No	1546
Davis	100	100	Variable	No	1568
Goldberg	100	100	Fixed	No	1585
Goldberg	100	100	Variable	No	1605
Davis	100	100	Fixed	No	1649

Table 2: Parameters that did Badly

Interestingly, the three direct representation systems, which gave the best results with the right parameters also gave the worst results when the parameters were inappropriate. This is probably because the indirect representations are more brittle; making the desired evolution more difficult, but making premature convergence less likely.

7. Conclusions and Further Work

The problem has been solved using evolutionary methods, and the results have been of real practical use. Each of the systems produced schedules which were good, and considerably better than those which could be hand crafted, or found through random search. Some useful data has been collected concerning the algorithm parameters for each of the systems used. These may be valuable to other researches using similar systems for different applications. Replacement strategies which replace only a small number of individuals at a time have been shown to markedly reduce premature convergence with this application. Decreasing the crossover rate and increasing the mutation rate through the course of evolution has proved to be useful, as has the linear normalisation of fitnesses.

A great deal of further work could be done in the area. Other systems, operators and methods of varying operator rates could be experimented with, and further investigation of the best parameter set for each system could be carried out. Experiments with other selection and replacement policies would also be useful. Further investigation of the effect of linear normalisation is necessary in order to clearly explain the benefits of this policy.

Investigations could also be made into the recombination of the chromosomes sets of parents without always involving crossover, so that a child can be composed of some whole chromosomes from one parent and the remainder from the other parent.

8. Acknowledgements

I would like to thank Nick Radcliffe for pointing me in the right direction for permutation representations and for his useful comments and ideas. I would also like to thank Henri Luchian for providing a stimulating research environment in Romania and for persuading me to publish these results. Andrew Cumming, Mihai Petriuc, George Paechter, Dave Corne, Richard Macfarlane and Linda Black have also contributed ideas and support.

9. References

[1] Paechter B., Luchian H., Cumming A., and Petriuc M., "Two Solutions to the General Timetable Problem Using Evolutionary Methods", to appear in The Proceedings of the IEEE Conference of Evolutionary Computation, 1994.

[2] Michalewicz, Z., Genetic Algorithms + Data Structures = Evolution Programs, Springer-Verlag, London, 1992.

[3] Davis, L., Handbook of Genetic Algorithms, van Nostrand Reinhold, London, 1992.

[4] Oliver, I. M., Smith, D. J. and Holland, J. R. C. "A Study of Permutation Crossover Operators on the Travelling Salesman Problem", in The Proceedings of the Second International Conference on Genetic Algorithms, Lawrence Erlbaum Associates, Hillsdale, New Jersey, 1987.

[5] Goldberg, D. E. and Lingle, R., "Alleles, loci, and the Travelling Salesman Problem", Proceedings of an International Conference on Genetic Algorithms and their Applications, 1985.

[6] Goldberg, D. E. Genetic Algorithms in Search, Optimisation and Machine Learning, Addison Wesley, Reading, 1989.

[7] Corne, D., Fang, H-L and Mellish C., "Solving the Modular Exam Scheduling Problem with Genetic Algorithms", Proceedings of the Sixth International Conference of Industrial and Engineering Applications of Artificial Intelligence and Expert Systems, Edinburgh, 1993.

Genetic algorithms and flowshop scheduling: towards the development of a real-time process control system

Hugh M. Cartwright and Andrew L. Tuson

Physical Chemistry Laboratory, Oxford University,
South Parks Road, Oxford, England OX1 3QZ

e-mail: hugh@physchem.ox.ac.uk, altuson@physchem.ox.ac.uk

Abstract

Scheduling in chemical flowshops is one of a number of important industrial problems which are potentially amenable to solution using the genetic algorithm. However the problem is not trivial: flowshops run continuously, and for efficient operation those controlling them must be able to adjust the order in which products are made as new requests are received. In addition, there are in principle efficiency gains available if the topology of the flowshop can be varied, but the determination of a suitable topology is also a demanding problem. In this paper we discuss how a genetic algorithm can be implemented to handle an industrial flowshop, taking account of these requirements.

1 Introduction

The range of application of genetic algorithms has grown very rapidly in recent years. It is significant that many of these applications relate to industrial or practical problems, such as the design of aircraft (Bramlette and Bouchard, 1991), the routing of water mains (Murphy et al, 1993), or the construction of timetables (Colorni et al, 1993). In these applications, genetic algorithms show real promise, and already full-scale genetic algorithms are being brought into use in industry and elsewhere (Unicom, 1994). This is the clearest sign that the method is coming of age, and that a further rapid growth in applications in the next few years is likely.

This growth is occurring even though the mathematics which underpins the method is still far from complete. Indeed the use of genetic algorithms is in a very real sense experimental, since the guidelines used by researchers when work begins on a new problem remain largely empirical. Much 'experimentation' may be needed to ensure both that the method chosen to code solutions to the problem is appropriate, and that a suitable set of adjustable GA parameters is employed. Despite these complications, genetic algorithms have shown considerable promise when applied to large-scale industrial problems.

As a consequence, the focus of the GA community is gradually moving away from model problems, such as the Travelling Salesman Problem (TSP), towards real-world applications. This is a more complex shift than one might at first imagine. A model problem may be deliberately formulated in such a fashion that it fits snugly into the GA paradigm, and thus promises results which can provide some insight into the way in which the algorithm works; the problem may in this way be moulded to fit the algorithm. By contrast, in an industrial application the situation is reversed: the genetic algorithm must tackle the problem according to the exact formulation in which it arises in industry, not in some stripped-down form which may be well-suited to solution by the algorithm, but is of limited industrial use. With increasing experience in the use of the algorithm, and understanding of the mathematics which underlies it, this is becoming an increasingly attractive option.

2 The Chemical Flowshop

It has been recognised almost from the start of work on genetic algorithms that their structure is particularly well-suited to scheduling and ordering problems. This doubtless accounts for the special place which the Travelling Salesman Problem seems to have in the hearts of many GA workers. The TSP is a *static scheduling problem* in the sense that, once the set of cities and their locations is specified, the problem constraints are entirely fixed; the cities do not start to roam about once the salesman begins his journey.

Scheduling problems are common in industry, in which one of the most important examples is the chemical flowshop (Ku *et al*, 1987). This is a particular example within the large class of job-shop problems, which are widespread in industry due to the adoption of production line methods. In a flowshop high-value chemicals are produced on a continuous-flow production line (fig. 1), in which chemicals pass in succession through a string of reactors joined in serial fashion.

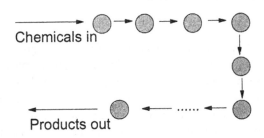

Fig. 1. A serial chemical flowshop.

During synthesis of any chemical in bulk, economies of scale are important, since both the cost per kilogram and the percentage yield of a product may be affected by the size of the batches in which it is made. However, continuous large-scale production of chemicals such as paint dyes, which are required in only small amounts, is clearly counterproductive, rapidly leading to over-stocking and financial loss. The conflict between the efficiency improvements to be gained by manufacturing on a large scale, and the problems of selling large quantities into a limited demand can be resolved in two ways:

a) sporadic batch processing; or
b) continuous flowshop processing.

Both methods employ large-scale production, so benefit from economies of scale; but the flowshop, as a continuous process, enjoys significant operating advantages, because continuous use of plant reduces 'dead time'.

Flowshops generally operate in a many-product fashion. Different chemicals can be synthesised by feeding the appropriate precursors into the first reactor in the line, and then adding suitable reagents as appropriate. Because each chemical has its own characteristic residence time in the various units, the efficiency of the line is strongly influenced by the order in which chemicals are produced. If a poor selection of order is made, the line may become blocked temporarily while a chemical finishes processing in a reactor in the middle of the line (fig. 2).

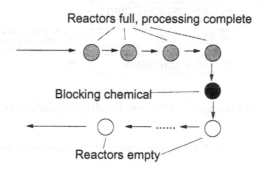

Fig. 2. Blocking of a flowshop by a single chemical.

Chemicals begin to build up in reactors behind the blockage, waiting to move on, while beyond the blockage reactors may empty, and therefore be unemployed. This has an obvious detrimental effect on throughput.

Determination of the optimum order in which chemicals should enter the line is a classic scheduling problem, and papers have already appeared discussing how the genetic algorithm might be used to derive suitable orders (Cartwright and Long, 1993; Reeves, 1994). However, although these papers provide persuasive evidence of the power of the algorithm to generate solutions of high quality, they only address the static problem, in which a single set of chemicals is to be scheduled.

In practice, real flowshops are *dynamic*. The most efficient order in which to synthesise products might be determined, but that order will remain optimum only for as long as no fresh requests for chemicals arrive. When new requests do reach the flowshop, the original optimum production order will almost certainly require modification. To further complicate the situation, product requests may have time constraints associated with them (e.g. "synthesise 50kg of red dye by 1400 hrs"). These constraints impose additional limitations on what constitutes an acceptable order, since requests must be inserted into the existing order in such a position that the constraints will be satisfied, while the highest possible operating efficiency is maintained. Thus an algorithm is required which not only is capable of establishing a good order for the requests known when the flowshop begins production, but which can also efficiently adjust processing orders as new requests are received. This constitutes a real-time process control system, which needs to be fully integrated with systems controlling the line itself (fig. 3).

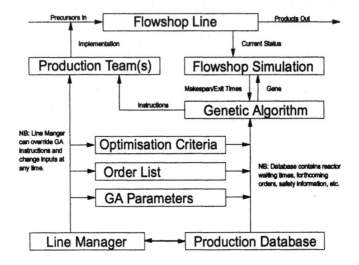

Fig. 3. A schematic for a flowshop controlled by a genetic algorithm.

This paper discusses some early studies into the feasibility of such a real-time system, and also introduces preliminary results of work on the effect of adding extra processing units to the flow-line to give an enlarged line of 'unrestricted' topology.

3 Variable Topology Flowshops

The determination of the optimum schedule for a static flowshop has already been discussed in the literature, and there is good evidence that the genetic algorithm is a powerful tool for finding high-quality orders. However, real flowshops differ in at least two important respects from the static linear flowshop shown in fig. 1. We have already commented on the dynamic nature of the flowshop, a consequence of the need to accommodate the flow of new requests. Empirical studies show that significant improvements in efficiency are possible if a few additional reactors are added in parallel to the serial line. These extra units require the addition of outlets to existing units, in the manner illustrated in fig. 4.

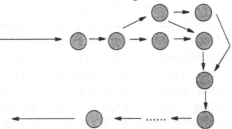

Fig. 4. A 'restricted' topology: a maximum of two reactors is available at each station.

Temporary blockages may arise in the line, as the processing of one chemical in a reactor holds up progress of other chemicals waiting to move on. The addition of an extra reactor can provide an 'escape route' by which chemicals can be diverted around such a blockage, and thus continue processing. This adjustment of the topology of the flowshop has the potential to substantially improve line efficiency. However, it also increases the demand upon the scheduling algorithm, which needs now to be able to deal not only with the dynamic nature of the flowshop but also the possibility that efficiency can be enhanced by adjusting the placement of the extra reactors.

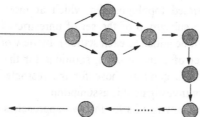

Fig. 5. An 'unrestricted' topology: more than two reactors may occupy one station.

Preliminary results have already been published (Cartwright and Long 1993) dealing with flowshops of 'restricted' topology; that is, in which at most one extra unit can be placed at each station (fig. 4). We discuss now the 'unrestricted' situation, in which extra units may be placed in any topology, including configurations in which several additional reactors occupy the same station in parallel (fig. 5).

4 Results

4.1 Restricted Versus Unrestricted Topologies

The results discussed here were generated on Hewlett Packard 9000/700 series workstations using the parameters shown below.

Number of chemicals (n)	20
Number of mainline reactors (m)	20
Number of extra reactors (e)	variable
Number of additional outputs (o)	variable
Population size	50
Number of generations	2,000
Crossover probability	0.8 per string per generation
Order mutation probability	0.002 x n per string per generation
Topology mutation probability	0.002 x e per string per generation

Table 1. Genetic Algorithm parameters.

A processing times matrix consisting of randomly-generated values between zero and ten hours was read into the program at the start of each run. For brevity we will write a flowshop system in the form $n/m/e/o$. Main implementation details are discussed elsewhere (Tuson, 1994a and 1994b).

The objective of the genetic algorithm is to find the product order and flowshop topology yielding the minimum makespan or flowtime, which is the total time required to process every chemical. All makespans quoted, unless stated otherwise, were the average over 6 runs with identical genetic algorithm parameters, but different random number seeds.

The set of restricted topologies (in which at most one extra reactor is permitted per station) is a sub-group of the set of unrestricted topologies (in which any number of reactors can be added to each station). In view of this we would expect that, given a fixed number of extra reactors, solutions for the unrestricted topology should always be at least as good as those for the restricted topology, and often superior. In this section we investigate this assumption.

Table 2 illustrates how results for the two approaches compare for a system in which 20 chemicals pass through a flowshop consisting of 20 reactor stations, with ten extra reactors available.

Number of extra outlets (o)	1	2	3	4
Restricted topology makespan (hrs.)	213	209	208	206
Unrestricted topology makespan (hrs.)	213	205	204	204

Table 2. Average makespans for restricted and unrestricted topologies.

The makespans are identical for $o=1$ as in this case the systems are equivalent. It is evident from Table 2 that the addition of one new output ($o=2$) yields a significant reduction in makespan, whether or not the topology is restricted. As we would expect, unrestricted solutions are superior to restricted solutions. The addition of further outputs leads to smaller, though still useful, improvements in makespans.

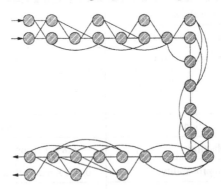

Fig. 6. The best topology found by the GA for four extra outlets.

Makespans can be reduced by increasing the number of outlets from reactors because some chemicals have zero processing time in a reactor; these chemicals can make use of a bypass link (one which connects two non-adjacent reactors) if such a link is available. As extra reactor outlets become available this provides additional flexibility. The best solution for the set of GA runs with $o=4$ (makespan = 202 hr) is found by the unrestricted calculation and has the topology depicted in fig. 6.

It is evident that this topology is permitted both in the restricted and the unrestricted calculation, so one might reasonably ask why it is that the unrestricted genetic algorithm has found it while the restricted algorithm has not.

An explanation may lie in the way that the GA handles the topology string during crossover and mutation. Crossover of the topology in a restricted system will frequently lead to an illegal topology, in which more than one additional reactor is in place at a station. The restricted topology representation must then effect a `repair` on the string to restore legality. This repair inevitably disrupts the string to an extent and hence potentially useful schemata may be lost. The unrestricted topology representation is much less prone to this, as the topologies which crossover generates are always legal and so the disruptive repair is not required.

4.2 Dependence of Topology on the Number of Extra Reactors

The GA was run for both restricted and unrestricted 20/20/e/4 flowshop systems with a varying number of extra reactors, e. A comparison of the best times found for restricted and unrestricted systems is shown in table 3. It is immediately apparent that there is little difference between solutions discovered by the two approaches. Indeed, comparison of the topologies that they generate leads to the conclusion that they frequently converge to identical orders and topologies, though at differing speeds.

Number of extra reactors (e)	3	5	7	10	13	15	17	20
Restricted topology makespan (hrs.)	227	225	217	206	195	188	177	161
Unrestricted topology makespan (hrs.)	228	223	218	204	195	185	176	162

Table 3. The variation of makespan with number of extra reactors.

This is a little disconcerting, since it seems that the unrestricted calculation is taking no advantage of the opportunity to reduce the makespan by stacking reactors, but is instead merely finding restricted-type solutions. Indeed, the unrestricted algorithm converges routinely to restricted solutions, even when the number of extra reactors approaches the total number of reactors in the main serial line.

However, these results are neither anomalous nor unreasonable. The unrestricted system is reluctant to stack reactors because of the uniformity of reactor residence times along the flowline. Temporary blockages may form anywhere along the line, though certain reactors are 'black spots' (see section 4.6). However, the addition of one extra reactor at a station relieves congestion so effectively that, when a further extra reactor is available, it is almost always more efficient to place this at a station at which no extras are present, than to choose a station where they are.

This interpretation can be confirmed by modifying the reactor times matrix so that the residence times are less uniform. For example, if we assign a large reactor time (30 hrs) for five chemicals in reactor ten, the system immediately responds by stacking reactors at this station in preference to spreading them out. For this amended matrix of residence times, the average makespan achieved for the unrestricted topology was 176 hours, compared with an average makespan of 192 hours for the restricted topology representation. This demonstrates that for reactor residence times which do not lie uniformly within the chosen residence times interval, the unrestricted representation has the potential for superior solutions and shows that the GA has the ability to find these solutions, when they exist.

4.3 Distribution Analysis

To assess the quality of GA solutions, distribution analyses were performed for both the restricted and unrestricted topology representations. 100,000 random solutions to the flowshop system were generated and the makespans determined, to give a distribution against which GA solutions could be compared.

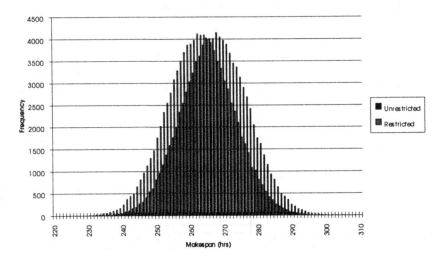

Fig. 7. The makespan for 100,000 randomly-chosen solutions.

Fig. 7 shows the distribution for the 20/20/10/4 system. This has a mean $(\bar{x})$ of 267.6 hours and a standard deviation (σ) of 9.6 hours for the unrestricted topology and $\bar{x} = 262.6$ hours and $\sigma = 9.6$ hours for the restricted topology. (The average makespan for the unrestricted topology is poorer that that for the restricted, since with only a small number of extra reactors available, virtually all stacked configurations, many of which the unrestricted algorithm will generate, will be of poor quality.)

It is evident that the solutions provided by the GA (restricted: 206 hrs, unrestricted: 204 hrs) are superior to the lowest makespan in the distribution shown above. To gain a quantitative idea of the efficiency of the GA the data in fig. 7 were fitted to a pair of Gaussian functions, which well represents the data. The probability that a randomly generated solution will have a makespan equal to or less than that calculated by the GA can then be determined to be of the order of 2×10^{-11} for the unrestricted system and 2×10^{-9} for the restricted system.

These results demonstrate the power of the GA to search these large and rather intractable spaces. It is notable that the GA can find high-quality solutions for the unrestricted system although the density of such solutions in the search space is low, and the quality of the average solution is significantly poorer.

4.4 The Use of Genetic Algorithms to Handle Flowshop Constraints

Real flowshop schedules must not only maximise the rate of chemical production but also ensure that production deadlines are met. This work demonstrates that a GA can find solutions that meet order deadlines without significantly compromising productivity.

We will start by considering a 20/20/0/3 flowshop system. Without order deadlines the GA finds a solution with a best makespan of 234 hours with the chemical order shown below:

20 5 19 17 9 15 8 6 16 4 7 1 18 2 10 14 12 3 13 11

The effect on the makespan of applying increasingly strict order deadlines on the last chemical to be processed (chemical 11) was investigated, with results shown in fig. 8.

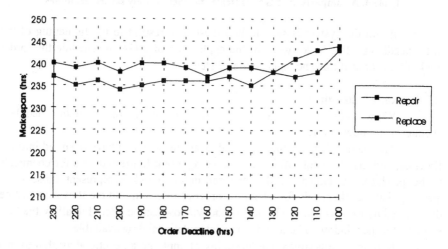

Fig. 8. The variation of makespan with order deadlines for different GA strategies.

Strings generated by the manipulations of the algorithm which do not satisfy the deadline constraints can be repaired (by moving forward in the order the chemical whose constraints are broken, until the deadline is met) or replaced by a new random string. For both methods there is little change in the makespan found by the GA until the deadline becomes quite strict. Surprisingly, the *Replace* strategy leads to better solutions than the unconstrained GA (240 hrs), which suggests that the introduction of new random solutions into the GA could improve performance by giving the GA fresh schemata to process, hence avoiding premature convergence. We are now investigating the use of a string fridge (Costello, 1993) to see whether the broadening of the genetic base that such a mechanism entails will help the algorithm.

The ability of the GA to observe order deadlines whilst simultaneously optimising chemical order and flowshop topology was investigated by considering a 20/20/5/4 unrestricted topology system. With order deadlines the GA found a best makespan of 219 hours, with the chemical order shown below:

5 20 16 17 15 7 6 1 9 2 19 8 4 14 18 13 10 12 3 11

The effect on the makespan of applying increasingly strict order deadlines on the last chemical to leave (chemical 11) was investigated with results shown in the table below:

Deadline (hrs.)	220	210	200	190	180	170	160	150	140	130	120	110	100
Repair (hrs.)	223	221	224	225	224	224	224	224	223	224	225	227	233
Replace (hrs.)	222	223	222	223	224	221	222	223	223	225	225	227	229

Table 4. A comparison of GA strategies for increasingly severe deadlines.

Again the system is seen to be robust with respect to the tightening of the order deadlines - this bodes well for the application of a GA to schedule the 'order book' of a chemical flowshop and similar systems.

4.5 Dynamic Scheduling

We will consider the use of a 'scheduling window' of size w to schedule n chemicals, as shown in fig. 9.

The chemicals preceding the window are those which have been fed into the flowline, and the chemicals after the window represent future requests for chemicals to be produced. Chemicals within the scheduling window represent those on a 'pending list', i.e. chemicals for whom the order of insertion into the flowshop will be determined by the algorithm. The size of the window can be altered to allow the best solution for the window to be found in the computational time available.

In order to investigate the feasibility of applying a genetic algorithm to this problem, a simple serial flowline is considered. The GA is used to schedule a 20/20/0/0 system and the effect of reducing the size of the scheduling window, w, is investigated. This is an approximation to the situation where the scheduler has unknown future requests in addition to those already being processed in the flowline.

The genetic algorithm was run as before but with an order mutation rate of $0.002 \times w$ respectively. The results obtained are shown in table 5.

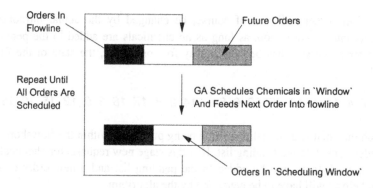

Fig. 9. Dynamic scheduling with a 'scheduling window'.

Size of 'scheduling window' (w)	20	18	16	14	12	10
Makespan for 20 chemicals (hrs)	244	246	246	247	249	248

Table 5. The effect of the 'scheduling window' size on the makespan.

As table 5 shows, there is some degradation in performance with decreasing window size; however the degradation is not severe. This suggests that, in principle, a windowing implementation of a dynamic scheduling system based upon the genetic algorithm is feasible.

However, there are other practical matters to be considered in using a genetic algorithm to control a dynamic flowshop. In particular, there is the question of the amount of computing time required for the calculations. When a chemical is inserted into the flowshop for processing, its place in the pending list may be filled by an incoming request. The set of chemicals whose processing time the algorithm is trying to minimise is thus changed, and in principle the algorithm must then be rerun from scratch to calculate an optimum order for the revised pending list. As optimisation of the pending list requires a not insignificant amount of computer time, it is conceivable that the algorithm might fail to keep up with the speed of operation of the flowshop. When the flowshop requests the identity of the next chemical to be fed in, the algorithm may not have completed its calculation and hence will have found only a sub-optimal order.

Fortunately, preliminary studies suggest that, when a chemical in the pending list is replaced by a new request, the optimum order for the new pending list often has sub-orders of chemicals in common with the old list. For example, suppose that the algorithm had determined the following order in which entries from the pending list should enter the flowshop.

7 4 11 1 19 18 20 6 5 14 2 8 17 16 3 9 12 15 10 13

This order will not, of course, be changed by the actual introduction of chemicals into the flowshop, as long as no chemicals are added to the pending list. Thus, after processing has been underway for some time, the state of the flowshop might be

7 4 11 1 19 18 20 6 5   *14 2 8 17 16 3 9 12 15 10 13*

in which the chemicals in italic are now being processed within the flowshop, and the remainder constitute the pending list. If at this stage new requests for chemicals 5 and 9 are received, these will be added to the pending list and a new order of this list (shown below) will have to be generated by the algorithm.

7 4 11 18 20 19 1 5 6 5 9   *14 2 8 17 16 3 9 12 15 10 13*

We notice that the processing order in the new pending list shown immediately above has segments in common with the original processing order. For example, the sequences {7 4 11}, {18 20} and {6 5} occur in both the revised list and the original list. This is related in a very direct manner to the interpretation of genetic algorithm behaviour in terms of the combination of 'building blocks'. Some of the schemata which were effective for the initial problem are still of value in handling the updated pending list. As a consequence, because schemata in the old problem can in effect be re-used for the new pending list, the calculation does not need to be started from scratch every time a new request arrives. Instead, the updated problem can be cast as a continuation of the old calculation, with consequent savings in computer time, and less likelihood that the computer will fall behind the flowshop whose processing order it is meant to be determining.

4.6 The use of a GA to Design Flowshop Systems with Long-term Fixed Topology

Previous work (Cartwright and Long, 1993) has demonstrated that simultaneous optimisation of chemical order and topology leads to better solutions than separate optimisation. However, it would be impractical to change the topology of the flowshop system dynamically as new orders are received - the topology can only be changed when the line is shut down, an expensive action which would be performed only for maintenance, or when calculations suggest that, for the present mix of products, the topology of the current line is far from optimum.

A topology is therefore required that is as good as possible for the range of orders the flowshop is likely to receive over a period of time. Assuming that it is possible to produce a sensible forecast of the required production of the flowshop over a period in the future, could the GA suggest a near-optimum long-term topology for the flowline?

To investigate this the GA was run 100 times for a 20/20/5/4 unrestricted topology flowshop system, optimising order and topology simultaneously. A random set of requests for products chosen from the chemicals the line could produce was chosen, and the number of reactors placed by the GA at each position was recorded and is shown fig. 10.

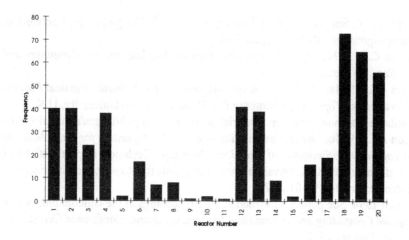

Fig. 10. Frequency of reactor placement for randomly-chosen sets of orders.

The data show that there is a strong preference for extra reactors to be placed at particular positions along the flowline. Thus, even if only a rough estimate is available of the future product requests, the algorithm can be valuable in predicting the most efficient topology for long-term operation. When the flowshop is shut down for maintenance the GA can be used to decide whether it is worthwhile to change the reactor topology in light of current production forecasts.

Acknowledgements

We are pleased to acknowledge the support of Hewlett Packard (UK) and Hewlett Packard (Europe) for this work.

References

Bramlette, M.F. and Bouchard, E.E., 1991. Genetic algorithms in parametric design of aircraft, in *Handbook of Genetic Algorithms*, L.Davis (ed.) Van Nostrand Reinhold, New York.

Cartwright, H.M. and Long, R.A., 1993. Simultaneous optimisation of chemical flowshop sequencing and topology using genetic algorithms. *Ind. Eng. Chem. Res.*, 32, 2706-2713

Colorni, A., Dorigo, M. and Maniezzo, V., 1993. A genetic algorithm to solve the timetable problem (submitted for publication).

Costello, R. 1993. Chemistry Part II thesis, Oxford University, U.K.

Ku H.M.A., Rajagopalan D., and Karimi I.A. 1987, Scheduling in Batch Processes. *Chem. Eng. Prog.*, 83, 8, 35.

Murphy, L.J., Simpson, A.R. and Dandy, G.C., 1993. Design of a pipe network using genetic algorithms, *Water*, August 40-42.

Reeves, C.R. 1994. A genetic algorithm for flowshop sequencing. *Computers and Ops. Res.*, (in press).

Tuson A.L. 1994a. *The Use of Genetic Algorithms To Optimise Chemical Flowshops of Unrestricted Topology*, Chemistry Part II thesis, Oxford University, U.K. [Available via anonymous ftp at muriel.pcl.ox.ac.uk in pub/theses/altuson1994.tar.Z].

Tuson A.L. 1994b. *The Implementation of a Genetic Algorithm for the Scheduling and Topology Optimisation of Chemical Flowshops.* Technical Report TRGA94-01, Oxford University, U.K. [Available via anonymous ftp at muriel.pcl.ox.ac.uk in pub/techreports/trga94-01.ps.Z].

Unicom 1994. See papers published in the Proceedings of the Unicom conference on *Adaptive Computing and Information Processing*, Brunel Conference Centre, London, January 25-27.

Genetic Algorithms for Digital Signal Processing

Michael S. White and Stuart J. Flockton

Physics Department, Royal Holloway (University of London), Egham, Surrey
TW20 OEX, UK

Abstract. Recursive digital filters are potentially less computationally expensive than their non-recursive counterparts. However, algorithms for adjusting the coefficients of recursive filters may produce biased or suboptimal estimates of the optimal coefficent values. In addition, recursive filters may become unstable if the adaptive algorithm updates a feedback coefficient so that one of the poles remains outside the unit circle for any length of time. This paper details an adaptive algorithm for optimizing the coefficients of recursive digital filters based on the genetic algorithm. Stability considerations are addressed by implementing the population of adaptive filters as lattice structures which allows the entire feasible, stable coefficient space to be searched whilst ensuring that crossover and mutation do not produce invalid (unstable) filters. Results are presented showing the application of this technique to the tasks of system identification and adaptive data equalization.

1 Introduction

Adaptive filtering techniques have been applied to many important areas of digital signal processing. Recursive digital filters offer potential computational savings over non-recursive filters but conventional adaptive algorithms for recursive filters suffer a number of drawbacks. Algorithms utilizing the equation-error formulation may converge on biased estimates of the optimal filter coefficients whilst output-error approaches can produce multi-modal error surfaces, leading to the possible entrapment of gradient-based searches in local minima. This paper reports on the use of genetic algorithms for adapting recursive digital filters for a variety of different tasks.

A brief introduction to digital signal processing and adaptive filtering is given in Section 2. This is followed by a review of the application of genetic algorithms to the problem of modelling unknown systems and Section 4 details work carried out by the authors in this same area. Section 5 introduces the task of data equalization and simulation results of a genetic algorithm approach to this problem are presented in Section 6. The paper concludes with a summary and details areas for further study.

2   Digital Signal Processing

Digital Signal Processing (DSP) is used to transform and analyze data and signals that are either inherently discrete or have been sampled from analogue sources. With the availability of cheap but powerful general-purpose computers and custom-designed DSP chips, digital signal processing has come to have a great impact on many different disciplines from electronic and mechanical engineering to economics and meteorology. In the field of biomedical engineering, for example, digital filters are used to remove unwanted 'noise' from electrocardiograms (EKG) while in the area of consumer electronics DSP techniques have revolutionised the recording and playback of audio material with the introduction of compact disk and digital audio tape technology.

Any DSP algorithm or processor can be reasonably described as a filter. Digital filters may be divided into recursive and non-recursive categories depending on their use of feedback. The response of non-recursive, or FIR filters is dependent only upon present and previous values of the input signal. Recursive, or IIR filters, however, depend not only upon the input data but also upon one or more previous output values. As a consequence of this feedback, recursive filters with just a few coefficients are often able to obtain similar output characteristics to non-recursive filters requiring (say) 100 or more coefficients. This potentially greater computational efficiency of recursive filters over their non-recursive counterparts is tempered by several possible shortcomings. Firstly, recursive filters may become unstable, that is, their output may grow without limit, if the feedback coefficients are chosen incorrectly. In addition, recursive filters are, in general, unable to produce the linear-phase responses achievable by non-recursive filter implementations and the presence of feedback may have an adverse effect on the accuracy to which the filter coefficients need to be specified.

The design of a conventional digital signal processor requires a priori knowledge about the statistics of the data to be processed. When this information is inadequate or when the statistical characteristics of the input data are known to change with time, adaptive filters are used. Adaptive filters have the property of self-optimization. They consist, primarily, of a time-varying filter, characterised by a set of adjustable coefficients and a recursive algorithm which updates these coefficients as more information concerning the statistics of the relevant signals is learned. Most current applications of adaptive signal processing (the modelling of unknown systems, echo cancellation and the digital representation of speech etc) utilise non-recursive digital filters. Since non-recursive filters do not have feedback, the output is a linear function of the coefficients and this greatly simplifies the derivation of gradient-based adaptive algorithms.

The greater computational efficiency that recursive filters offer has led researchers to try and develop reliable algorithms for their adaptation. Fundamentally, two approaches to the problem of adaptive IIR filtering have been investigated (Shynk, 1989). In both cases, after each iteration of the adaptive algorithm the performance of the digital filter is assessed on the basis of some optimization criterion, commonly some function of the total or mean squared error. The two approaches differ in the formulation of this prediction error. The equation-error formulation effectively transforms the recursive filter into two coupled non-recursive filters, allowing well-understood FIR adaptive algorithms to be used. Unfortunately, in the presence of noise, adaptive algorithms based on this formulation can converge to biased estimates of the filter coefficients. The second approach, known as the output-error formulation adjusts the coefficients of the time-varying digital filter directly in recursive form. The output of a recursive filter is a non-linear function of the coefficients. Consequently, the prediction error is not a quadratic function and may have multiple local minima. Adaptive algorithms based on gradient-search methods, such as the widely used LMS, may then converge to sub-optimal estimates of the filter coefficients. This paper seeks to introduce a class of adaptive algorithms based on the natural processes of evolution and population genetics which can overcome some of these problems.

3 Genetic Algorithm Approaches to System Identification

Many problems in the areas of adaptive control and signal processing can be reduced to that of system identification. In this task, a system with adjustable coefficients is used to model the dynamics of an unknown system also known as the plant. The model is frequently a non-recursive filter in order that conventional adaptive algorithms based on least-squares or gradient techniques may be used to produce the best estimate of the unknown system. Approaches to system identification which employ an IIR filter system model seek to benefit from the computational economy that they offer. In (Etter et al. 1982) a genetic adaptive algorithm was used to adjust the coefficients of a population of 11 recursive adaptive filters. Each system model was represented as a binary string by the genetic algorithm. Every generation, 10 pairs of these filters were probabilistically chosen to undergo single-point crossover on the basis of their ability to model the dynamics of the unknown system. A single filter was similarly chosen to undergo mutation. This scheme was shown to be able to correctly identify simple systems (first or second order) whose response surfaces were either uni- or bimodal.

Nambiar et al (1992) demonstrated similar performance for their GA-based adaptive algorithm. In this case, a parallel filter realization was implemented in order that the stability of the system models could be monitored. Several 'unknown' systems, from fourth, to tenth order were used and the effect of varying some of the genetic algorithm parameters (population size, crossover and mutation rates) was investigated. Kristinsson and Dumont (1992) applied a genetic algorithm to the identification of both discrete and continuous time systems. Second-order systems with minimum and non-minimum phase characteristics were modelled by IIR adaptive filters realized as cascade structures. Their results indicated that in some cases the genetic adaptive algorithm converged on solutions which were biased from those of the system and they attributed this to an insensitivity to changes in the biased coefficients. A population size of 100 was implemented and the single-point crossover and mutation operators were applied at rates of 0.8 and 0.01 respectively. During a run, if the percent involvement (the proportion of the current population producing offspring) declined significantly, as would be the case when a few 'super' individuals were receiving most of the opportunities to reproduce, a fitness ranking scheme was implemented. This limited the number of offspring that any individual model could produce and helped to maintain population diversity.

4 A Genetic Adaptive Lattice Algorithm

Earlier work by the authors (1993) addressed the problem of maintaining the stability of the genetically-derived adaptive filters by realizing them as IIR lattice structures. Monitoring the stability of conventional, direct-form filter implementations is computationally expensive and generally not robust. Alternative filter realizations such as the parallel and cascade forms used in the work detailed in Section 3 simplify stability monitoring by reducing the problem to one of ensuring that the coefficients of the cascaded second-order filter sections lie within the 'stability triangle'. The lattice realization enables stability to be controlled even more simply as it can be shown that a necessary and sufficient condition for the filter to remain stable, is for the feedback coefficients to have magnitude less than unity. The structure of a lattice filter is illustrated below in Figure 1.

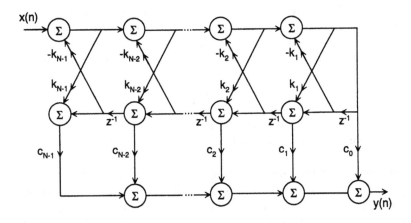

Fig. 1. Structure of a lattice filter

This genetic adaptive algorithm differs from conventional adaptive algorithms in its population-based approach. Rather than adjusting the filter coefficients of a single filter, the genetic algorithm operates on a number of points in the search space simultaneously (at least in principle — simulation on a serial machine means that all evaluations have to take place sequentially). Each lattice filter is represented as a bit-string. These strings are constructed by quantizing the feedback and feedforward coefficients (the k's and c's in Figure 1) of a single filter and concatenating them to form a binary code. The initial population of filters is randomly generated. Each filter is then evaluated according to its ability to accurately model the plant. On the basis of this fitness value, selection probabilities are generated, with filters corresponding to smaller error values being assigned a proportionally higher selection probability. Subsequent generations are generated by selecting members of the current population according to their assigned probability and applying genetic operators to a proportion of these filter structures in order to introduce variation.

The lattice structure is ideally suited to adaptation by a genetic algorithm. The maintenance of filter stability is, in essence, achieved 'for free' as the GA necessarily specifies a range onto which the quantized coefficients are mapped (decoded) when evaluation of the filter structures takes place. This is in marked contrast to other, non-lattice structures which require factorization of polynomials and schemes to regain stability or unduly restrict the range of values which the feedback coefficients can take. Because the stability criterion is built into the coefficient decoding mechanism the mutation and crossover operators are unable to generate invalid (unstable) filters, consequently, there are no constraint violations to be dealt with.

The two-point crossover mechanism used in these simulations exchanges portions of the 'genetic material' (binary strings) of two randomly selected parents in order to create a pair of new filter structures. This is accomplished by randomly selecting two cut-points on the parent bit strings and crossing-over the bits in between these points. In the real filter coefficient space this has one of two possible effects. Should the cut-points fall between the binary codes for two coefficients, the child structures each receive some of the coefficients of the two parent structures. If, however, one or both of the cut-points falls within the code for a coefficient then a child receives the most significant part of the binary-encoded coefficient of one parent and the least significant part from the other. This can be viewed as a combining of coefficient values from the two parents along with a perturbation of the coefficient within which a cut-point falls. The mutation operator is used to find new points in the search space to evaluate and acts by flipping randomly chosen bits in the binary strings. In the real coefficient space this has the effect of perturbing the coefficient within which the mutation occurs, the size of perturbation relating to the bit or bits mutated. This evaluation-selection-variation cycle is repeated either for a fixed number of iterations (generations) or until the unknown system has been modelled to the desired accuracy.

Initial experiments sought to investigate the suitability of lattice filters for genetic adaptation. The example given in (Johnson and Larimore 1977) uses an adaptive filter model of lower order than the plant, to produce an error surface which has both a local and a global minimum. This example was originally conceived to demonstrate the inability of the recursive LMS algorithm to locate the optimal coefficients of the plant and has since been used to highlight the inadequacies of true gradient algorithms when started from within the basin of attraction of a local minimum. Figure 2 shows the results of a simulation run using a genetic adaptive algorithm to identify the system described above. In this graph, the normalised squared error in decibels (the mean of ten runs) is plotted against the generation number. The dB reference level was taken as the the mean squared error generated by an adaptive filter with all coefficients set to zero. From the randomly initialized population of 40 lattice filters, the genetic algorithm requires three generations to concentrate its efforts about the two minima. Two generations later the local minimum has been abandoned and the entire population is clustered around the global minimum.

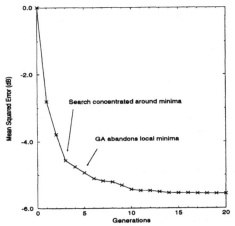

Fig. 2. Identification of example from Johnson and Larimore (1977)

Two methods have been used to characterize the dynamics of the unknown system. In the first, a unit impulse, which is finite at time $n = 0$ but zero elsewhere, is applied to the input of the unknown system. The output of the unknown system is then compared to that of each time-varying filter in turn in order to determine its fitness. As the 'excitation' is confined to the instant $n = 0$, any output signal observed after this time is characteristic of the system itself. This output is known as the impulse response and is of finite length for non-recursive digital filters (hence non-recursive filters are also known as finite impulse response or FIR filters). Conversely, the impulse response of a recursive or infinite impulse response (IIR) filter never decays exactly to zero. Although the impulse response uniquely characterizes a digital filter, certain practical problems arise in the production of a unit impulse for input to a real continuous-time system. As a consequence, simulations have also been undertaken using pseudo-random gaussian noise as the input to the plant and time-varying filters. These runs take significantly more generations for the genetic adaptive algorithm to converge on the optimal adaptive filter coefficients.

In an attempt to provide a comparison with other non-traditional adaptive algorithms some attempt was made at minimizing the number of impulse response samples that were used to characterize the plant and adaptive filters in this first simulation. Results seem to indicate that for the unknown system detailed above, impulse responses of less than approximately 40 samples cause the genetic adaptive algorithm to converge to sub-optimal values of the filter coefficients. This is because the unknown system is a recursive filter and thus has an infinitely long impulse response and cannot be uniquely characterized by a very short burst of this output signal. However, with 40 samples of impulse response being processed by each of the 30 time-varying lattice filters in the population, 1200 time samples are evaluated every generation. From Figure 2 above, 13 generations are required for convergence, on average, resulting in a total of 15,600 time samples being processed in each run. In comparison the Stochastic Learning Automata (SLA) approach of Nambiar et al. (1992) requires about 11,000 time samples to be processed in order to identify the same system. Whilst this means that on this problem the genetic adaptive algorithm performs no better than the SLA, genetic algorithms are thought to be able to tackle problems with high-dimensional search spaces which are difficult for the stochastic learning automata approach to solve. Additionally, the genetic algorithm requires only 40 different time samples as the same output signal is used in the evaluation of every generation. This could give a GA-based approach an advantage if the adaptation could be accomplished off-line.

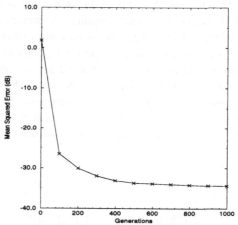

Fig. 3. Identification of sixth order plant

Subsequent simulations sought to test the ability of the genetic adaptive algorithm to optimize higher order adaptive filters. The plant in this case was a sixth order, low-pass Butterworth filter, providing a thirteen dimensional space for the genetic algorithm to search. Mutation and crossover rates were as in the previous example but a much larger population size (720) was found to be necessary. Figure 3 shows the mean squared error in dB (the dB reference level was derived as in the previous experiment) of ten independent runs plotted against the generation number. Convergence to a mean squared error (MSE) of $-30\,$dB was accomplished in just 200 generations but after this point the improvement slowed down dramatically, resulting in a final MSE value of −40 dB after 10000 generations had been evaluated. The minimum mean squared error achievable with the 10 bit precision used to encode each filter coefficient is −51 dB and this was achieved by one run out of the ten.

5 Intersymbol Interference and Data Equalization

In an analogue communications system information from analogue sources is transmitted directly over the communications channel using one of the conventional modulation techniques. In a digital communications system, analogue signals are converted into digital form prior to transmission. The most commonly employed technique for transmitting this digital information is known as pulse code modulation or PCM (Stremler 1990). This pulse modulation technique represents the amplitude of the analogue signal at regular sampling intervals as digital words in a serial bit stream. The transmission characteristics of most real communications channels are usually far from perfect. Twisted pairs of wires, coaxial cable or radio channels all have nonideal frequency response characteristics and may introduce noise or interference which will corrupt the signal transmitted through the channel. A result of the amplitude and delay distortion caused by the nonideal channel frequency response characteristic is intersymbol interference (ISI). Digital pulses subject to intersymbol interference are elongated so that a pulse corresponding to any one bit will smear into adjacent bit slots. The effect of intersymbol interference on a stream of randomly generated polar encoded (±1) digital pulses is illustrated in Figure 4. The discrete-time channel model is represented as a second order transversal filter taken from an example in (Proakis 1983). On this graph, the channel input and output are superimposed in order to show the extent of the distortion caused by ISI.

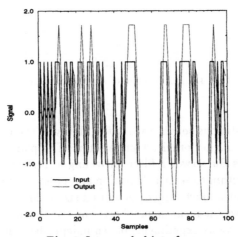

Fig. 4. Intersymbol interference

The effects of intersymbol interference can be reduced by equalization. An equalizing filter is a structure designed to compensate for the imperfect transmission characteristics of the channel. In practical communications systems the frequency response of the channel is usually not known with sufficient accuracy to enable a time-invariant equalizer to be constructed. Similarly, in communication systems operating over switched telephone lines the variation in channel characteristics from one line to another may be so great that the equalizer has to adjust its response to each individual channel. Consequently, an adaptive equalizer is often used, with a response which can be adjusted to meet specific measured channel characteristics. A widely implemented form of adaptive equalizer is the linear transversal equalizing filter. This non-recursive filter is realized as a tapped delay-line with tap weights adjusted by some recursive adaptation algorithm. The criterion used to optimize the equalizing filter coefficients is usually some function of the mean squared error or peak distortion (worst case intersymbol interference). The next Section details preliminary work in the genetic adaptation of recursive equalizing filters undertaken by the authors.

6   A Genetic Adaptive Algorithm for Data Equalization

The results of simulation experiments are presented here in order to illustrate the capabilities of a genetic adaptive algorithm used for data equalization. Two different channels are modelled, producing intersymbol interference at low and high degrees of severity. In each case a traditional bit-string genetic algorithm was used to optimize the coefficients of a population of time-varying recursive filters. The defining parameters of the GA were, for these initial experiments, those identified by Grefenstette (1986) as producing optimal performance with respect to the online performance measure, the on-line performance being defined as the average performance of all tested structures over the course of the search. Thus, the population size was kept small at 30 time-varying filters and the genetic operators of mutation and two-point crossover were applied at rather high frequencies, 0.01 and 0.95 respectively. A sequence of 100 pseudo-random, polar (± 1) encoded, values was used as the input to the channel throughout each run and the genetic algorithm was set the task of minimizing the mean squared error, generated by subtracting the output of the channel from the desired output value at each sample time. Use of this minimization criterion assumes that the adaptive filter has prior knowledge of the transmitted information sequence in order for it to form the error signal . Such information (known variously as a learning sequence or preamble) is typically made available during a short training period before the data is transmitted. The coefficients of time-varying filters may be continuously updated if a decision-directed mode of operation is implemented, in which decisions on the output of the equalizer are assumed to be correct and used in place of the desired output sequence. As long as the receiver is operating at low error rates adaptive algorithms are able to converge on the optimal equalizing filter coefficients.

The results of the first equalization problem are shown in Figure 5. The

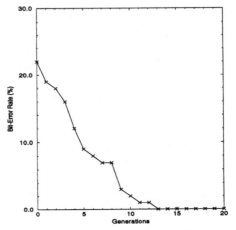

Fig. 5. Equalization of data channel

percentage of bit-errors (the mean of 10 runs) is plotted against the generation number. In this simulation, a channel model with a low-pass frequency and linear phase characteristic is implemented. The transfer function of this channel is:

$$H(z) = 0.407 + 0.815z^{-1} + 0.407z^{-2} \qquad (1)$$

and the distorting effect that it has on an input stream of polar encoded binary digits was illustrated previously in Figure 4. The effect of the intersymbol interference is such that 22% of the bits arriving at the receiver are incorrect. After having evaluated only 13 generations the genetic adaptive algorithm is able to find a set of equalizing filter coefficients which reduce the bit-error rate to zero.

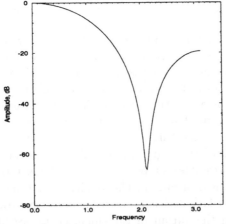

Fig. 6. Frequency response of channel with severe intersymbol interference

The next simulation aims to demonstrate the ability of the genetic adaptive algorithm to compensate for a severely distorting channel. The frequency response of the fourth order discrete-time channel model is shown in Figure 6, illustrating its poor spectral characteristics. The intersymbol interference generated by the channel as a result of its imperfect frequency and phase characteristics distort the input signal such that 40% of the data bits arriving at the receiver are erroneous.

Using a recursive equalizer to compensate for the severe intersymbol interference of the channel, the genetic adaptive algorithm requires, on average, 160 generations to reduce the bit-error rate to zero. This performance was compared to that of a non-recursive equalizer using the same number of filter coefficients (nine). The results of this comparison are given in Figure 7. On this graph, the bit-error rate (mean of ten runs) is plotted against the number of generations evaluated. Within the length of the simulation run the bit-error rate never drops below 15% when using the non-recursive equalizer. This simulation then, clearly illustrates the improved performance that can be obtained from a recursive equalizing filter over a non-recursive equalizer of similar complexity.

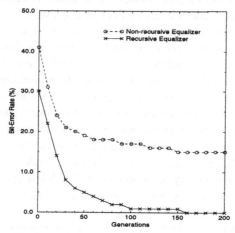

Fig. 7. Equalization of channel with severe intersymbol interference

7 Summary

The results of the simulation experiments presented in this paper demonstrate that a genetic adaptive algorithm can be used to optimize the coefficients of recursive time-varying digital filters. By choosing to implement the adaptive filters as lattice structures the entire feasible coefficient space can be searched without there being any risk of the coefficient set becoming unstable. Additionally, lattice filters are known to be less sensitive to the effects of coefficient round-off. Since one of the major problems of recursive output-error adaptive filters is their potentially multimodal error surfaces, the ability of the genetic algorithm to search spaces of this type is a significant advantage. Other optimization algorithms have also demonstrated this capability. However, when higher-order filters are adapted using the SLA approach of Nambiar et al. the number of actions of the automata becomes increasingly large thus slowing the speed of convergence. In contrast, genetic algorithms have been shown to be capable of tackling very high dimensional problem spaces (Mühlenbein and Schlierkamp-Voosen 1993).

In the system identification configuration, the genetic adaptive algorithm has demonstrated its ability to converge to the optimal filter coefficient values. Using a standard bit-string genetic algorithm, the precision to which these values can be determined is dependent on the number of bits used to encode each coefficient value. For very high orders of adaptive filter, this would require a correspondingly large string representation for each filter structure. Consequently, research is currently being undertaken by the authors into applying real-coded genetic algorithms to the task of filter adaptation. This type of genetic algorithm manipulates vectors of floating-point numbers rather than strings of binary digits and and has been shown to be faster and more consistent from run to run in certain problem domains (Janikow and Michalewicz 1991). The mutation and recombination operators of a real-coded genetic adaptive algorithm may be tailored for the particular task in hand and act at the level of the filter coefficients themselves rather than on a binary representation of them.

The comparison between recursive and non-recursive equalizers (Figure 7) highlights the need for algorithms which can reliably adapt recursive digital filters. Results from the data equalization simulations demonstrate that a GA-driven recursive equalizing filter is able to compensate for the imperfect frequency and phase characteristics of a digital communications channel. Future work in this area will investigate the use of genetic adaptive methods in equalizing time-varying channels. Previous simulations indicate that a conventional genetic algorithm using a strong selection policy and small mutation rate quickly eliminates population diversity as it seeks out the global optimum. Whilst this may be desirable in problems where the global optimum remains static, the performance of the standard genetic adaptive algorithm is adversely affected when the channel characteristics alter with time. In essence, this type of algorithm is often unable to track a moving optimum. Recent studies (Grefenstette 1992; Cobb and Grefenstette 1993), have explored the effectiveness of various mutation-based schemes in enhancing the performance of genetic algorithms operating in changing environments. In addition to these mechanisms, the authors are currently investigating the integration of local search (individual learning) heuristics into the genetic adaptive algorithm framework.

8   Acknowledgements

This work was supported by the UK Science and Engineering Research Council and the Defence Research Agency under the CASE scheme.

References

Cobb, H. G., Grefenstette, J. J.: Genetic Algorithms for Tracking Changing Environments. Proceedings of the Fifth International Conference on Genetic Algorithms. (1993) 523–530

Etter, D. M., Hicks, M. J., Cho, K. H.: Recursive Adaptive Filter Design Using an Adaptive Genetic Algorithm. Proceedings of the IEEE International Conference on Acoustics, Speech and Signal Processing (ICASSP 82). 2 (1982) 635–638

Flockton, S. J., White, M. S.: Pole-Zero System Identification Using Genetic Algorithms. Proceedings of the Fifth International Conference on Genetic Algorithms. (1993) 531–535

Grefenstette, J. J.: Optimization of Control Parameters for Genetic Algorithms. IEEE Trans. Systems, Man, and Cybernetics. 16 (1986) 122–128

Grefenstette, J. J.: Genetic Algorithms for Changing Environments. Parallel Problem Solving from Nature 2. (1992) 137–144

Janikow, C. Z., Michalewicz, Z.: An Experimental Comparison of Binary and Floating Point Representations in Genetic Algorithms. Proceedings of the Fourth International Conference on Genetic Algorithms. (1991) 31–36

Johnson, C. R., Larimore, M. G.: Comments on and Additions to 'An Adaptive Recursive LMS Filter'. Proceedings IEEE. 65 (1977) 1399–1401

Kristinsson, K., Dumont, G. A.: System Identification and Control Using Genetic Algorithms. IEEE Trans. Systems, Man, and Cybernetics. 22 (1992) 1033–1046

Mühlenbein, H., Schlierkamp-Voosen, D.: Predictive Models for the Breeder Genetic Algorithm: I. Continuous Parameter Optimization. Evolutionary Computing. 1 (1993) 25–49

Nambiar, R., Tang, C. K. K., Mars, P.: Genetic and Learning Automata Algorithms for Adaptive IIR Filtering. Proceedings of the IEEE International Conference on Acoustics, Speech and Signal Processing (ICASSP 92). 5 (1992)

Oppenheim, A. V., Schafer, R. W.: Discrete-Time Signal Processing Prentice-Hall (1989)

Proakis, J. G.: Digital Communications. McGraw-Hill (1983)

Shynk, J. J.: Adaptive IIR Filtering. IEEE ASSP Magazine. April (1989) 4–20

Stremler, F. G.: Introduction to Communication Systems, Third edition. Addison-Wesley (1990)

White, M. S., Flockton, S. J.: A Genetic Adaptive Algorithm for Data Equalization. Proceedings of the IEEE World Congress on Computational Intelligence. (to appear)

Complexity Reduction using Expansive Coding

David Beasley[1], David R. Bull[2] and Ralph R. Martin[1]

[1] Department of Computing Mathematics,
University of Wales College of Cardiff, Cardiff, CF2 4YN, UK.
[2] Department of Electrical and Electronic Engineering,
University of Bristol, Bristol, BS8 1TR, UK.

Abstract. This paper describes a new technique for reducing the complexity of algorithms, such as those used in digital signal processing, using a genetic algorithm (GA). The method, referred to as *expansive coding*, is a representation methodology which makes complicated combinatorial optimisation tasks easier to solve for a GA. Using this technique, the representation, operators and fitness function used by the GA become more complicated, but the search space becomes less epistatic, and therefore easier for the GA to tackle. This reduction in epistasis (interaction between parameters) is essential if the difficult task of complexity reduction is to be successfully achieved. Expansive coding *spreads* the task's complexity more evenly among the operators, fitness function and search space. We demonstrate how this technique can be applied to two cases of reduction of complexity of algorithms: a multiplier for quaternion numbers, and a Walsh transform computation. We suggest why the technique is more successful on the former task than the latter.

1 Introduction

Expansive coding is a new representation technique which we have devised to help solve complexity reduction tasks using a genetic algorithm (GA).

There have been several applications of GAs to the design of signal processing algorithms [4, 8, 13, 15]. These have mostly concentrated on the design of digital filters to satisfy particular frequency response templates. In some cases, the GA evolves a set of coefficients for a filter of fixed topology, while in other cases the GA must determine the complete configuration of the filter (i.e. coefficients and topology).

The topic of the work presented here is a variation on the conventional algorithm design task. Conventionally, the function an algorithm must perform is specified, and the task of the GA is to find a suitable configuration of processing elements which realises this function. Our research has concentrated on *algorithm simplification*. In this, we assume that both the function *and* a suitable configuration (perhaps an "obvious" one) are known. The task of the GA is to find *alternative* configurations which perform the same function, but are of lower complexity. Lower complexity may be measured in any desired way, for example: time complexity, component count, total area required, or a combination of these.

The ease with which a GA can solve such a task depends critically on the *representation* used. The most obvious, direct representations suffer from a high degree of interaction among the parameters (or *genes*). This interaction, known as *epistasis*, presents difficulty for any search algorithm, including GAs [5]. Our expansive coding technique introduces some isolation between the different parts of the task to be performed, and thereby reduces the interaction between them. By lowering the epistasis in this way, the task is made more tractable for a GA.

This paper describes the application of the expansive coding method to selected tasks in algorithm optimisation. We first describe the task domain, and then explain the expansive coding technique itself. We then show how the technique has been applied to two practical tasks: simplifying an algorithm to perform quaternion multiplication, and the derivation of the fast Walsh transform. Finally we give our conclusions.

2   The Task Domain

Many algorithms common in digital signal processing take the form of Eqn. 1 where, for a set of N inputs, x_1 to x_N, and a set of M outputs, y_1 to y_M, the output vector, $\mathbf{y} = (y_1 \ldots y_M)$, is generated by the product of a coefficient matrix, $\mathbf{a}$, with the input vector $\mathbf{x}$.

$$y_k = \sum_{j=1}^{N} a_{jk} x_j \tag{1}$$

The direct implementation of such an algorithm requires $N \times M$ multiplications, and $(N-1) \times M$ additions. In many instances, however, elements of the coefficient matrix, $\mathbf{a}$, may be related in some way. In this case, some equivalent computations may be duplicated, giving scope for algorithm simplification. For example, the multiplication of two complex numbers, $(g + \mathrm{i}h)$ and $(x + \mathrm{i}y)$, simplistically requires four real number multiplications to form the real and imaginary coefficients of the result, $[(gx - hy) + \mathrm{i}(gy + hx)]$. However, this may be achieved with only *three* real multiplications [12] by computing the product as, for example, $[(gx - hy) + \mathrm{i}((g + h)(x + y) - gx - hy)]$.

Here, the four multiplications $\{gx, hy, gy, hx\}$ are replaced by the three multiplications $\{gx, hy, (g + h)(x + y)\}$. In most cases, however, finding an improved computational arrangement is non-trivial. It is difficult because of the high degree of interaction between different elements of the coefficient matrix (i.e. if one element is changed, then several other elements may also need changing in order to compensate).

3   Expansive Coding

Expansive coding [2] involves splitting a large, epistatic, task into a number of *sub-tasks*, so that even though high epistasis may remain *within* each sub-task, epistasis *between* sub-tasks is lower. Because the regions of high epistasis are localised, they are easier to deal with.

The steps involved in designing the representation and the GA can be described as follows:

- **Splitting.** The task is split into sub-tasks, so that any combination of valid sub-solutions gives a valid overall solution. Sub-task representations are concatenated together to form a chromosome.
- **Local constraints.** These must be placed on each sub-task, to ensure that the sub-solutions represented are always valid. They can be enforced by using, instead of conventional crossover and mutation, task-specific operators which always maintain the validity of a sub-task.
- **Merging algorithm.** The fitness is calculated by attempting to merge the sub-solutions into a global solution. Methods for doing this will be task-specific. The more merging that is successfully carried out, the fewer distinct sub-solutions will remain, and the higher the fitness.

The GA is left with the greatly simplified task of juxtaposing relatively weakly-interacting sub-solutions to find the overall solution of highest fitness. The search is simplified because the search space is restricted to the space of *valid* algorithms.

To prevent *crossover* from disrupting the validity of any sub-solution, it will often be necessary to restrict crossing sites to lie between sub-tasks. This means that the effective number of symbols in the chromosome will be equal to the number of sub-tasks. Similarly, *mutation* operators will normally work on one sub-task at a time.

4   The Quaternion Multiplier Task

Quaternions [11, 14] have *four* components, and may be written as $q \equiv a + ib + jc + kd$, where a, b, c and d are real numbers, and i, j and k are the three *quaternion operators*. Each of these is analogous to the complex number operator, i. If two quaternions, $q_1 \equiv (a + ib + jc + kd)$ and $q_2 \equiv (p + iq + jr + ks)$ are multiplied to give $(w + ix + jy + kz)$, then the components to be computed are:

$$w = (ap - bq - cr - ds), \qquad x = (aq + bp + cs - dr), \qquad (2)$$
$$y = (ar - bs + cp + dq), \qquad z = (as + br - cq + dp). \qquad (3)$$

Quaternion multiplication may be viewed as 16 input–output mappings. For example, the mapping $a \rightarrow w$, is achieved by multiplying by the value p. This can be represented as a linear signal flow graph, as shown in Fig.1.

This method requires sixteen real-number multiplications. Our task is to devise an algorithm which uses fewer than this.

In Fig. 1, the result of each multiplication is used only once. However, because of redundancy, it is possible to *reuse* results, allowing inputs and/or outputs to *share* multiplications. A generic circuit arrangement for this is shown in Fig. 2.

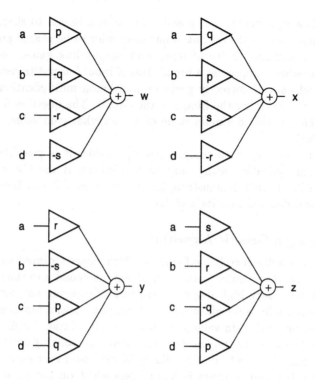

Fig. 1. Simple quaternion multiplication

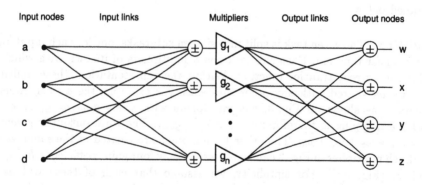

Fig. 2. Generic circuit for quaternion multiplier

Here there are n multipliers, which multiply by factors $g_1, g_2, g_3 \cdots g_n$. Each of these gains, g_i, can be specified by four coefficients, $h_p(g_i), h_q(g_i), h_r(g_i)$ and $h_s(g_i)$, such that $g_i = h_p(g_i)p + h_q(g_i)q + h_r(g_i)r + h_s(g_i)s$. Each of the input nodes is connected via a set of *input links* to the input of each multiplier. Each input link represents a *potential* connection between an input node, and an adder/subtracter node at the input to a multiplier. Each input link therefore represents a connection with a gain in $\{-1, 0, 1\}$. Similarly, the output of each multiplier is connected to each output node via an *output link*, with a gain also in $\{-1, 0, 1\}$.

With this arrangement, it is possible for several inputs to share a common multiplier, and also for the output of one multiplier to be shared among several outputs. By a suitable choice of input and output link gains, and multiplier gains, it is possible to represent a broad class of input/output transfer functions, a subset of which will correctly perform quaternion multiplication. The lower the value of n, the higher the fitness of the circuit. The question for the GA to answer is: What is the minimum value of n, and what gain values are required to achieve this?

Clearly, by analogy with complex multiplication, it is possible to perform quaternion multiplication using only 12 multipliers. If the GA can find this solution, it will at least demonstrate its competence; if it can improve on this solution, it will demonstrate its usefulness.

4.1 Applying a Genetic Algorithm

If we were to use a direct coding scheme, in which values for input link gains, output link gains, and multiplier gains were all simply coded into the chromosome, there would be a very high degree of epistasis. Changing any multiplier gain value can, potentially, alter *all* of the input/output transfer functions. Similarly, changing just one link gain value can make a valid solution invalid. So interdependent are the gains, it is impossible to improve a reasonably good chromosome by making small changes to it. No building blocks could ever form, since the fitness of any sub-group of genes is highly dependent on the values of most of the other genes. Expansive coding can be used to overcome these problems, as detailed below.

Splitting. The given task is split into sixteen sub-tasks, one for each input/output transfer function. Each sub-task has S multipliers of its own with which to fulfil the correct transfer function. Within the representation, these multipliers are *not* shared with any other sub-tasks. For example, the $a \rightarrow w$ transfer function, as shown in Fig. 3, has multipliers with gains $g_{aw1}, g_{aw2}, \cdots g_{awS}$, and the mapping is therefore: $w_a = a \sum_{i=1}^{S} g_{awi}$. The total output is given by $w = w_a + w_b + w_c + w_d$. The other 15 mappings are treated in the same way.

Each multiplier in Fig. 3 can be represented by four gain coefficients, h_p, h_q, h_r and h_s (Fig. 5(c)). For simplicity, we assume that each of these will be in $\{-1, 0, 1\}$. Input and output link gains do not need to be represented explicitly; they can be deduced during the merging phase (see *Merging Algorithm*, below).

Local Constraints. Having split the task into sixteen sub-tasks, we now consider what local constraints should be applied to each of these. For a sub-task mapping input $u \in \{a, b, c, d\}$ to output $v \in \{w, x, y, z\}$, the transfer function is

$$v_u = p \sum_{i=1}^{S} h_p(g_{uvi}) + q \sum_{i=1}^{S} h_q(g_{uvi}) + r \sum_{i=1}^{S} h_r(g_{uvi}) + s \sum_{i=1}^{S} h_s(g_{uvi}) \quad (4)$$

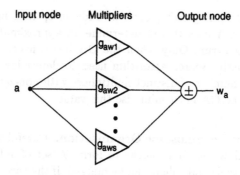

Fig. 3. Circuit for one sub-task

Eqns. (2) and (3) give the total gains required in each of the sixteen cases. For example, for the $a \rightarrow w$ mapping, we require $\sum h_p(g_{uvi}) = 1$, $\sum h_q(g_{uvi}) = 0$, $\sum h_r(g_{uvi}) = 0$, $\sum h_s(g_{uvi}) = 0$. This means that within each sub-task, the sums of the gain coefficients, h_p, h_q, h_r and h_s, must be maintained at specific values. To achieve this, chromosomes in the initial population are set up with valid sums, and the operators used are designed to maintain this validity (see *Operators*, below).

Merging Algorithm. The merging algorithm must bring together multipliers which have equal gains and compatible input/output connections. An example is shown in Fig. 4.

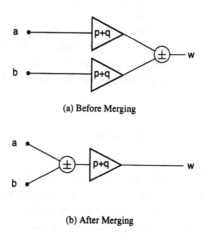

(a) Before Merging

(b) After Merging

Fig. 4. Illustration of merging

In general, there will be several different ways of merging a set of multipliers, so to find the optimum merging pattern, an exhaustive search must be done. This is a slow process, but fortunately we can use an *approximate* fitness evaluation method [9, pp138,206]. A *greedy algorithm* finds an optimal merging pattern in most cases, so our GA uses this to determine an approximate fitness for each chromosome during a run. Only when the GA has converged, and a solution found is the exhaustive search algorithm used to determine the exact fitness. After merging, the number of distinct multipliers with non-zero gain is taken as the fitness value. The GA must minimise this value.

Two-Dimensional Chromosome Organisation. Careful organisation of the chromosome will allow building blocks to form. A set of sub-solutions is well adapted if many of their multipliers can be merged. If they are also close together on the chromosome, they can form a building block. Merging can only take place between multipliers which share common input or output connections. So, a chromosome organisation is needed where sub-tasks are close together if they share common inputs, *or* if they share common outputs. This cannot be achieved with a conventional 1-dimensional chromosome, but is easily arranged on a 2-dimensional chromosome.

The most natural organisation is therefore a 4×4 array of the sub-tasks, (Fig. 5(a)), where each sub-task is represented by S multipliers, (Fig. 5(b)). Each *row* of the array contains sub-tasks relating to the same output node. Conversely, each *column* contains sub-tasks relating to the same input node. Since merging can only take place between multipliers in the same row or column, it is possible for building blocks to form as coherent rows or columns evolve.

Operators. To avoid creating invalid chromosomes, crossover points are only allowed at sub-task boundaries. A 2-D analogy of 2-point crossover is used, with the chromosome treated as a torus.

Mutation is performed by making a small change to a coefficient value of a randomly chosen multiplier. Then, an equal and opposite change is made to the corresponding coefficient of another multiplier within the same sub-task. The effect is to maintain the sums of the gain coefficients, h_p, h_q, h_r and h_s, at their correct values.

4.2 Quaternion Multiplier Results

The expansive representation scheme was tested using a simple generational replacement GA, based on Goldberg's SGA [9], implemented in Pop-11 [1]. We used a form of linear fitness scaling, similar to *sigma truncation* [9, p124]. The number of reproductive opportunities for the most fit individual in each generation was 2.0. We used a crossover probability of 0.8, and mutation probability of 0.064 per sub-task. A run was terminated when the population average fitness, measured over a moving window of a fixed number of generations, stopped increasing.

We used $S = 8$ for most runs. This gives a chromosome of 1024 bits, and a search space of approximately 10^{150} valid chromosomes. At the outset we were aware of a solution to the task requiring only 12 multipliers. Our approach

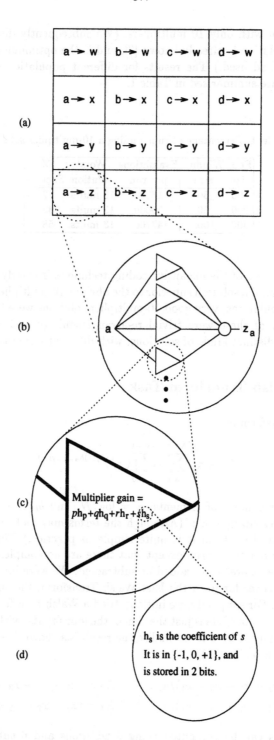

Fig. 5. Chromosome organisation: (a) 4 × 4 sub-task array; (b) a single sub-task; (c) a single multiplier; (d) a single gain coefficient.

found a solution with only 10 multipliers. (We subsequently discovered that a 10-multiplier solution is already known [7], and is the optimum within the task constraints we had used.) The results for different population sizes, averaged over 100 runs, are summarised in Table 1.

Table 1. Percentage of runs finding a 10-multiplier solution

Pop. size	Window size	Evaluations per run	Worst solution	% Success
100	30	16100	15 mults.	2
200	60	56600	13 mults.	33
400	100	147700	13 mults.	58

The effectiveness of the expansive coding technique is clearly demonstrated. Every run produced a solution superior to the obvious 16-multiplier arrangement. With larger populations, very good 10-multiplier solutions were found regularly. Although many of these were trivial re-arrangements of each other, our GA discovered two distinct kinds of solutions, with different structures.

5 The Walsh Transform Task

The Walsh transform is:

$$X_u = \frac{1}{N} \sum_{i=0}^{N-1} x_i \prod_{j=0}^{n-1} (-1)^{b_j(i) b_{n-1-j}(u)} \tag{5}$$

where N is the number of elements, $n = \log_2 N$, and $b_k(z)$ is the kth bit of the binary representation of z. (Although the terms may look complicated, the $\prod$ term expands to ± 1, so it is quite simple in practice.) This is similar to the discrete Fourier transform, except that there are no complex roots of unity involved—elements are either added or subtracted. We were interested to see if a GA could discover for itself the Fast Walsh Transform, the equivalent of the FFT algorithm. For simplicity, we initially tried a Walsh transform with $N = 4$.

Each of the four outputs is just the sum of the four inputs, with some negated, and the total divided by 4 to normalise the result, as shown below (division by 4 will henceforth be ignored):

$$X_1 = (x_1 + x_2 + x_3 + x_4)/4, \qquad X_2 = (x_1 + x_2 - x_3 - x_4)/4, \tag{6}$$
$$X_3 = (x_1 - x_2 + x_3 - x_4)/4, \qquad X_4 = (x_1 - x_2 - x_3 + x_4)/4 \tag{7}$$

Clearly, this can be computed using 6 additions and 6 subtractions. The optimum arrangement is well known, and requires only 4 additions and 4 subtractions, as shown in Fig. 6.

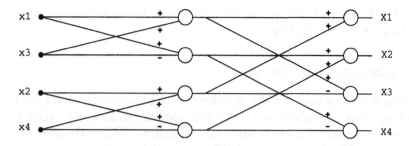

Fig. 6. Optimum solution for the Walsh transform task

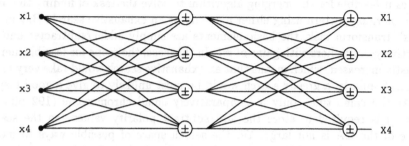

Fig. 7. Generic circuit for the Walsh transform task

A suitable generic circuit is shown in Fig. 7. Here we assume that we know that two stages of addition will be required.

As with the quaternion multiplier task, we can split this into sub-tasks, where each sub-task consists of one input–output mapping, with a number of *paths* connecting the input node to the output node. One such sub-task with four *paths* $(S = 4)$ is shown in Fig.8.

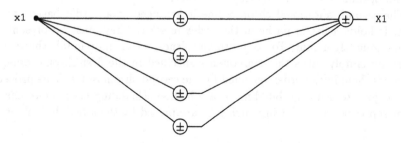

Fig. 8. One of the Walsh transform sub-tasks

Each *path* passes through two stages of addition/subtraction, each of which may effectively multiply the signal by $+1$ or -1. Not all paths need be used. There are thus five possibilities for each path, which may be represented as: $\{+, +\}, \{+, -\}, \{-, +\}, \{-, -\}$ or "not present". Thus each path can be represented in 3 bits, each sub-task in $3 \times S$ bits, and the whole generic circuit of 16 sub-tasks in $3 \times S \times 16$ bits.

5.1 Difficulties with Merging

In some ways this task is much simpler than the quaternion multiplier task—which involves multiplication by different coefficients. However, when the expansive coding technique is applied, this "simplicity" makes the merging phase much harder. The large number of different multiplier gain values in the quaternion task restricts the scope for the merging algorithm to a reasonable size. This makes it feasible for the merging algorithm to solve the task of finding the optimum merging pattern using either a greedy or an exhaustive search. But in the Walsh transform task, the basic elements are simply adder/subtracter units—effectively multipliers with gains of ± 1. In this case, the scope for feasible merges is vastly increased, which means that an exhaustive search would take very much longer. Similarly, a greedy search would be most unlikely to give a good result.

At the same time, only a comparatively small chromosome (192 bits for $S = 4$) is required to store the add/subtract polarity values—so the search space of the GA is not large. Yet the search space of possible ways to merge these sub-tasks is much larger. Clearly, the division of effort between the GA and the merging algorithm is out of balance—the GA has a simple task, while the merging algorithm has a difficult one. To solve this task effectively, the balance must be tipped, giving the GA more of the burden of search.

5.2 A Two-Chromosome Representation

The difficult task of merging the sub-tasks can be given to the GA by using a *two-chromosome* representation for each individual. The GA then has the task of searching in *two* search spaces, to find both a suitable set of path polarities, *and* an optimal way to merge them together.

The second, additional chromosome is arranged as an order-based chromosome. It holds a list of paths in the order in which they must be presented to the merging algorithm. To compute the fitness of an individual, the merging algorithm simply takes the components specified in the first chromosome, and processes them in the order specified by the second chromosome. This procedure is analogous to schedule building in a job-shop scheduling GA, where chromosomes represent the order in which jobs are placed by the schedule builder [16].

5.3 The Merging Process

As with the merging carried out in the quaternion multiplier task, two paths may be merged if they share common input or common output nodes, and if they have compatible polarities.

The merging algorithm takes the list of paths specified by the second chromosome, and considers the path at the head of the list for merging. An attempt is made to merge it with the following path in the list. If this merge is not possible, then the path after that is tried. A maximum number of merge attempts, known as the *lookahead distance*, is made. If no merge is possible, then the first path is moved to the end of the list. If a merge *is* possible, the merged paths are placed at the end of the list. The new path at the head of the list is then processed. This continues until the whole list is traversed without any merges being made.

After the merging process is complete, we can compute the number of additions and subtractions required by the resultant graph. The (un)fitness of the individual is computed by forming a weighted sum. For our investigations, we used the following (arbitrary) weights: additions, 3, subtractions, 4.

5.4 Reproduction

Mating involves two parents, each having two chromosomes. One is a two-dimensional, value-based chromosome, the other is an order-based chromosome. The two chromosomes are crossed over and mutated quite independently. Different crossover and mutation routines are required for the two chromosomes, since they are have completely different structures. There are separate crossover probability (*pcross*) and mutation probability (*pmutation*) parameters for each chromosome.

The first chromosome has a structure similar to that used by the quaternion multiplier, and the crossover and mutation operators work in the same way. For the second chromosome we use uniform order-based crossover [6]. For mutation we use a type of scramble sub-list mutation [6, p81]. In this, we choose a contiguous sub-section of the chromosome, and randomly scramble the order of its genes.

5.5 4-Input Walsh Transform Results

We experimented with a variety of crossover probabilities, mutation probabilities and population sizes. Like many GA researchers in the past, we found that no parameters are critical, so the GA performs well over a range of parameter values [10]. In our tests we used the performance measuring methods used for previous work [3].

With $S = 4$ (4 paths per sub-task), and a *lookahead distance* of 2, we found that the best results were obtained with a population size of 32, $pcross = 0.5$ and $pmutation = 0.03$ for the first chromosome, and $pcross = 0.5$ and $pmutation = 0.2$ for the second chromosome. In 100 runs, we obtained an optimum solution in an average of 1942 evaluations (±297 with 95% confidence margin, standard deviation 1379). By comparison, a random search of the same task space found only one solution in 1 200 000 evaluations.

The results we obtained all require only one path per sub-task. Starting with 4 paths per sub-task, it is clear that the GA must do a significant amount of work simply to eliminate the redundant paths. When we ran the GA with fewer paths per sub-task, we found that it converged much more quickly: in only 603 evaluations ($\pm$ 139) with $S = 1$.

We also tried increasing the lookahead distance to 4. This gave a further speed improvement to 242 evaluations ($\pm$ 51). An extension to 6 gave a smaller improvement, to 204 ($\pm$ 55).

Such a large speedup is worrying. The performance is now so good that solutions often appear in the initial population! It therefore seems that little credit can be given to the GA. It seems that our task representation is so attuned to the task, that little searching is required. This may mean that the GA is in a stronger position to tackle more difficult tasks of the same type.

5.6   8-Input Walsh Transform Results

We ran the GA on a similar Walsh Transform task, but with 8 inputs, and 64 sub-tasks. Each path has *three* stages of addition/subtraction, making the merging much more complicated. With 4 paths per sub-task ($S = 4$), we found that the average fitness steadily improved during each run, eventually reaching the fitness of a typical, randomly generated individual with $S = 1$. But starting with $S = 1$, and using a variety of mutation and crossover rates, population sizes up to 512 and lookahead distances up to 8, we found that the population entirely failed to converge.

The uniform order-based crossover used for the second chromosome was found to be largely responsible. When this was replaced with PMX [9], the average fitness of the population converged, but still, little improvement in the maximum fitness of the population was observed. The "best" individuals of each run showed only trivial merges, with little evidence of a coherent structure emerging.

We also tried fixing the first chromosome of all members of the population, such that each contained an optimum set of genes throughout each run. In this case, the GA only had to search over the second chromosome. However, the performance was indistinguishable from before.

6   Conclusions

At first sight, expansive coding seems counter-intuitive, since it makes the search space larger. The increase in the number of parameters is, however, offset by the restriction of the search space to only *valid* chromosomes. Furthermore, the task is made simpler, since the interaction (epistasis) between the elements which the GA has to manipulate (the sub-tasks) is reduced.

With appropriate representation and operators, the inherent complexity of a task may be shifted, so that although the fitness decoding function becomes more complicated, the GA finds the task easier. In theory, this allows any task to be made trivially easy to solve, from the point of view of the GA [17]. This was the approach adopted with the quaternion multiplier task—a significant amount of the search effort was performed by the merging algorithm.

With the Walsh transform task, however, the same approach would have relied too heavily on the merging algorithm, and not enough on the GA. We therefore transferred much of the effort involved in the merging process *back* to the GA. By using *two* chromosomes, the idea is that the GA solves two interrelated tasks (a value-based task and an order-based task) at the same time.

Our initial results with the 4-input Walsh transform task seemed to show that this approach had been successful. But further work shows that, primarily, all the GA is doing is eliminating the redundancy introduced by having more than one path per sub-task. Once the GA has solved this part of the task, the representation ensures that a solution is found easily.

Much the same was found with the 8-input Walsh transform task. The GA can successfully eliminate redundant multiple paths. But it has difficulty finding solutions much better than a randomly generated population given one path per sub-task as a starting point.

It therefore appears that the introduction of the second chromosome has not, in fact, enabled the GA to solve the Walsh transform task. The merging task, encoded in the second chromosome, is, *by itself*, too difficult for our simple GA to solve.

We suspect that this is because the structure of the Walsh transform task is, in certain respects, too "simple". In the quaternion multiplier task, there are many constraints on which paths can be merged, since there are many possible multiplier gain values. This significantly reduces the area of the search space (of the merging task) which must be explored—to the extent that even an exhaustive search is feasible. However, in the Walsh task, each node is effectively a multiplier with a gain of ± 1. The constraints on merging are therefore minimal, and almost the whole search space the merging task must be considered. This is a large search space. For the 8-input Walsh task, with 1 path per sub-task, the first chromosome has 256 bits, while the second is twice this size. Consequently, whether the merging is performed by a conventional search procedure (as in the quaternion task), or by the GA (using a second chromosome), it is a difficult task to solve.

It therefore seems that expansive coding is more useful where each sub-task is specified by a large number of bits. This allows a significant proportion of potential merges to be ruled out quickly, thereby reducing the amount of searching which must be done in the merge space. For example, in the quaternion task, each path is specified by 8 bits, whereas in the 4-input Walsh task, each path requires only 3 bits to specify it. Hence, for a randomly-distributed set of sub-task parameters, there will only be a 1 in 256 chance that two quaternion paths would have the same set of parameters, while for the Walsh task, the figure is 1 in 8. So the search space which needs to be considered while merging the quaternion task is of the order of 32 times smaller.

A further difficulty of the Walsh task is that the known, optimal solution provides such a large improvement over the simple solution—from $O(n^2)$ to $O(n \log n)$. This is obtainable because of the high degree of regularity in the task. A relatively "weak" search method, however, which knows nothing about this regularity, would be unlikely to find the optimal solution.

We are currently investigating the application of expansive coding to tasks which are more closely related to the quaternion multiplier task, yet are scalable in nature. Our initial investigations into manipulations involving symmetrical Toeplitz matrices show that simplifications, which are far from obvious, can be obtained. We are looking into ways of maintaining good performance as task size is scaled up.

GAs are good for tasks of intermediate epistasis [5]. On highly epistatic tasks, therefore, a suitable representation and operator set must be found which sufficiently reduces the epistasis. The expansive coding technique is one such approach. Complexity, in terms of epistasis in the original task, is traded for complexity in terms of an increased chromosome size, a more complicated fitness function, and the need for task-specific operators. We have therefore split one large dose of complexity into three smaller doses—making the task easier to tackle. It is important, however, to maintain a balance of complexity among these three areas. If too much complexity is transferred to the fitness function (e.g. in the merging), the GA will not be successful.

Our practical applications of this method show that it can be effective for certain types of tasks, so long as the complexity of merging does not become too great.

References

1. A. Barrett, A. Ramsay, and A. Sloman. *POP-11: a Practical Language for Artificial Intelligence*. Ellis Horwood, Chicester, 1985.
2. D. Beasley, D.R. Bull, and R.R. Martin. Reducing epistasis in combinatorial problems by expansive coding. In S. Forrest, editor, *Proceedings of the Fifth International Conference on Genetic Algorithms*, pages 400–407. Morgan Kaufmann, 1993.
3. D. Beasley, D.R. Bull, and R.R. Martin. A sequential niche technique for multimodal function optimization. *Evolutionary Computation*, 1(2):101–125, 1993.
4. C-H.H. Chu. A genetic algorithm approach to the configuration of stack filters. In J.D. Schaffer, editor, *Proceedings of the Third International Conference on Genetic Algorithms*, pages 219–224. Morgan Kaufmann, 1989.
5. Y. Davidor. Epistasis variance: Suitability of a representation to genetic algorithms. *Complex Systems*, 4:369–383, 1990.
6. L. Davis. *Handbook of Genetic Algorithms*. Van Nostrand Reinhold, 1991.
7. H.F. de Groote. On the complexity of quaternion multiplication. *Information Processing Letters*, 3(6):177–179, 1975.
8. D.M. Etter, M.J. Hicks, and K.H. Cho. Recursive adaptive filter design using an adaptive genetic algorithm. In *IEEE Int Conf Acou Speech Sig Proc*, pages 635–638, 1982.
9. D.E. Goldberg. *Genetic Algorithms in search, optimization and machine learning*. Addison-Wesley, 1989.

10. J.J. Grefenstette. Optimization of control parameters for genetic algorithms. *IEEE Trans SMC*, 16:122–128, 1986.

11. W.R. Hamilton. *Elements of Quaternions*. Cambridge University Press, 1899.

12. D.H. Horrocks and D.R. Bull. The synthesis of complex signal multipliers. In V. Cappellini and A.G. Constantinides, editors, *Proc. Int. Conf. Digital Signal Processing*, pages 207–210. North-Holland, 1987.

13. S.J. Louis and G.J.E. Rawlins. Designer genetic algorithms: Genetic algorithms in structure design. In R.K. Belew and L.B. Booker, editors, *Proceedings of the Fourth International Conference on Genetic Algorithms*, pages 53–60. Morgan Kaufmann, 1991.

14. R.R. Martin. Rotation by quaternions. *Mathematical Spectrum*, 17(2):42–48, 1983.

15. D. Suckley. Genetic algorithm in the design of FIR filters. *IEE Proc-G*, 138:234–238, 1991.

16. G. Syswerda. Schedule optimization using genetic algorithms. In L. Davis, editor, *Handbook of Genetic Algorithms*, chapter 21, pages 332–349. Van Nostrand Reinhold, 1991.

17. M. Vose and G. Liepins. Schema disruption. In R.K. Belew and L.B. Booker, editors, *Proceedings of the Fourth International Conference on Genetic Algorithms*, pages 237–242. Morgan Kaufmann, 1991.

The Application of Genetic Programming to the Investigation of Short, Noisy, Chaotic Data Series

E. Howard N. Oakley

Institute of Naval Medicine, Alverstoke, Gosport, Hants PO12 2DL, UK

Abstract. Techniques to investigate chaotic data require long noise-free series. Genetic programming allows fitting of arbitrary functions to short noisy datasets. Conventional genetic programming was used to fit Lisp S-expressions to a known chaotic series (the Mackey-Glass equation, discretized to a map) with added noise. Embedding was performed by including previous values in time in the terminal set. Prediction intervals were 20–1065 steps into the future, based upon near-minimal 35 'training' points from the series.

Fittest S-expressions yielded useful structural information. Semilogarithmic plots of normalised root mean squared error of the fittest forecasts against the length of forecast showed two dominant slopes. Noise led to a small exponential increase in this error. Genetic programming appears useful, as it compares favourably with established techniques, is robust to noise, and easily avoids overfitting.

1 Background

Unpredictability is a characteristic of chaotic systems, and has been studied in terms of summary measures, such as Lyapunov exponents, and attempts to develop methods of prediction. Although there is controversy over the very discrimination of chaos (e.g. Ruelle [13] and Farmer and Sidorowich [4]), there is now an extensive literature devoted to the efficacy of different prediction techniques. Casdagli [1] has succinctly cast the problem thus:

> "The standard problem in dynamical systems is, given a nonlinear map, describe the asymptotic behavior of iterates. The inverse problem is, given a sequence of iterates, construct a nonlinear map that gives rise to them. This map would then be a candidate for a predictive model."

Initial studies of chaos in physical systems, in which precise measurements can be made into large datasets, have encouraged similar work in other fields, such as biological systems. In the latter, the requirements of tests for chaos – including such long series of noise-free data – are usually very difficult to satisfy, although many such systems are likely to be chaotic [5]. Prediction techniques may also be unsuitable for the data available. Early work employed radial basis functions [1] for local prediction, and has been succeeded by other techniques including piecewise linear approximation [3] (which used 40 000 observations in the fitting set), neural networks, and the genetic algorithm. Among neural network approaches, Stokbro and Umberger [14] used weighted maps and training series of 500 to 5000 observations to predict just 6 steps into the future, and Mead et al. [9] used a modified radial basis function neural network (termed a "connectionist normalised local spline network") containing 35 or more nodes with 500 points in the learning series. Some, such as Nychka et al. [11], have also used neural networks to estimate the most positive Lyapunov exponent in short, noisy datasets. An alternative approach is exemplified by Meyer and Packard [10], who employed the genetic algorithm on series embedded in phase space, to forecast limited regions of attractors.

Koza's monograph on genetic programming [7] includes two demonstrations of simple non-linear systems, one of which is also chaotic, which have been fitted with predictive equations using this new technique. Although the chaotic example is the logistic map, one of the simplest instances of chaos, his limited exploration of short term prediction is promising, and consistent with Casdagli's suggestion that genetic programming may be useful in forecasting chaotic series [3].

The work reported here is the second part of a study eventually intended to develop quantitative techniques for the investigation of non-linearities in the control of human peripheral blood flow. The first step [12] was to examine genetic programming in this role using a standard system which is known to be chaotic; this second report describes the results from the application of the same techniques to noisy theoretical data, and the third will extend this to experimental data.

The chaotic system chosen as being similar in visual appearance to that of the physiological data was the Mackey-Glass equation [8]:

$$\frac{dx_t}{dt} = \frac{bx_{t-\Delta}}{1 + (x_{t-\Delta})^c} - ax_t \ , \tag{1}$$

where a is usually taken as 0.1, b as 0.2, c as 10.0, and Δ as 30.0. Expressed as a flow, this is not only non-trivial to approximate, but is not explicitly soluble as a Lisp S-expression. For the purposes of this study, it was therefore discretized to a map:

$$x_{t+1} = x_t + \frac{bx_{t-\Delta}}{1 + (x_{t-\Delta})^c} - ax_t \ , \tag{2}$$

which retains its chaotic characteristics (see Fig. 1) whilst being expressible as an S-expression. It therefore satisfies Koza's criterion [7] as being sufficient. For comparison, Fig. 6 shows one example of the physiological data which is the inspiration for this work.

Rather than use very long training series, it was decided to restrict the training data close to that just sufficient to predict further members of the series, which if Δ is 30, would be 31 points. It is also of particular interest, as it is at the limit of the $D \leq 2\log_{10} N$ rule (D being the correlation dimension, and N the number of observations in the time series) expounded by Ruelle [13] and

Mackey–Glass Map

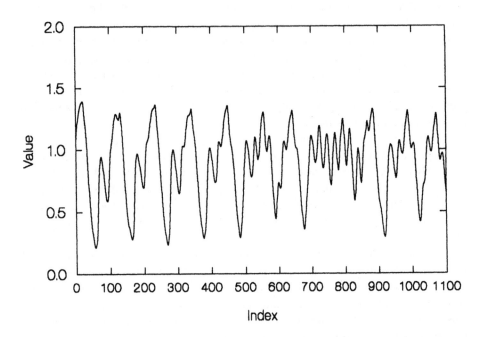

Fig. 1. Members of the uncontaminated Mackey-Glass map used in the experiments

derived from the Grassberger-Procaccia algorithm [6]. Given that the correlation dimension of the Mackey-Glass series is of the order of 3 [1], the expected minimum series from which useful information could be extracted is 32.

2 Methods

In the first phase, a large number of runs (several hundred in all) were undertaken to attempt to predict the values of the Mackey-Glass mapping 20–1065 time steps into the future, given the same 35 sequential exemplars of the mapping. The exemplars were generated using double-precision floating point calculation, seeding the series with 35 random numbers between 0.5 and 1.5, and applying (2). The first 1000 in the generated series were discarded, and only data following those were used experimentally; these are illustrated in Fig. 1. Genetic programming was performed using the Simple Lisp implementation [7] with local performance enhancements. A typical tableau is shown in Table 1.

Time delay embedding was effectively achieved by the inclusion in the terminal set of previous values in time, with an upper limit to dimension of 10.

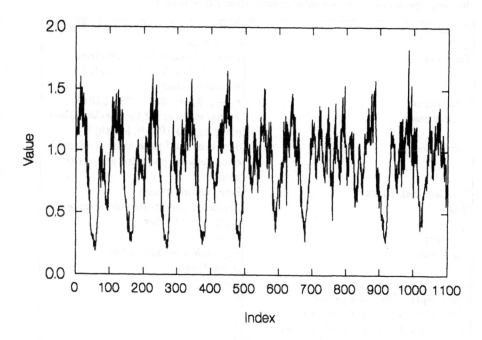

Fig. 2. Members of the most noisy Mackey-Glass map used in the experiments, contaminated with 50% noise

Unusually, because previous values were deliberately not interspersed equally in time, it was possible for S-expressions to effectively incorporate multiple dimensional embedding, with unequal time delays.

Koza's [7] criterion of sufficiency required that the terminal set included each of the variables in (2), and that the function set included each of the arithmetic functions in (2). Additional terminals were added in order to produce a set which was more generally applicable (it has since been used on series for which underlying equations are unknown), similarly with functions. Although a priori knowledge introduces inevitable bias, only 2 of the 10 terminal variables actually appear in (2), and aside from the basic 4 arithmetic operators, only 1 of the 3 functions is used in (2). Thus, the actual search space was very much larger than one confined to the variables and functions appearing in (2).

Many runs were also augmented with fitness functions which included comparisons between Fourier power spectra of the predicted and real data series. These were used as the sole fitness measure, and in combination with the squared error measure given in Table 1. Fitness was computed as the squared error be-

Table 1. Typical tableau for fitting chaotic series. % is the protected divide operation. **EXP10** is exponentiation to the power of 10 (x^{10}) protected from numeric overflow such that any operand of absolute value greater than 2.0 returns 2^{10}

Objective	Predict next N, 1–10 times in Mackey-Glass map
Terminal set	Time-embedded data series from $t = 1, 2, 3, 4, 5, 6, 11, 16, 21, 31,$ $\Re$ (real constants)
Function set	+, −, %, *, SIN, COS, EXP10
Fitness cases	Actual members of the Mackey-Glass mapping (20–1065 in length)
Raw fitness	Sum over the fitness cases of squared error between predicted and actual points
Standardized fitness	Same as raw fitness
Hits	Predicted and actual points are within 0.001 of each other
Wrapper	None
Parameters	Population 50–5000; 51 or 101 generations
Success predicate	None
Max. depth of new individuals	6
Max depth of new subtrees for mutants	4
Max depth of individuals after crossover	17
Fitness-proportionate reproduction fraction	0.1
Crossover at any point fraction	0.2
Crossover at function points fraction	0.7
Selection method	Fitness-proportionate
Generation method	Ramped half-and-half

tween estimates of power across predicted and actual spectra following Fourier analysis. When used in combination with the untransformed squared error, the latter was estimated first, and only if it exceeded a given value (40.0, half the fitness achieved by the best linear extrapolation) was the power spectral fitness measure used by adding both power spectral and untransformed summed squared errors together. In some other runs, tournament selection was used (with groups of size 3–15), rather than fitness proportionate methods.

The second series of runs examined prediction periods of 20, 30, 40 and 100 into the future, which were applied over 10 (periods 20–40) or 7 (period 100) successive non-overlapping periods of the Mackey-Glass series. In other words, each S-expression was 'trained' against 7 or 10 separate windows of 35 real data points, before being used to predict the following 20–100 points. These repeated forecasts were performed using the original data series, and the same series contaminated with varying amounts of multiplicative (proportionate) random noise. According to normalised root mean squared error estimates, the amount of noise was 0.12, 0.25 and 0.50, and will be referred to as 12%, 25%, and 50% respectively. Figures 1 and 2 illustrate the contrast between uncontaminated and 50% noisy series.

Forecasting error was computed as the normalised root mean squared error of the fittest S-expression when assessed against the uncontaminated Mackey-Glass data series [4]. Thus, measurement of the effectiveness of prediction was made against the standard of the original data series, even though prediction may have been based on 'training' data containing noise. This is similar to the situation pertaining when using real data.

3 Results

The predicted results from a selection of fitter and more interesting S-expressions are shown in Fig. 3. Those illustrated are:

GFFT1, long-term fit using the FFT fitness function,

```
(COS (* (- Y31 0.6168952889819135) (- (- Y31 Y1)
                                  -1.0653563570535542))))
```

which is equivalent to

$$x_{t+1} = \cos\left((x_{t-\Delta} - 0.616895\cdots) \times ((x_{t-\Delta} - x_t) + 1.065356\cdots)\right) \ . \qquad (3)$$

GFFT2, long-term fit using the FFT fitness function,

```
(% (EXP10 Y2) Y31)
```

which is equivalent to

$$x_{t+1} = \frac{(x_{t-1})^{10}}{x_{t-\Delta}} \ . \qquad (4)$$

GST1, long-term fit without the FFT fitness function,

```
(COS (SQRT% (SQRT% (- Y31 Y1))))
```

which is equivalent to

$$x_{t+1} = \cos\left((x_{t-\Delta} - x_t)^{0.25}\right) \ . \qquad (5)$$

GST2, long-term fit without the FFT fitness function,

```
(COS (- Y31 (COS Y31)))
```

which is equivalent to

$$x_{t+1} = \cos(x_{t-\Delta} - \cos(x_{t-\Delta})) \ . \qquad (6)$$

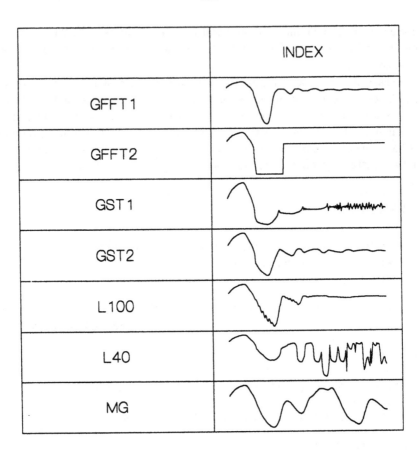

	INDEX
GFFT 1	
GFFT 2	
GST 1	
GST 2	
L 100	
L 40	
MG	

Fig. 3. Plots of a selection of S-expressions and the Mackey-Glass mapping. In each case, the first 35 points are common, and represent the 'training series'. A total of 200 points are shown for each series, and S-expressions are given in the text. From Oakley [12]

L100, short-term fit over 100 points

```
(SIN (* (* (COS Y31)
           (+ (SIN (* (COS Y31)
                      (+ (COS Y31)
                         (* (SIN (SIN (COS Y31)))
                            (SIN (* (COS Y31)
```

```
                  (+ (+ (COS Y31) Y21) Y21)))))))
        Y21))
    (+ (* (COS Y31) (COS Y31)) Y21)))
```

L40, short-term fit over 40 points

```
(COS (- (* (COS Y4)
           (+ (* (COS (COS (- (* Y31
                                 (+ (* (COS (COS (COS
        (+ (COS Y2) Y31)))) Y31) Y31)) (COS (COS Y2))))) Y31) Y31))
    (COS (COS (COS (COS (COS (* (+ (* (COS Y4) Y31) Y31) Y31)))))))))
```

MG, original Mackey-Glass mapping

```
(+ Y1 (- (% (* 0.2 Y31) (+ 1.0 (EXP10 Y31))) (* 0.1 Y1)))
```

which is equivalent to (2).

The relationship between forecasting period and fitness was investigated by varying the number of fitness cases between 20 and 1065, between different series of runs. Normalized root mean squared errors [4] of results are shown in Fig. 4, which demonstrates that forecast accuracy falls with increasing length of forecast (increasing number of fitness cases).

Given the a priori knowledge of the Mackey-Glass equation, a number of interesting S-expressions emerged from experimental runs. Surprisingly, the terminal set member at a time delay of 30 (corresponding to $x_{t-\Delta}$ in the Mackey-Glass mapping) was featured in the majority of fittest S-expressions, although it did not appear in the form $1/(x_{t-31})^{10}$ found in (2). Of 22 runs to perform long-term prediction, this delay appeared in 20 of the fittest S-expressions (χ^2 test significant at $p \ll 0.001$), and of 22 runs to perform short-term prediction, the same delay appeared in 17 of the fittest S-expressions (χ^2 test significant at $p \ll 0.001$).

Forecasting error increased with increasing amounts of noise. This relationship is shown in Fig. 5, where it is apparent that, with the exception of remarkably low error when forecasting 30 steps ahead with the noise-free dataset, the results closely follow exponential curves, rising little even with large amounts of noise.

4 Discussion

4.1 Deception and Fitness

Although these results confirm that genetic programming appears to be of value in short-term prediction of short and noisy chaotic series, it may appear disappointing that no S-expression came close to that for the Mackey-Glass mapping which generated the data. However, this upholds the contention that the series studied is chaotic, as it possesses the quality of unpredictability. In this sense, the noise-free data series is deceptive, and the addition of noise will only compound

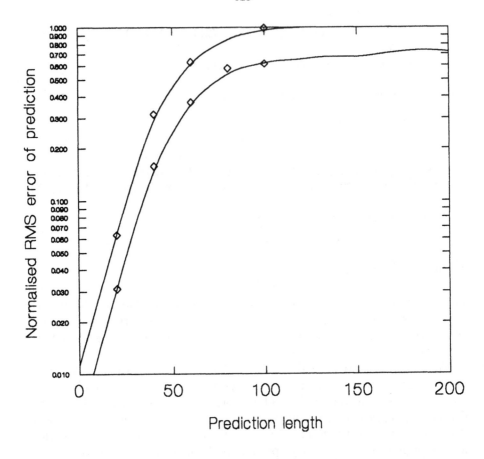

Fig. 4. The effect of forecast length on error. This is a semilogarithmic plot of the normalised root mean squared error of the fittest forecasting attempt, against the number of steps in the forecast period. The upper line represents data from Casdagli [1], the lower line the single forecast series from this study

the problem of prediction. Performance could be improved by modification of the genetic programming technique; Oakley [12] has already discussed the use of compound or 'steered' fitness measures, segregation of the population into demes, and hierarchical automatic function definition [7]. Employment of local Lyapunov exponents and dimensions as part of the estimation of fitness may also prove worthwhile, although computationally very intensive.

4.2 Effectiveness of Prediction

Given the very short 'training' data series, both with and without noise, predictions appear surprisingly good. For example, comparisons with data esti-

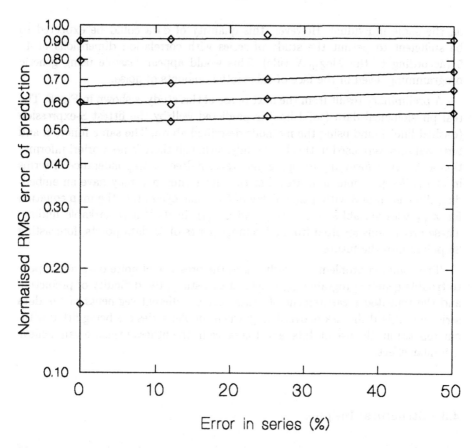

Fig. 5. The effect of noise on forecasting error, for different lengths of forecast. This is a semilogarithmic plot of the normalised root mean squared error of the fittest forecasting attempts obtained when forecasting 20 (lowermost line), 30, 40 and 100 (uppermost) steps ahead, in relation to the amount of noise (%) in the data series. All results are derived from forecasts repeated 7 or 10 times along the data series

mated from Casdagli's iterative radial basis function predictions [1] – the best-performing technique reported in his study, and still [14] one of the best techniques of all – are shown in Fig. 4.

Most physiological time series are brief and noisy, relative to those derived from purely physical systems. For instance, a rheometer measurement system typically records blood flow at 40 Hz from each of two sites, but this is reduced to approximately 1 Hz when analyzed in cardiac cycles; periods of up to 10 min are usually collected, providing series of 100–600 data points, and subject to ±10% observational error in addition to other noise. These would be insufficient for neural network techniques [9, 14], particularly as few experiments are repeated on the same individual. However, this quantity of data could be expected to be sufficient to permit the study of series with correlation dimensions of 4–5 (according to the $2 \log_{10} N$ rule). This would appear feasible using genetic programming, even in the face of substantial amounts of noise.

A preliminary result from the next phase of this study is shown in Fig. 6. The real physiological data (solid line) is predicted well by the fittest S-expression (dashed line) found using the methods described above. The same function and terminal sets were used in this later study, although there is no a priori information as to what form any mapping may take. Indeed, a preponderance of terms in fittest S-expressions indicates that the underlying map may have an embedding dimension of 3 with a time delay of 5 cardiac cycles (i.e. the map equation for x_{t+1} uses variables of x_t, x_{t-5} and x_{t-10}), in itself a remarkable finding. These predictions resulted from a 'training' series of 33 data points, forecasting 36 points into the future.

The common problem of overfitting in the presence of noise does not appear to trouble genetic programming much. In this study, the difficulty of prediction and the repeated measurement of fitness across different segments of the data series minimised the risk of overfitting; when simpler series are being fitted without repetition, the use of 'hits' and tolerance in the fitness measure can achieve a similar effect.

4.3   Structural Insights

A number of characteristics of the fittest S-expressions found may provide structural insights into a non-linear system, which could in turn help in the construction of a model. In the first instance, the form of the relationship in Fig. 4 will be influenced by the presence or absence of chaos, and gradients reflect the resulting unpredictability of the system. Casdagli [1] has pointed out that there are two likely underlying exponential equations which express scaling laws. The two prevalent gradients in the original data are, using natural logarithms, 0.08121 ($N < 60$) and 4.655×10^{-4} ($N > 80$). These compare with a computed metric entropy (equal to the sum of positive Lyapunov exponents) of approximately 0.01 for the Mackey-Glass flow [10]. The form of the curves in Fig. 4 may also be used to infer the type of model most appropriate to the underlying data [2].

Another clue to structure may lie in the preponderance of S-expressions incorporating a time delay Δ of 30; in the absence of a priori knowledge of the structure of a non-linear system, it could be a useful starting point for model construction, as has been suggested in the preliminary results from the third phase of this study, referred to above. Few other techniques appear capable of this, particularly as they may hide empirical relationships discovered in the data series. In contrast, genetic programming encapsulates them in S-expressions which are rich ground for further examination.

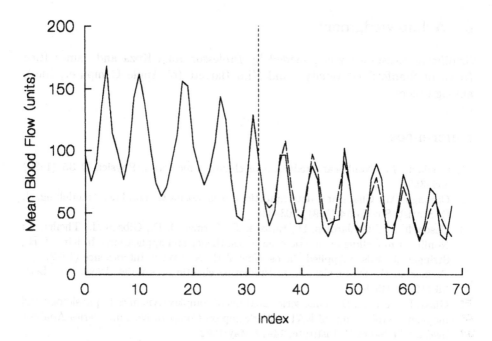

Fig. 6. Preliminary results from one experiment in the third phase of the study. A real physiological dataset is shown, of mean blood flow plotted against cardiac cycle (solid line), together with the fittest predicted series from genetic programming (dashed line). The vertical marker indicates the division between 'training' and predicted points for the latter. (Work by Oakley, submitted for publication)

5 Conclusions

Investigation of a well-studied chaotic system by means of prediction using genetic programming has resulted in relatively accurate forecasts, even when 'trained' on datasets of nearly minimal length (only slightly above the number predicted by the $2\log_{10} N$ rule), and in the presence of considerable amounts of noise. Applied carefully, this technique appears useful in the forecasting of chaotic series.

Furthermore, the S-expressions evolved appear capable of providing structural information about the non-linear system which can be useful to its investigation and modelling. The graph of normalized root mean squared prediction error against the length of forecast, and examination of the composition of fitter S-expressions, may be of particular use.

6 Acknowledgments

Significant assistance was provided by Professor John Koza and James Rice (both of Stanford University), and Kim Barrett (of Apple Computer, Inc.), among others.

References

1. Casdagli, M.: Nonlinear prediction of chaotic time series. Physica D **35** (1989) 335–356
2. Casdagli, M.: Chaos and deterministic versus stochastic non-linear modelling. J. R. Statist. Soc. B **54** (1991) 303–328
3. Casdagli, M., des Jardins, D., Eubank, S., Farmer, J. D., Gibson, J., Theiler, J.: Nonlinear modelling of chaotic time series: theory and applications. In Kim, J. H., Stringer, J. (Eds.): Applied Chaos. New York, NY: Wiley-Interscience (1992)
4. Farmer, J. D., Sidorowich, J. J.: Predicting chaotic time series. Phys. Rev. Lett. **59** (1987) 845–848
5. Glass, L., Kaplan, D.: Time series analysis of complex dynamics in physiology and medicine. Proceedings of NATO Workshop on Comparative Time Series Analysis held at the Santa Fe Institute, 14–17 May 1992
6. Grassberger, P., Procaccia, I.: Measuring the strangeness of strange attractors. Physica **9D** (1983) 189–208
7. Koza, J. R.: Genetic Programming. On the Programming of Computers by Means of Natural Selection. Cambridge, MA: MIT Press (1992)
8. Mackey, M. C., Glass, L.: Oscillation and chaos in physiological control systems. Science **197** (1977) 287–289
9. Mead, W. C., Jones, R. D., Lee, Y. C., Barnes, C. W., Flake, G. W., Lee, L. A., O'Rourke, M. K.: Prediction of chaotic time series using CNLS-net – example: the Mackey-Glass equation. In Casdagli, M., Eubank, S. (Eds.): Nonlinear Modelling and Forecasting. Redwood City, CA: Addison-Wesley (1992)
10. Meyer, T. P., Packard, N. H.: Local forecasting of high-dimensional chaotic dynamics. In Casdagli, M., Eubank, S. (Eds.): Nonlinear Modelling and Forecasting. Redwood City, CA: Addison-Wesley (1992)
11. Nychka, D., Ellner, S., Gallant, A. R., McCaffrey, D.: Finding chaos in noisy systems. J. R. Statist. Soc. B **54** (1992) 399–426
12. Oakley, E. H. N.: Two scientific applications of genetic programming: the development of stack filters, and the fitting of non-linear equations to chaotic data. In Kinnear, K. (Ed.): Advances in Genetic Programming. Cambridge, MA: MIT Press (1994)
13. Ruelle, D.: Deterministic chaos: the science and the fiction. Proc. Roy. Soc. Lond. A **427** (1990) 241–248
14. Stokbro, L., Umberger, D. K.: Forecasting with weighted maps. In Casdagli, M., Eubank, S. (Eds.): Nonlinear Modelling and Forecasting. Redwood City, CA: Addison-Wesley (1992)

Springer-Verlag
and the Environment

We at Springer-Verlag firmly believe that an international science publisher has a special obligation to the environment, and our corporate policies consistently reflect this conviction.

We also expect our business partners – paper mills, printers, packaging manufacturers, etc. – to commit themselves to using environmentally friendly materials and production processes.

The paper in this book is made from low- or no-chlorine pulp and is acid free, in conformance with international standards for paper permanency.

Lecture Notes in Computer Science

For information about Vols. 1–792
please contact your bookseller or Springer-Verlag

Vol. 830: C. Castelfranchi, E. Werner (Eds.), Artificial Social Systems. Proceedings, 1992. XVIII, 337 pages. 1994. (Subseries LNAI).

Vol. 831: V. Bouchitté, M. Morvan (Eds.), Orders, Algorithms, and Applications. Proceedings, 1994. IX, 204 pages. 1994.

Vol. 832: E. Börger, Y. Gurevich, K. Meinke (Eds.), Computer Science Logic. Proceedings, 1993. VIII, 336 pages. 1994.

Vol. 833: D. Driankov, P. W. Eklund, A. Ralescu (Eds.), Fuzzy Logic and Fuzzy Control. Proceedings, 1991. XII, 157 pages. 1994. (Subseries LNAI).

Vol. 834: D.-Z. Du, X.-S. Zhang (Eds.), Algorithms and Computation. Proceedings, 1994. XIII, 687 pages. 1994.

Vol. 835: W. M. Tepfenhart, J. P. Dick, J. F. Sowa (Eds.), Conceptual Structures: Current Practices. Proceedings, 1994. VIII, 331 pages. 1994. (Subseries LNAI).

Vol. 836: B. Jonsson, J. Parrow (Eds.), CONCUR '94: Concurrency Theory. Proceedings, 1994. IX, 529 pages. 1994.

Vol. 837: S. Wess, K.-D. Althoff, M. M. Richter (Eds.), Topics in Case-Based Reasoning. Proceedings, 1993. IX, 471 pages. 1994. (Subseries LNAI).

Vol. 838: C. MacNish, D. Pearce, L. Moniz Pereira (Eds.), Logics in Artificial Intelligence. Proceedings, 1994. IX, 413 pages. 1994. (Subseries LNAI).

Vol. 839: Y. G. Desmedt (Ed.), Advances in Cryptology - CRYPTO '94. Proceedings, 1994. XII, 439 pages. 1994.

Vol. 840: G. Reinelt, The Traveling Salesman. VIII, 223 pages. 1994.

Vol. 841: I. Prívara, B. Rovan, P. Ružička (Eds.), Mathematical Foundations of Computer Science 1994. Proceedings, 1994. X, 628 pages. 1994.

Vol. 842: T. Kloks, Treewidth. IX, 209 pages. 1994.

Vol. 843: A. Szepietowski, Turing Machines with Sublogarithmic Space. VIII, 115 pages. 1994.

Vol. 844: M. Hermenegildo, J. Penjam (Eds.), Programming Language Implementation and Logic Programming. Proceedings, 1994. XII, 469 pages. 1994.

Vol. 845: J.-P. Jouannaud (Ed.), Constraints in Computational Logics. Proceedings, 1994. VIII, 367 pages. 1994.

Vol. 846: D. Shepherd, G. Blair, G. Coulson, N. Davies, F. Garcia (Eds.), Network and Operating System Support for Digital Audio and Video. Proceedings, 1993. VIII, 269 pages. 1994.

Vol. 847: A. L. Ralescu (Ed.) Fuzzy Logic in Artificial Intelligence. Proceedings, 1993. VII, 128 pages. 1994. (Subseries LNAI).

Vol. 848: A. R. Krommer, C. W. Ueberhuber, Numerical Integration on Advanced Computer Systems. XIII, 341 pages. 1994.

Vol. 849: R. W. Hartenstein, M. Z. Servít (Eds.), Field-Programmable Logic. Proceedings, 1994. XI, 434 pages. 1994.

Vol. 850: G. Levi, M. Rodríguez-Artalejo (Eds.), Algebraic and Logic Programming. Proceedings, 1994. VIII, 304 pages. 1994.

Vol. 851: H.-J. Kugler, A. Mullery, N. Niebert (Eds.), Towards a Pan-European Telecommunication Service Infrastructure. Proceedings, 1994. XIII, 582 pages. 1994.

Vol. 852: K. Echtle, D. Hammer, D. Powell (Eds.), Dependable Computing – EDCC-1. Proceedings, 1994. XVII, 618 pages. 1994.

Vol. 853: K. Bolding, L. Snyder (Eds.), Parallel Computer Routing and Communication. Proceedings, 1994. IX, 317 pages. 1994.

Vol. 854: B. Buchberger, J. Volkert (Eds.), Parallel Processing: CONPAR 94 – VAPP VI. Proceedings, 1994. XVI, 893 pages. 1994.

Vol. 855: J. van Leeuwen (Ed.), Algorithms – ESA '94. Proceedings, 1994. X, 510 pages.1994.

Vol. 856: D. Karagiannis (Ed.), Database and Expert Systems Applications. Proceedings, 1994. XVII, 807 pages. 1994.

Vol. 857: G. Tel, P. Vitányi (Eds.), Distributed Algorithms. Proceedings, 1994. X, 370 pages. 1994.

Vol. 858: E. Bertino, S. Urban (Eds.), Object-Oriented Methodologies and Systems. Proceedings, 1994. X, 386 pages. 1994.

Vol. 859: T. F. Melham, J. Camilleri (Eds.), Higher Order Logic Theorem Proving and Its Applications. Proceedings, 1994. IX, 470 pages. 1994.

Vol. 860: W. L. Zagler, G. Busby, R. R. Wagner (Eds.), Computers for Handicapped Persons. Proceedings, 1994. XX, 625 pages. 1994.

Vol: 861: B. Nebel, L. Dreschler-Fischer (Eds.), KI-94: Advances in Artificial Intelligence. Proceedings, 1994. IX, 401 pages. 1994. (Subseries LNAI).

Vol. 862: R. C. Carrasco, J. Oncina (Eds.), Grammatical Inference and Applications. Proceedings, 1994. VIII, 290 pages. 1994. (Subseries LNAI).

Vol. 863: H. Langmaack, W.-P. de Roever, J. Vytopil (Eds.), Formal Techniques in Real-Time and Fault-Tolerant Systems. Proceedings, 1994. XIV, 787 pages. 1994.

Vol. 864: B. Le Charlier (Ed.), Static Analysis. Proceedings, 1994. XII, 465 pages. 1994.

Vol. 865: T. C. Fogarty (Ed.), Evolutionary Computing. Proceedings, 1994. XII, 332 pages. 1994.

Vol. 866: Y. Davidor, H.-P. Schwefel, R. Männer (Eds.), Parallel Problem Solving from Nature-Evolutionary Computation. Proceedings, 1994. XV, 642 pages. 1994.

Vol 867: L. Steels, G. Schreiber, W. Van de Velde (Eds.), A Future for Knowledge Acquisition. Proceedings, 1994. XII, 414 pages. 1994. (Subseries LNAI).

Vol. 868: R. Steinmetz (Ed.), Advanced Teleservices and High-Speed Communication Architectures. Proceedings, 1994. IX, 451 pages. 1994.

Vol. 869: Z. W. Raś, Zemankova (Eds.), Methodologies for Intelligent Systems. Proceedings, 1994. X, 613 pages. 1994. (Subseries LNAI).

Vol. 870: J. S. Greenfield, Distributed Programming Paradigms with Cryptography Applications. XI, 182 pages. 1994.